THE SMITHSONIAN HISTORY OF SPACE EXPLORATION

THE SMITHSONIAN HISTORY OF SPACE EXPLORATION

From the Ancient World to the Extraterrestrial Future

ROGER D. LAUNIUS

Smithsonian Books
Washington, DC

Page 2: American astronaut Bruce McCandless conducting the first untethered space walk on February 7, 1984, during the Challenger IV Space Shuttle mission. © Time Life Pictures/NASA/The LIFE Picture Collection/Getty Images

This book was designed and produced by

White Lion Publishing
The Old Brewery
6 Blundell Street
London N7 9BH

Senior Editors	Hannah Phillips, Elspeth Beidas, Carol King
Editor	Rob Cave
Senior Designer	Isabel Eeles
Designer	Michelle Kliem
Production Manager	Rohana Yusof
Editorial Director	Ruth Patrick
Publisher	Philip Cooper

Published in North America by Smithsonian Books

This book may be purchased for educational, business, or sales promotional use. For information, please write: Special Markets Department, Smithsonian Books, P.O. Box 37012, MRC 513, Washington, DC 20013

Library of Congress Cataloging-in-Publication Data
Names: Launius, Roger D., author.
Title: The Smithsonian history of space exploration: from the ancient world to the extraterrestrial future / Roger D. Launius
Other titles: History of space exploration
Description: Washington, DC: Smithsonian Books, [2018]
Identifiers: LCCN 2018017014 | ISBN 9781588346377 (hardcover)
Subjects: LCSH: Outer space—Exploration—History. | Astronautics—History. | Manned space flight—History. | Space flight to Mars.
Classification: LCC TL790 .L3275 2018 | DDC 629.409—dc23
LC record available at https://lccn.loc.gov/2018017014

Manufactured in China, not at government expense

22 21 20 19 18   5 4 3 2 1

Contents

Introduction

It all began with dreams. Throughout human history, we have been constantly fascinated with our natural universe, leading to a desire to learn more about it. The early spaceflight pioneers relentlessly worked to make their dreams of exploring the solar system a reality. Enthusiasts around the globe considered possibilities, worked to build a theoretical framework, and experimented with primitive rockets in a quest to demonstrate humanity's potential to reach beyond Earth. World War II served as a watershed; every belligerent nation developed some type of rocket technology as a weapon. The German V-2 was easily the most advanced of these efforts, and every nation venturing into space afterward studied or adapted its design and acknowledged the pioneering work of Wernher von Braun and his fellow rocketeers.

After the war, scientists and engineers in the United States and the Soviet Union realized that spaceflight was something that was suddenly within their grasp. The two nations were soon engaged in a head-to-head race to get humans into space and later to land them on the Moon.

Thereafter, spacefaring nations—and there were seemingly more every year—followed on with increasingly complex satellite launches, robotic missions to the planets, and further human operations in Earth orbit. By the year 2000, the established space powers of the United States and Russia still played the leading role in space exploration, but the whole enterprise had become a more international pursuit, particularly after the establishment of the International Space Station. Earth orbit has become a normal realm of human activity, but the exploration has not ended there. Mighty space telescopes have probed the depths of the universe and brought us knowledge about its very origins and evolution.

It might be tempting to take all this for granted, but this is something we should never do. As the great American science fiction writer Ray Bradbury once commented: "Too many of us have lost the passion and emotion of the remarkable things we've done in space. Let us not tear up the future, but rather again heed the creative metaphors that render space travel a religious experience. When the blast of a rocket launch slams you against the wall and all the rust is shaken off your body, you will hear the great shout of the universe and the joyful crying of people who have been changed by what they've seen." The power of a big rocket launch is indeed daunting. Impressive enough when viewed on television, in person it is overwhelming and uniquely magical. No one who witnesses a launch firsthand leaves unchanged. The experience is thrilling and transformative.

President John F. Kennedy captured this sense of wonder well in 1962 when he remarked in a speech at Rice University: "We choose to go to the Moon in this decade and do the other things, not because they are easy, but because they are hard, because that goal will serve to organize and measure the best of our energies and skills, because that challenge is one that we are unwilling to postpone, and one which we intend to win, and the others, too." He also said that "we set sail on this new sea because there is new knowledge to be gained and new rights to be won, and they must be won and used for the progress of all people." As well as being an immense feat of precision engineering, spaceflight is a deeply spiritual experience, one that has the power to reach into the depths of the human psyche.

From the beginning of human spaceflight in the 1950s, the most eloquent advocates of space exploration have been those who have journeyed into space themselves. Yuri Gagarin, the first person to go into orbit, recalled, "The road to the stars is steep and dangerous. But we're not afraid.... Spaceflights can't be stopped. This isn't the work of one man or even a group of men. It is a historical process which mankind is carrying out in accordance with the natural laws of human development." John Glenn, the first American to orbit the Earth, summoned images of the American heritage of pioneering when he commented that "space represents the modern frontier for extending humanity's research into the unknown. Our commitment to manned programs must remain strong even in the face of adversity and tragedy. This is our history and the legacy of all who fly."

The first person on the Moon, Neil Armstrong, reflected in a similar manner: "The important achievement of Apollo was demonstrating that humanity is not forever chained to this planet and our visions go rather further than that and our opportunities are unlimited." Armstrong's sentiments have been echoed by recent NASA astronaut Franklin Chang-Diaz, who said, "My vision is a future for humanity where we will be completely free to pursue activities outside of our planet."

Even those who merely witnessed the opening age of space travel remain moved by the endeavor. Legendary American journalist Walter Cronkite reflected at the turn of the twenty-first century that "yes, indeed, we are the lucky generation.... [We] first broke our earthly bonds and ventured into space. From our descendants' perches on other planets or distant space cities, they will look back at our achievement with wonder at our courage and audacity and with appreciation at our accomplishments, which assured the future in which they live."

The story of space exploration was motivated by fantastic dreams, the spirit of discovery, and the thrill of voyaging into the unknown. Properly conducted, space exploration can provide a hopeful future. It can help teach us all important lessons about what humans can achieve when we work together and how we can learn to better live together on our small and precious world before we leave it for another.

The story that follows is a recounting of a global pursuit of space exploration that stretches back to when humans first gazed up at the sky with wonder in their hearts. It looks back at the dawn of human spaceflight in the mid-twentieth century, when we first dipped our toes into the cosmic ocean, and on to a future of space exploration filled with boundless opportunities. The story is far from complete; indeed, it has only just begun.

1

Laying the Foundations for Space Exploration

Throughout human history, people have been awed and fascinated by the heavens. The magnificence and movement of the Sun, the stars, and the planets have repeatedly inspired wonder, study, and a search for meaning among the various civilizations that beheld them. While the astronomers of the ancient world may have viewed the cosmic firmament through a religious lens, identifying its various elements with a rich tapestry of gods, heroes, and beasts of many kinds, they nonetheless managed to establish the size of the Earth and chart the rising and setting of constellations, the movement of the planets, and distances around the globe, even going as far as to build structures, such as Stonehenge in England, that aligned with the movement of particular celestial bodies.

Gradually, humanity uncovered the science that explains the basic workings of the solar system. From the start of the scientific revolution of the sixteenth century, the work of astronomers, such as Copernicus and Galileo (p. 14) and the experimental physicist Isaac Newton (p. 27), did much to transformation humanity's perspective on the universe. In turn, their scientific achievements helped inspire generations of science fiction writers, from Cyrano de Bergerac to Jules Verne (p. 17), and even early filmmakers, such as Georges Méliès and Fritz Lang (p. 32), to imagine and explore their ideas of what space travel might be like.

But it wasn't until the opening years of the twentieth century that the great pioneers of space travel—the Russian Konstantin Tsiolkovsky (p. 22), the American Robert Goddard (p. 24), and the German Hermann Oberth (p. 30), together with numerous amateur rocket societies around the globe (p. 34)—began to take the first steps toward making space travel, and space exploration, a reality.

OPPOSITE The heavens and the Milky Way fascinated the ancients, who gave it religious and mythical meaning.

Speculations on Life beyond Earth

RIGHT This 1791 aquatint depicts the French author and astronomer Bernard Le Bovier de Fontenelle with a telescope, contemplating life on other planets.

BELOW Aztec calendar stones detail the recurring 52-year period of time that was more significant within Aztec culture than the annual cycle of the seasons.

Belief in life beyond our immediate frame of reference has been a constant feature of human history. When traveling through new territories, explorers often encountered unfamiliar "alien" creatures. They would later return home with tales of all manner of wondrous beasts, both real and imagined. Understandably, this led many people and civilizations to anticipate, both in seriousness and in jest, that among the stars and planets they saw in the night sky were worlds of great influence and importance that doubtlessly hosted a variety of strange creatures.

The ancient Egyptians viewed the Milky Way as a heavenly version of the River Nile, which sustained their civilization. This comparison helped them explain some of the forces of nature that affected their daily existence. Later, around 700 BCE, astronomers in Babylon (located in what is now Iraq) charted the paths of several planets and compiled observations of fixed stars (astronomical objects that do not appear to move in relation to each other), assigning them religious meaning and influence over people's lives. Around 400 BCE, they devised the zodiac, the first mechanism to divide the year into lunar periods and ascribe significance to a person's date of birth for foretelling the future.

In around 165 CE, Lucian of Samosata (ca. 125 CE–ca. 180 CE) wrote his *True History*. Widely viewed as the first work of science fiction, its narrative involves a vessel of 50 Greek soldiers sailing to the Moon, where they encounter its native inhabitants and a range of bizarre beasts. Later still, in what became the Americas, the Incan and Aztec cultures developed their own astronomical observatories. The Aztecs drew upon their astronomical observations to develop the 365-day *xiuhpohualli* (year count) calendar, as well as a 260-day *tonalpohualli* (day count) ritual cycle. Together they formed a 52-year unit of time whose spiritual significance was more important to the Aztecs than a calendar based on seasons.

The scientific revolution of the sixteenth century saw the gradual adoption of an increasingly rigorous and systematic approach to science that challenged all previous notions of the universe and humanity's place in it (p. 344). These new ideas served to increase speculation that, as Bernard Le Bovier de Fontenelle (1657–1757) wrote in his 1686 book *Conversations on the Plurality of Worlds*, life might exist elsewhere in the universe.

In the seventeenth century, astronomer and mathematician Johannes Kepler (1571–1630) even suggested that the craters on the Moon might actually be city walls constructed by its indigenous population, although he did so in the form of a science fiction story, "The Dream," first published in 1634. Two centuries later, the Munich-based astronomer Franz von Paula Gruithuisen (1774–1852) claimed to have actually observed these walled lunar cities. Belief in the habitability of the Moon continued well into the nineteenth century, when telescope technology improved sufficiently to disprove the theory. Attention then shifted to Venus and Mars as possible homes to extraterrestrial life. All this speculation only served to further fuel humanity's desire to find a means to reach and explore these worlds.

> *"When ships to sail the void between the stars have been built, there will step forth men to sail these ships."*
>
> JOHANNES KEPLER (1571–1630), GERMAN ASTRONOMER

BELOW LEFT This satellite image shows Angkor Wat in northwestern Cambodia. The buildings visible at the center were constructed with knowledge of the Earth's orbital motion that long predated modern scientific advancements and discoveries.

BELOW RIGHT Around 1,000 years ago, the Anasazi people built several groups of structures in what is now northwestern New Mexico. Many of these buildings are thought to have been constructed to reflect or align with solar and lunar cycles.

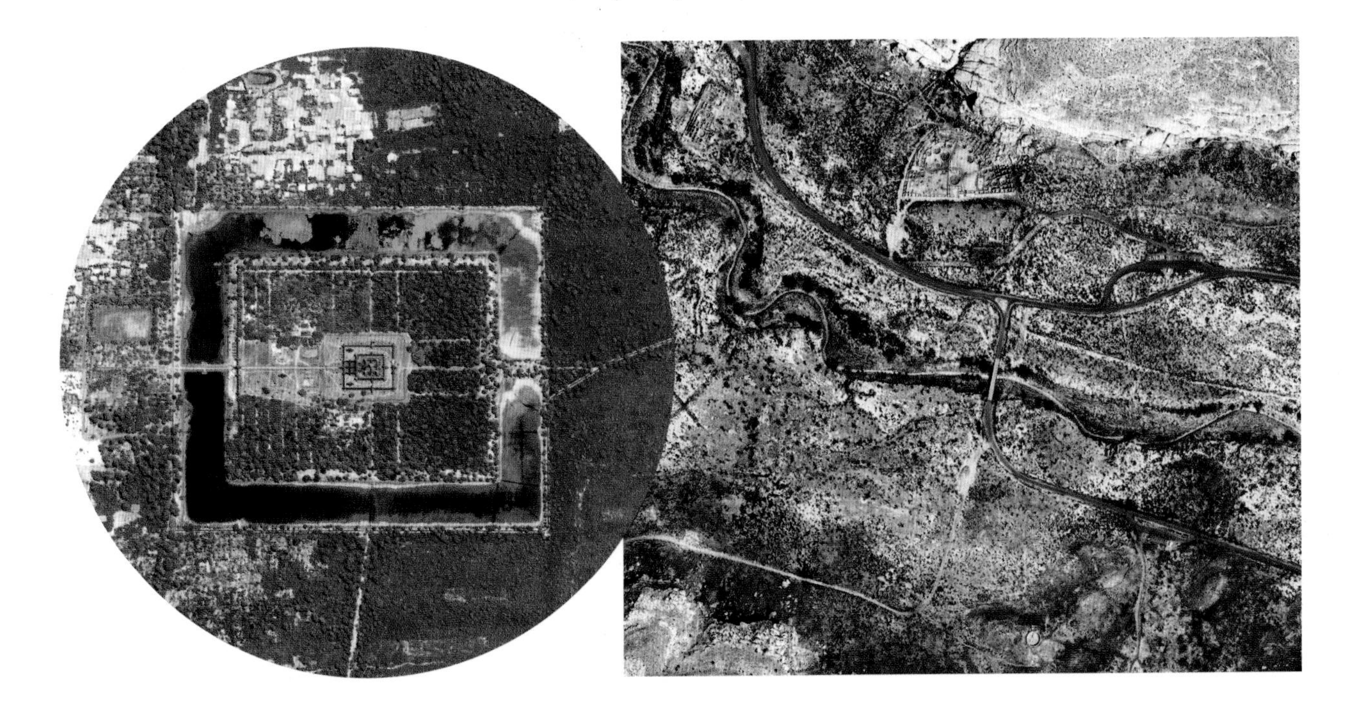

The Shifting Center of the Universe

This depiction of the Ptolemaic universe, from Andreas Cellarius's *Harmonia Macrocosmica* (1660), shows the signs of the zodiac and the solar system, with the Earth at the center.

Humans have always looked up at the sky to help them understand their place in the universe. In antiquity, ancient astronomers used their observations of the heavens to devise a number of different models to explain how the universe worked. The Greek astronomer Aristarchus of Samos (ca. 310–ca. 230 BCE) was among the first to propose that Earth revolves around the Sun. However, this heliocentric (Sun-centered) notion failed to gain much widespread acceptance.

The model of the universe that would come to dominate European thought until well into the sixteenth century took its name from one of its most prominent advocates, the Roman astronomer and theorist Claudius Ptolemy (87–150 CE). As set out in Ptolemy's great astronomical and mathematical compilation *The Almagest*, the Ptolemaic model describes a geocentric plan of the universe that positions the Earth at its core, surrounded by nine crystalline spheres in which the Sun, stars, and planets are embedded. Ptolemy's model declared the geometric and static harmony of the universe and placed humanity squarely at its center.

The Roman philosopher and statesman Marcus Tullius Cicero (106–43 BCE) summarized the Ptolemaic system this way: "The Universe consists of nine circles, or rather of nine moving globes. The outersphere is that of the heavens, which embraces all the others and under which the stars are fixed. Underneath this, seven globes rotate in the opposite direction from that of the heavens. The first circle carries the star known to men as Saturn; the second carries Jupiter, benevolent and propitious to humanity; then comes Mars, gleaming red and hateful; below this, occupying the middle region, shines the Sun, the chief, prince, and regulator of the other celestial bodies, and the soul of the world which is illuminated and filled by the light of its immense globe. After it, like two companions, come Venus

The Geocentric Model

Ptolemy ranked the celestial spheres as seen below, with the Earth placed unmoving at the center of his concept of the universe. The outermost layer was called the Primum Mobile, which was identified as the cause of the daily revolutions observed in the movements of the inner spheres.

Earth
Moon
Mercury
Venus
Sun
Mars
Jupiter
Saturn
Fixed Stars
Primum Mobile

Claudius Ptolemy was a philosopher and scientist from the Roman province of Egypt who gave his name to a model of the universe that placed the Earth and humanity at its center.

Astronomy before Telescopes

Before the advent of the telescope in 1608, astronomers mainly used their naked eyes to conduct their astronomical observations. Many astronomical observatories built before this period, including the Maragheh observatory, used murals on their walls to serve as quadrants to help people measure the altitudes of various astronomical objects in the sky as they passed overhead.

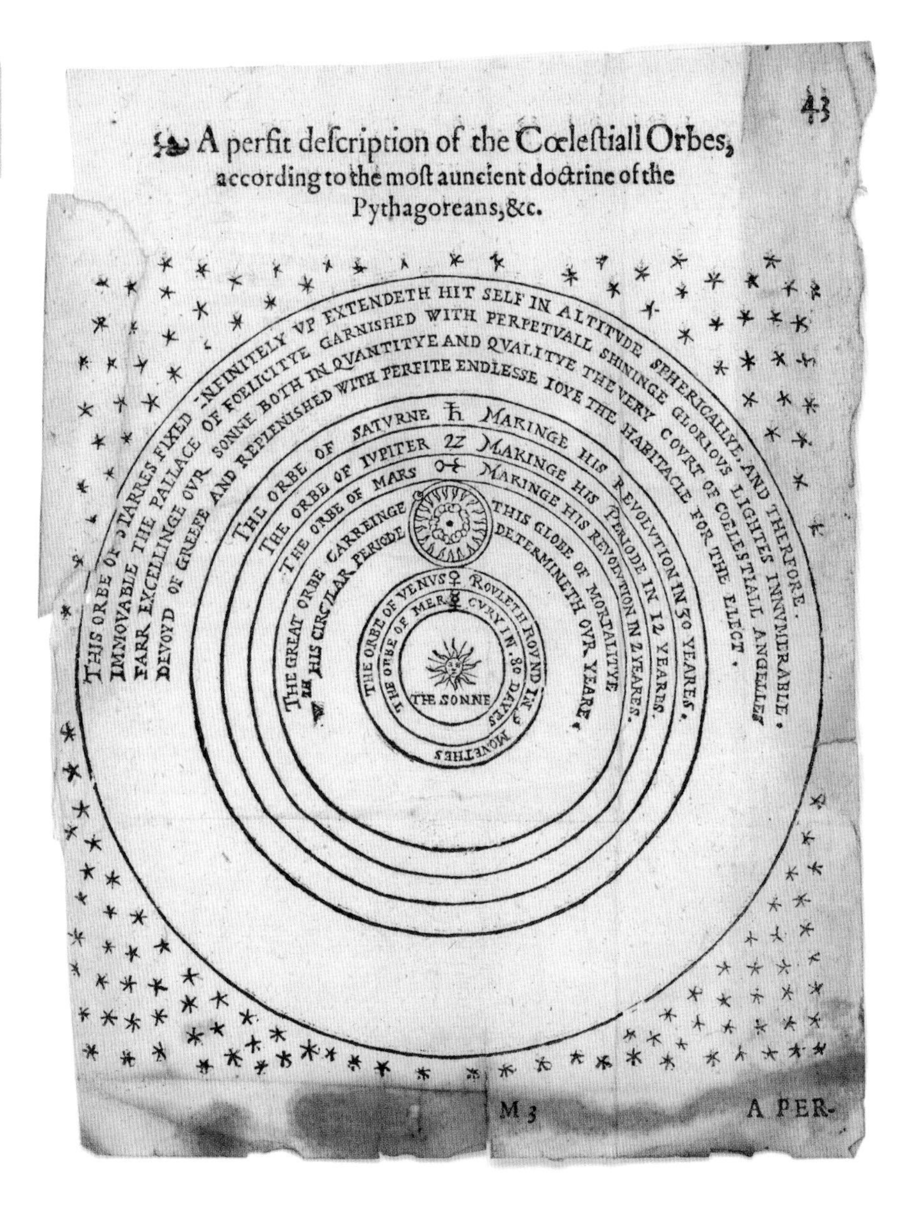

This depiction of the Copernican universe is from Thomas Digges, *A Perfit Description of the Coelestiall Orbes* (1576).

and Mercury. Finally, the lowest orb is occupied by the Moon, which borrows its light from the Sun. Below this last celestial circle there is nothing by what is mortal and perishable except for the minds granted by the gods to the human race. Above the Moon, all things are eternal. Our Earth, placed at the center of the world, and remote from the heavens on all sides, stays motionless and all heavy bodies are impelled towards it by their own weight."

Despite the Ptolemaic model's popularity and acceptance across the Roman Empire and among the followers of the emerging Christian and Islamic faiths, many astronomers found it progressively difficult to reconcile with the evidence of their astronomical observations. The movement of some planets just did not seem to match the model. Over the years, numerous people attempted to devise ways to account for these discrepancies, and among them was the Persian astronomer Nasir al-Din Tusi (1201–74), who served as the director of the Maragheh observatory in what is today the East Azerbaijan Province of Iran. Working with a group of fellow stargazers that included Chinese astronomer Fao Munji, al-Din Tusi identified adjustments to the Ptolemaic model that helped produce more accurate predictions of planetary movements.

Between 1514 and 1542, the Polish mathematician Nicolaus Copernicus (1473–1543) gradually developed his own solution to the problem in the form of a new heliocentric model in which all the planets, including the Earth, circle around in the Sun's orbit. Copernicus wrote a detailed six-volume explanation of his work under the title *On the Revolutions of the Heavenly Spheres* but refused to publish it during his lifetime. He feared contradicting the Roman Catholic Church, which was committed to a view that placed the Earth, and therefore humanity, at the center of the universe, in agreement with the Ptolemaic model.

After Copernicus's death, and the subsequent publication of his book, support for his heliocentric model slowly began to grow across Europe. This spread further increased once the Italian polymath Galileo Galilei (1564–1642) began to provide evidence in support of the heliocentric model. Using the newly invented technology of the telescope, he identified four of the largest moons orbiting Jupiter and also Saturn's rings, which he initially mistook for moons. The movement of the moons swiftly demonstrated that some celestial objects did not revolve around the Earth.

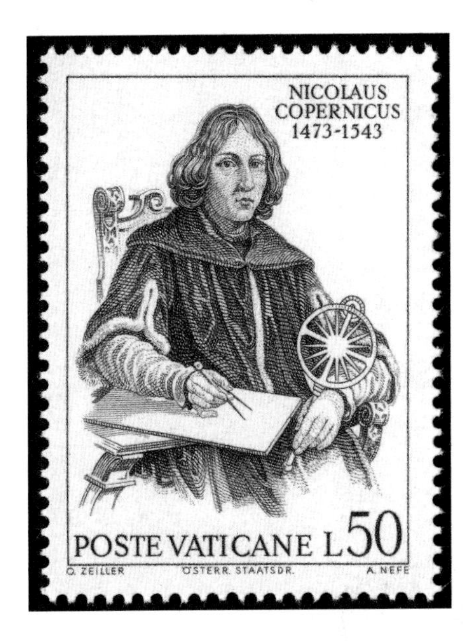

On June 19, 1973, Vatican City issued a set of four stamps commemorating the work of Nicolaus Copernicus, but it would be almost two more decades before the Catholic Church officially admitted Galileo Galilei had been right in supporting Copernicus's ideas.

The Trials of Galileo Galilei

The Catholic Church twice put Galileo Galilei on trial for promoting the idea that the Earth was not the center of the universe, contrary to the Church's interpretation of the Bible. In 1616, he was brought before the Inquisition in Rome, where he confronted Monsignor Francesco Ingoli in a debate over the merits of the Copernican model of the universe. Despite a spirited discussion, Pope Paul V declared that the Ptolemaic heliocentric model was "physically true" and instructed Galileo to completely abandon the teaching and defense of Copernican ideas. By agreeing not to teach the Copernican model publicly, Galileo initially avoided any direct punishment. However, in 1632 Galileo was brought to trial again for violating the prohibitions of the Church. He was compelled to recant under threat of torture, and was then forced into retirement and house arrest at his home in Florence. Nonetheless, it seems that Galileo remained quietly defiant, maintaining his belief in the Copernican system to the end of his days.

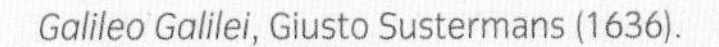
Galileo Galilei, Giusto Sustermans (1636).

Reaching Wider Audiences

The first proposal for a space station appeared in Edward Everett Hale's short story "The Brick Moon," published in the *Atlantic Monthly* in 1869.

When Galileo Galilei published his first collection of telescopic observations of the solar system in *Sidereal Messenger* in 1610, he helped spark a flood of speculative science fiction writings about the possibilities of space exploration. Most of these texts focused on adventure and the exciting idea of contact with extraterrestrial beings and were far more accessible to general readers than the scientific texts they drew from. A few even wrote sci-fi works to comment satirically upon contemporary issues.

Among the earliest in this phase of sci-fi authors was the French novelist and playwright Cyrano de Bergerac (1610–55), who penned *The Voyage to the Moon*, published in 1649. In his book, de Bergerac describes several attempts made by his hero, also named Cyrano, to travel to the Moon. First, the fictional Cyrano ties a string of bottles filled with dew around himself, hoping that when the heat of the Sun evaporates the bottled liquid, he will be similarly drawn upward. When this fails, the hapless explorer jumps into a craft powered by firecrackers that successfully launches him to the Moon. Consequently, he becomes the first fictional flyer to reach the Moon by rocket.

In an 1869 edition of *Atlantic Monthly*, the American writer Edward Everett Hale (1822–1909) published the first known proposal for a space station around Earth in a short story entitled "The Brick Moon." In his tale, Hale describes how an artificial moon, literally made out of bricks and intended as a navigational aid to oceangoing vessels, is constructed. The story goes on to describe how an accident sends the brick moon off into space with 37 people on its surface. Although the author notes the absence of atmosphere in space, the travelers nevertheless manage to survive their journey into the vacuum.

While much early science fiction suffered from its authors' poor understanding of science, over time an increasing number of writers began to demonstrate a far firmer understanding of the scientific underpinnings of space exploration in their work. Among these more scientifically literate writers were Jules Verne (1828–1905) and Herbert George Wells (1866–1946), although Verne in particular still relied on a hefty amount of artistic license in his stories. For example, in his *From*

the Earth to the Moon*, first published in 1865, and *Around the Moon*, first published in 1866, Verne's characters set off on their voyage by being fired out of a cannon—something that would be extremely inadvisable to try in real life.

Wells was on more scientifically solid ground with his great work, *The War of the Worlds*. First published in 1897, the story explores the mixture of fascination and horror with which humanity greets the arrival of aliens from Mars. These aliens then swiftly reveal both their hostile intent and their technological superiority, which enables them to decimate London, then capital of the British Empire and at the very pinnacle of its global might. Using their three-legged vehicles, which are armed with deadly heat rays, the aliens soon lay waste to Britain's military forces. Ultimately, it is not humanity's resistance that leads to the defeat and downfall of the invaders, but the aliens' own lack of immunity to terrestrial bacteria.

Interestingly, through their works of fiction, many of these science-inspired authors would in turn go on to provide invaluable inspiration to many of the scientists and engineers who would make space exploration a fact in the twentieth century.

ABOVE LEFT Jules Verne published his first science fiction novel, *From the Earth to the Moon*, in 1865. In this illustration, passengers in Verne's ship enjoy their first taste of weightlessness.

ABOVE RIGHT In this scene from the book *The War of the Worlds*, by the science fiction pioneer H. G. Wells, the technologically advanced Martians use their large three-legged vehicles to overcome an artillery attack by the British military.

The Origins of Rocketry in War

Humanity first experimented with explosives and developed primitive flying projectiles long before the scientific laws explaining how they work were first codified in the seventeenth century (p. 10). While it remains unclear exactly who invented the forerunner of the rockets used today, we do know that the Chinese were using self-propelled projectile weapons in warfare by the time of Genghis Khan (ca. 1162–1227) in the twelfth century CE. Some sources have speculated that such rockets could be used to transport individuals high above the Earth, but there is no firm evidence that anyone attempted such a feat. In his book *On the Wonders of the World*, the German Dominican friar Albertus Magnus (1193–1290) gave a recipe for making gunpowder and suggested that it could be used either "for flying or for making thunder." The crude Chinese gunpowder-fueled rockets, which bear some similarity to modern fireworks, served as a form of artillery, delivering both combustible material and a nasty surprise to enemy infantry.

Chinese forces are first recorded to have employed early rockets in warfare around 1000 CE. Here a Chinese soldier aims a rocket-powered lance at an enemy village.

Armies from China's Song dynasty employed similar weapons, known as "fire lances," extensively in military operations during the twelfth-century Jin-Song Wars. These fire lances consisted of a bamboo tube packed with gunpowder, a slow match to act as a fuse, and a spear or similar long, thrusting weapon. When ignited, the gunpowder would eject the spear from the bamboo with a brief jet of flames, like a bullet from a gun. Later versions of the weapon mixed iron pellets or pottery shards with the gunpowder and even employed tubes constructed of thick paper instead of bamboo. The Chinese also used gunpowder rockets against the Mongols at the siege of Kai Fung Foo in 1232. Arrows placed within tubes packed with gunpowder produced a hail of "flying fire arrows." The Mongol attackers swiftly dispersed when facing these terrifying weapons, despite the inaccuracy of their aim.

Around 1400, the use of rockets in combat had spread to the battlefields of Europe, but they were soon largely superseded by more powerful cannon technology. Then, in the late eighteenth century, rocket-based weapons experienced a brief resurgence when the British East India Company was dealt a series of defeats by the forces of the Sultanate of Mysore using so-called Mysore rockets. Thought to have descended from the earlier Chinese rockets, these devices consisted of a tube of soft, hammered iron that was sealed at one end and packed with gunpowder. The rockets produced greater thrust and had longer ranges that the rockets known in Europe at the time. The deadly efficacy of the Mysore rockets in battle prompted the British army to develop improved gunpowder rockets of their own in the early nineteenth century under artillery officer Sir William Congreve (1742–1814). Congreve's incendiary barrage missiles could be fired from either land or sea. They were just as accurate as conventional artillery cannon but were lighter and easier to transport, making them perfect for use in the farthest-flung corners of

ABOVE LEFT A Mysore soldier holding a flag that doubles as a rocket launcher.

ABOVE RIGHT This view of the British rocket bombardment of Fort McHenry shows some of the estimated 1,500 to 1,800 shells launched during British forces' attempted landing at Baltimore in 1814.

the globe. Congreve's rockets were used to great effect in several battles during the Napoleonic Wars, in the first decades of the 1800s. The British also employed them against the United States in the conflict known as the War of 1812. In "The Defense of Fort McHenry," an eyewitness poem recounting the British bombardment of the port of Baltimore written by Francis Scott Key (1779–1843), there are overt references to Congreve's rockets in the lines "the rockets' red glare, the bombs bursting in air, gave proof through the night that our flag was still there." Key's poem later provided the lyrics for the American national anthem, "The Star Spangled Banner."

By the mid-nineteenth century, further advances in conventional artillery technology, including improved aiming mechanisms for cannon, prompted the British military to largely abandon rockets once more. However, rocket technology was also adapted for civilian use around this time. By the 1820s, sailors were hunting whales using rocket-propelled harpoons. These were launched from shoulder-held tubes and were often equipped with circular safety shields. Such rocket harpoons were more powerful and could spear whales more effectively than hand-held harpoons, making whaling much more efficient. Rocket-powered rescue lines were also deployed in shipping to help pull a stricken vessel closer to the craft rendering assistance. It was only toward the end of the century that Konstantin Tsiolkovsky (p. 22) and other advocates of space exploration began to look at rocket technology as a means to travel beyond the Earth.

BELOW LEFT A nineteenth-century illustration of whalers using a rocket harpoon to kill their quarry.

BELOW RIGHT This whaling rocket harpoon of the 1870s was designed to be more effective than earlier hand-held harpoons, and it reduced the risk to whaling crews posed by flailing wounded whales.

William Congreve and the Development of Bombardment Rockets

The British military began research into developing its rocket artillery at the Royal Arsenal in Woolwich, South London, in 1801. There, Sir William Congreve developed powerful iron-cased gunpowder rockets with 16-foot (4.8-m) guide sticks. Congreve was the eldest son of Lt. General Sir William Congreve, a military officer credited with improving the strength of British artillery through his own gunpowder experiments. The younger Congreve's rockets were first demonstrated at the Royal Arsenal in 1805 and were subsequently tested in two naval engagements against Napoleon's forces at Boulogne, France, in 1806. By this time, Congreve was producing 32-pound (14.5-kg) rockets with a range of more than 3,000 yards (2,700 m). Congreve's innovations enabled artillery fire without the need for heavy guns; wherever a packhorse or an infantryman could go, the rocket could provide support.

The following year, Congreve witnessed 300 of his rockets successfully deployed at the Bombardment of Copenhagen, where they contributed to the burning of the city and the subsequent British victory. After this success, the British used Congreve's rockets more widely across various conflicts during the Napoleonic Wars, eventually forming a separate Rocket Brigade. This brigade fought effectively and to great acclaim alongside Swedish forces during the Battle of Leipzig in 1813. However, the rockets proved far less effective during the Anglo-American War of 1812, where they were deployed at the Battle of Baltimore in September 1814, and later at the Battle of New Orleans in January 1815. These failures may be attributed, in part, to the effectiveness of the Americans' defensive fortifications and their forces' ability to keep the British ships outside their most effective range. Despite these military setbacks, Congreve continued to experiment with rocket technology for the rest of his life. He even went on to establish a reputation for himself in providing fireworks for various celebrations in London.

ABOVE LEFT A model of two rockets designed by William Congreve sitting on their launcher.

ABOVE RIGHT Portrait of rocketeer Sir William Congreve, 2nd Baronet, by James Lonsdale (1777–1839).

Investigating Space with Rocket Devices

"The Earth is the cradle of humanity, but mankind cannot stay in the cradle forever."

KONSTANTIN EDUARDOVICH TSIOLKOVSKY, RUSSIAN SPACE THEORIST

TOP One of Konstantin Tsiolkovsky's many rocket designs. He examined fundamental scientific theories behind rocketry, and by the 1920s he had analyzed and mathematically formulated the technique for staged vehicles to reach escape velocities from Earth.

ABOVE Soviet model makers built this spacecraft based on Tsiolkovsky's designs and notes after the theorist's death. It was subsequently given to the Smithsonian as a gift from the Tsiolkovsky Russian National Space Museum, Kaluga, Russia.

Konstantin Eduardovich Tsiolkovsky (1857–1935) was one of the earliest scientific theorists to seriously consider the practical possibilities of space exploration. Born in Kaluga, a small town in imperial Russia, Tsiolkovsky contracted scarlet fever at the age of ten, which damaged his hearing. This impacted his ability to attend school and forced young Tsiolkovsky to pursue his early studies independently, using books on natural science and mathematics from his father's library. Enthralled with the idea of exploring the planets, he soon developed a passion for invention, constructing balloon-propelled carriages and various other experiments.

He went on to become a teacher at a local school but remained passionately interested in scientific matters, and, after a period in which he dabbled in writing his own science fiction tales, he eventually began writing theoretical papers on actual science in his spare time. In 1881, Tsiolkovsky submitted an article on the kinetic theory of gases for publication. This was rejected, but a second paper, "The Mechanics of a Living Organism," earned him election into the prestigious Society of Physics and Chemistry in St. Petersburg. Tsiolkovsky's subsequent articles,

"The Problem of Flying by Means of Wings" (1890–91) and "Elementary Studies of the Airship and Its Structure" (1898), showed his growing fascination with flight. He built the first wind tunnel in Russia in 1897 and even produced a monoplane design.

Tsiolkovsky first started writing about space in 1898, when he submitted an article to the Russian journal *Science Review.* "Investigating Space with Rocket Devices" described in depth the use of rockets for launching orbital spaceships and included calculations that established the relationship between a rocket's speed, the speed of the exhaust gas escaping from it, and the changing mass of the rocket's propellant. Now frequently referred to as the Tsiolkovsky formula or rocket equation, it opened the door to future writings on rocket-powered flight. Unfortunately, the journal's editors at the time viewed Tsiolkovsky as something of a crackpot, and he was finally able to persuade them to publish his analysis only in 1903.

After the Bolshevik Revolution of 1917 and the creation of the Soviet Union, Tsiolkovsky was formally recognized for his accomplishments in the theory of space flight. In 1921, he received a lifetime pension from the Soviet state that allowed him to retire from teaching at age 64. Thereafter, he was able to devote his full attention to further developing spaceflight theories. In the late 1920s, he went on to formulate the possibility of multistage rockets that could escape Earth's gravity, but he never launched any rockets based on his own designs and did not live to see his space travel theories realized with the first orbital launches, which occurred just a few short decades after Tsiolkovsky's death.

Although Tsiolkovsky's theoretical work greatly influenced later rocketeers in parts of Europe and across his native land, where he is known as "the father of theoretical and applied cosmonautics," Tsiolkovsky remained relatively unknown in the United States during his lifetime. It was only in the 1950s and 1960s, as American scientists sought to understand how the Soviet Union's rocket program had accomplished so much so swiftly, that his work became more universally recognized and studied.

Konstantin Tsiolkovsky's Key Accomplishments

Konstantin Tsiolkovsky is widely credited as being the first rocket theorist to derive and apply the equation that describes the ideal motion of a rocket if no other external forces are acting upon it.

Tsiolkovsky's rocket equation:

Δ_v	Maximum change of velocity (m/sec)
v_e	Effective exhaust velocity (m/sec)
ln	the natural logarithm function
m_0	Initial total mass, or "wet mass," including all propellant
m_1	Final total mass, or "dry mass"

$$\Delta_v = v_e \cdot \ln\left(\frac{m_{full}}{m_{empty}}\right)$$

Tsiolkovsky also studied the energies involved in vertical and horizontal launching, and considered the best overall shape for a rocket. He explored the properties of various types of propellant and determined that solid propellants could not provide the energy needed for interplanetary travel.

FAR LEFT Pins, or *znachki*, commemorating historic figures and achievements have a long and established tradition in Russia. During the Soviet period, collectible *znachki* were created as souvenirs of national celebrations. This example honors Konstantin Tsiolkovsky. Such space-themed items remain popular among collectors worldwide today.

LEFT Tsiolkovsky with some models based on his designs for space vehicles.

The Technology of Rockets

"It has often proved true that the dream of yesterday is the hope of today and the reality of tomorrow."

ROBERT GODDARD, AMERICAN ROCKETEER

Robert Goddard continued to experiment on different kinds of rocket-fuel mixes throughout his life.

Rocketry is the basic technology that underpins almost all space exploration. Rockets are the launch vehicles that place satellites in orbit (p. 90), put people on the Moon (p. 129), and send various probes out to the planets (p. 165) and beyond (p. 290). They work by harnessing the energy released from the controlled burning of a highly combustible fuel, usually with an oxidizing agent employed to increase the efficiency of this reaction.

Typically, rockets take the optimally aerodynamic shape of a long cylinder with a conical tip. This area is usually reserved for the rocket's payload, which could be anything from scientific equipment to an explosive warhead. They also contain a fuel tank, to house the combustible propellant, and an oxidizer tank, to hold the oxidizing agent. Each of these has its own valves and pumps that lead to the engine, comprising a combustion chamber and a nozzle through which the pressurized exhaust gases are expelled, driving the rocket forward. Rockets also require guidance systems, which can include adjustable tail fins and vanes, vernier thrusters (small additional rocket engines at the sides of the main rocket thrusters, used for fine adjustments in a rocket's attitude and velocity), and even thrust-vectoring technology that alters the direction of the engine's exhaust.

The best choice of fuel for a rocket was a key focus of the experiments of the American rocket pioneer Robert Goddard (1882–1945). Goddard made a systematic investigation of potential rocket propellants, including various kinds of solid fuel, in the early twentieth century. This eventually led him to realize the greater burn efficiency of liquid fuels (p. 29), but he also developed many other key technologies of modern rocketry, including turbopumps to drive the liquid propellant through the fuel lines, efficient engine nozzles, cooling systems, and launching apparatus. He also registered a total of 214 patents for rocket components over his lifetime.

BELOW LEFT An example of one of Robert Goddard's many rocket design patents; this one was filed in 1914.

BELOW RIGHT The general anatomy of a liquid-fueled V-2 rocket.

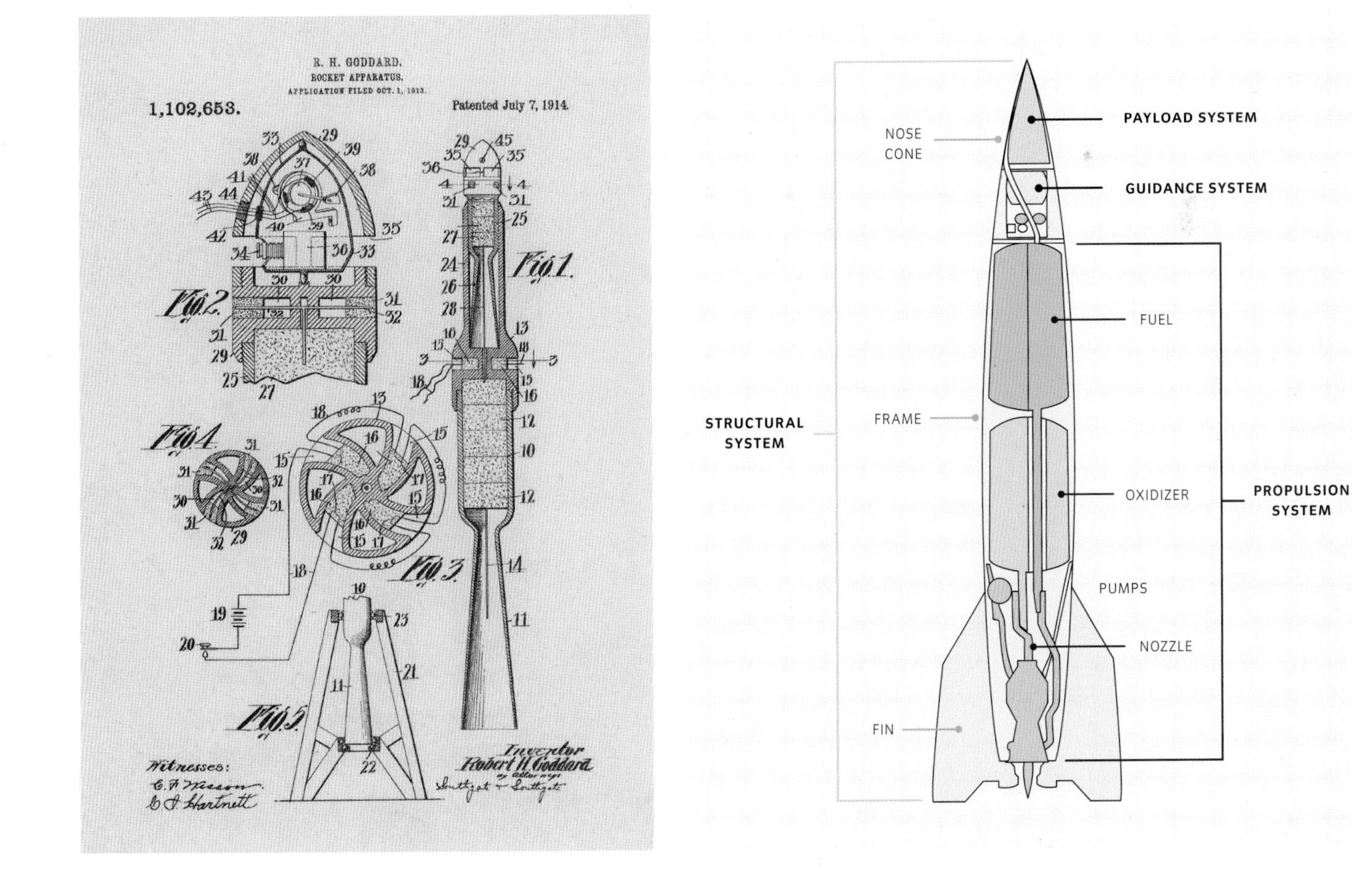

The Principles of Rocketry

Thrust is the force that moves the rocket through the air and into space. It is generated through the release of exhaust gases from the combustion of fuel and an oxidizing agent through a small aperture or nozzle. The release of the exhaust gases propels the rocket in the opposite direction in accordance with Newton's third law of motion.

Direction of the thrust is normally along the longitudinal axis of the rocket, through the rocket's center of gravity. But on some rockets the exhaust nozzle, and therefore the direction of the thrust, can be rotated, or gimbaled.

Magnitude of the thrust depends on the mass flow rate of the exhaust through the engine, and the exit velocity and pressure at which it exits the rocket's nozzle. The efficiency of the propulsion system is characterized by the specific impulse; the ratio of the amount of thrust produced to the mass flow of the propellants combusted to produce the exhaust.

Pressure and temperature of the exhaust gases are determined by the chemical properties of the fuel and the oxidizer used. Collectively, the fuel and oxidizer are termed a rocket's propellants.

Liquid-fueled rockets use liquid propellants, which are stored separately and pumped into and combusted in the combustion chamber, above the nozzle.

Solid rockets use fuel and oxidizer that have been mixed together into a solid propellant that is packed into the cylinder of the rocket. The propellant burns until all the propellant has been consumed.

The ability to get a rocket into space rests on a single mathematical equation. The product of the propellant mass flow (m) and its exhaust velocity (C) equals the thrust (T) generated by a rocket: $T = mC$. Thrust is the force that enables a rocket to overcome gravity, but what determines the vehicle's effectiveness in this goal is the efficiency or performance of its engines. Rocketeers often measure rocket engine performance in units of specific impulse (Isp). This is usually expressed in pounds of thrust generated per pound of propellant used per second.

Early rocket engines, including those developed by Goddard, were relatively inefficient. It wasn't until the end of World War II, when scientists began to work systematically on rocket technology in greater numbers, that engines capable of attaining the speeds of over 17,500 miles per hour (28,000 km/h) required to send a rocket into orbit became possible.

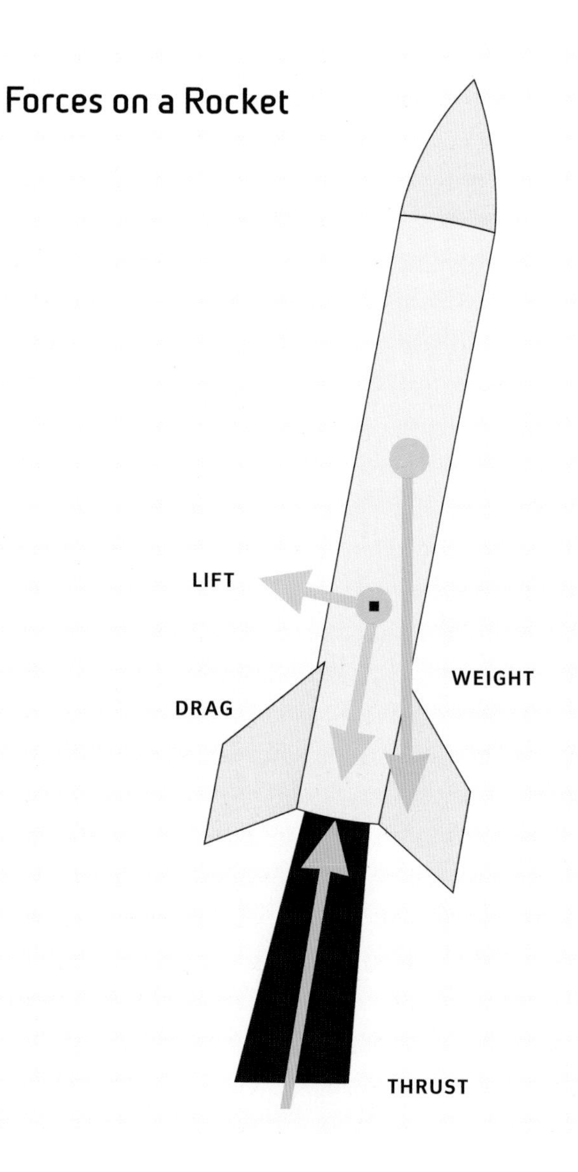

The forces at work in a rocket system.

DE MOTU CORPORUM

46 PHILOSOPHIÆ NATURALIS

PROPOSITIO X. PROBLEMA. V.

Gyretur corpus in Ellipsi: requiritur lex vis centripetæ tendentis ad centrum Ellipseos.

Sunto CA, CB semiaxes Ellipseos; GP, DK diametri conjugatæ; PF, Qt perpendicula ad diametros; Qv ordinatim applicata ad diametrum GP; & si compleatur parallelogrammum $QvPR$, erit (ex Conicis) PvG ad Qv *quad.* ut PC *quad.* ad CD *quad.* & (ob similia triangula Qvt, PCF) Qv *quad.* est ad Qt *quad.* ut PC *quad.* ad PF *quad.* & conjunctis rationibus, PvG ad Qt *quad.* ut PC *quad.* ad CD *quad.* & PC *quad.* ad PF *quad.* id est, vG ad $\frac{Qt\ quad.}{Pv}$ ut PC *quad.* ad $\frac{CDq \times PFq}{PCq}$. Scribe QR pro Pv, & (per Lemma XII.) $BC \times CA$ pro $CD \times PF$, nec non, punctis P & Q coeuntibus, $2PC$ pro vG, & ductis extremis & mediis in se mutuo, fiet $\frac{Qt\ quad. \times PCq}{QR}$ æquale $\frac{2BCq \times CAq}{PC}$. Est ergo (per Corol. 5 Prop. VI.) vis centripeta reciproce ut $\frac{2BCq \times CAq}{PC}$; id est (ob datum $2BCq \times CAq$) reciproce ut $\frac{1}{PC}$; hoc est, directe ut distantia PC. *Q. E. I.*

Idem aliter.

In PG ab altera parte puncti t posita intelligatur tu æqualis ipsi tv; deinde cape uV quæ sit ad vG ut est DC *quad.* ad PC *quad.* Et quoniam ex Conicis est Qv *quad.* ad PvG, ut DC *quad.* ad PC *quad:* erit Qv *quad.* æquale $Pv \times uV$. Unde quadratum chordæ

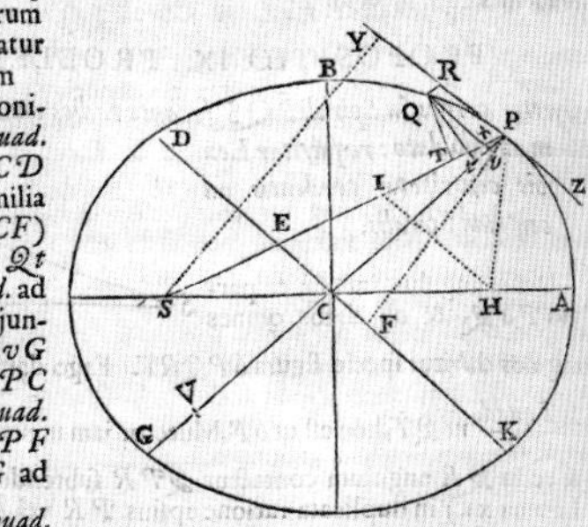

Isaac Newton and the Laws of Motion

1687

Newton first revealed his revolutionary laws of motion to the world in his book *Mathematical Principles of Natural Philosophy*.

The key scientific figure who performed the most important work to establish the theoretical and scientific foundations of space travel is the English polymath Isaac Newton (1646–1727). His theory of universal gravitation explained the movement of the planets, and his laws of motion set out the principles that underpin how rockets work. Newton's laws of motion stated:

1. An object at rest will remain at rest until some force acts on it. When a force does act, the object will remain in that state of uniform motion until another force acts on it.

2. The acceleration of an object is directly proportional to the net applied force and inversely proportional to the object's mass.

3. For every force, there is an equal and opposing force.

It is the third law, often simplified as "for every action, there is an equal and opposite reaction," that describes the core process at the heart of rocket-powered flight: exhaust gases from the internal combustion of fuel being forced out of a nozzle at one end of a rocket creates the thrust that propels the rocket in the opposite direction.

In his theory of universal gravitation, Newton also identified that all bodies in the universe, down to the very smallest, possess mass and exert gravitational attraction. The Earth's gravity pulls objects to its surface, but the gravity of other bodies also exerts attraction. Newton went on to show that the attraction of the Sun on a planet is directly proportional to the product of the two masses and inversely proportional to the square of the distance between them.

Together Newton's ideas gave the scientists who followed him the tools to understand how the universe works, and suggested that a sufficiently powerful flight technology would be able to overcome gravity. The search to find that technology eventually resulted in the development of powerful liquid-fueled rockets.

ABOVE LEFT A 1778 engraving of Isaac Newton, by William Sharp.

ABOVE RIGHT A page from Newton's *Mathematical Principles of Natural Philosophy*, in which he first explored the ideas that would become known as his laws of motion.

The First Liquid-Fueled Rocket

Robert Hutchings Goddard was the son of a machine-shop owner and was fascinated by flight from a young age. After excelling at school, he went on to pursue an academic career in science, working principally at Clark University in Worcester, Massachusetts, where he taught and continued to pursue his own research into rocketry.

A diagram showing the differences between more efficient liquid-fuel rockets used in space exploration and solid-fuel rockets that are used more frequently in suborbital ballistic missiles.

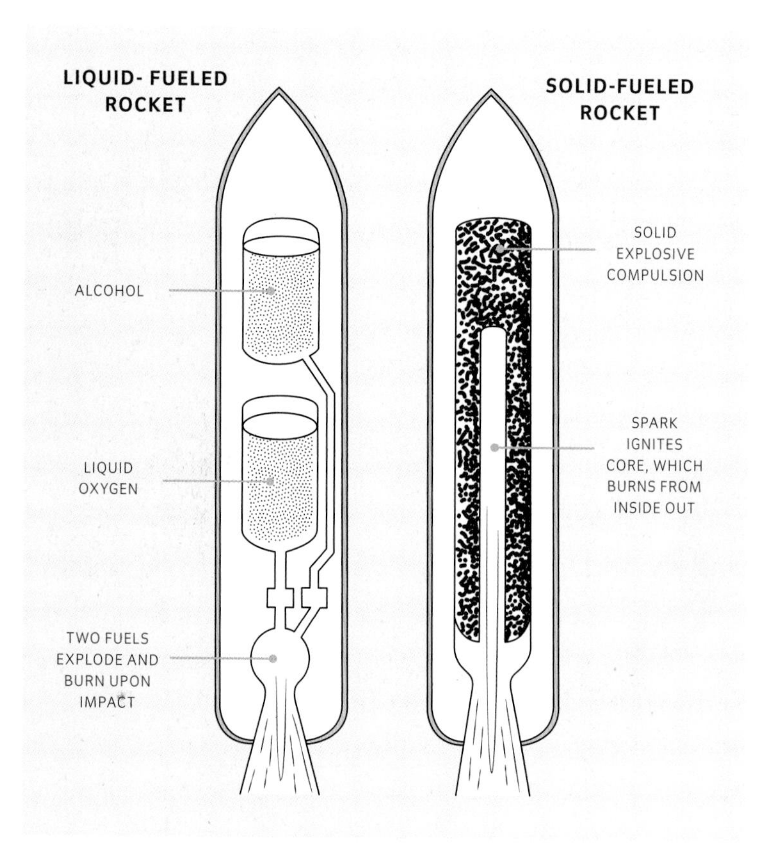

To support his experiments with gunpowder-based solid-fueled rockets (p. 38), Goddard applied to the Smithsonian Institution for assistance in 1916 and received a $5,000 grant from its Hodgkins Fund. This money helped facilitate the publication of his classic 1919 study "A Method of Reaching Extreme Altitudes," in which he detailed his "escape velocity" equation, stating that a flight speed of 6.95 miles per second is necessary to overcome Earth's gravity. Goddard also argued publicly that humans could reach the Moon using his rocketry techniques. His ideas were derided at the time in a *New York Times* opinion column, and it was not until humans finally landed on the Moon in 1969, 24 years after Goddard's death, that the *Times* finally issued a correction, effectively acknowledging that Goddard had been right all along.

Although the negative publicity prompted Goddard to become more secretive and reclusive, it did not stop his work. In September 1921, he began experimenting with liquid oxidizers that would eventually lead to his pioneering first liquid-fuel powered rocket flight in 1926. Goddard was subsequently awarded $50,000 from the Daniel Guggenheim Fund to continue his work. The money enabled him to move to Roswell, New Mexico, where he could more easily build and test even larger rockets, writing up the results of these experiments in his 1936 study "Liquid-Propellant Rocket Development."

It wasn't long afterward that the United States Navy began to express an interest in Goddard's research. It seemed the American military was about to get involved in developing its own rockets (p. 54).

FAR LEFT This is a bronze reproduction of the Congressional Gold Medal that was awarded posthumously to Robert Goddard in 1959 for his pioneering work in rocket development from the early 1900s until his death in 1945.

LEFT This four-chamber rocket is being prepared for test launch at Goddard's rocket launch site in Roswell, New Mexico, in 1940.

The First Liquid-Fueled Flight

1926

Robert Goddard pioneered the use of liquid fuels in rocketry with his historic test launch.

Goddard is credited with building the first liquid-fueled rocket, which he launched on his aunt's farm in Auburn, Massachusetts, on March 16, 1926. Its key innovation was its use of gasoline for fuel and liquid oxygen as an oxidizer to increase the efficiency of its combustion. This combination had the advantage of producing more thrust per unit rate of its fuel consumption than a solid-fueled rocket.

Goddard's small rocket was not particularly impressive to look at. Constructed mainly from metal tubes, it resembled a child's climbing frame, with its engine ignition chamber actually suspended some way above the fuel and oxidizer tanks. It didn't fly far or for a long time, reaching a height of just 41 feet (12.5 m) before crashing into a cabbage patch just 184 feet (56 m) from the spot it lifted off from 2.5 seconds earlier, but it was the rocketry equivalent of the Wright Brothers' first powered flight at Kitty Hawk in 1903.

This now-famous image shows Robert Goddard with his first liquid-fueled rocket just before its successful launch on March 16, 1926.

The Beginnings of German Rocketry

> *"The rockets...can be built so powerfully that they could be capable of carrying a man aloft."*
>
> HERMANN OBERTH, GERMAN ROCKET PIONEER, IN *THE ROCKET INTO INTERPLANETARY SPACE* (1923)

RIGHT First published in Germany in 1923 as *Die Rakete zu den Planetenräumen*, Hermann Oberth's landmark book is known in the English-speaking world as *The Rocket into Interplanetary Space*.

BELOW Portrait of Hermann Oberth, the great mentor of German rocketry, as a young man.

From a young age, Hermann Oberth had a great passion for science fiction and the works of Jules Verne in particular (p. 17). But it was not until after his World War I service in a German medical unit that he decided to study physics and pursue a career in science.

For his doctoral thesis, he chose the subject of rocketry, but his work was rejected by his doctoral supervisors. Undeterred, Oberth went on to pursue a career as a teacher and, in 1923, he had his thesis, *The Rocket into Interplanetary Space*, published as a book. Oberth's book deals with the mathematical theory of rocketry, and offers both designs for practical rockets and speculation on the potential for human spaceflight.

The success of this title put Oberth in contact with many others interested in rocketry, including Austrian spaceflight advocate Max Valier (1895–1930). With Oberth's blessing, Valier condensed and adapted Oberth's book into a much more accessible and popular format under the title *The Advance into Space*. Valier's book helped inspire a wave of rocket clubs to spring up all over Germany as rocket enthusiasts tried to put Oberth's theories into practice. The most widely influential of these new groups was undoubtedly the Society for Spaceship Travel.

Established on July 5, 1927, the society quickly built up a significant base of support, publishing a magazine and various scholarly studies as well as constructing and launching small rockets. Oberth continued to act as a sort of mentor for the group, but perhaps the society's key strength was its ability to publicize both its experiments and the notion of space exploration.

Under Valier's direction, the society developed and built an experimental rocket-powered race car in 1929. Working with the automobile designer and scion of the Opel car manufacturing dynasty Fritz von Opel (1899–1971), Valier continued to experiment on and race his rocket-powered vehicle. Both men shared an eye for publicity and rocket-powered cars. The spectacle of these cars racing at ever faster speeds helped maintain public interest in rocket technology, but such demonstrations were not without risk. In May 1930, Valier was killed in an explosion during the test of an alcohol-fueled rocket engine in Berlin.

Away from Valier's rocket car experiments, the rest of the society had continued with their non-automotive rocket tests. The group's next major success came on February 21, 1931, when it launched the liquid oxygen-methane-fueled rocket HW-1 near Dessau, Germany, to an altitude of approximately 2,000 feet (610 m). Among those present at this launch, the first for a liquid-fueled rocket in Europe, was an eager young rocketeer named Wernher von Braun (1912–77). Von Braun was both enthralled with the flight and impressed with the publicity it engendered, and he would later go on to gain plenty of fame and fortune himself as he assumed an even more prominent role in the development of rocketry in both Germany and eventually the United States (p. 54).

The Mirak Rocket

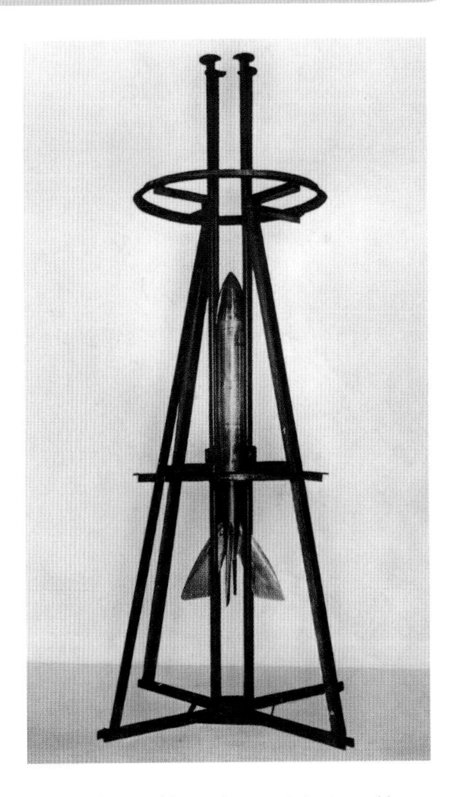

The rocket and launch stand designed by Hermann Oberth and Rudolf Nebel for their Mirak rocket. The Mirak rocket, which was flown over 100 times between 1931 and 1932, was instrumental in convincing the German military that rockets could serve as invaluable weapons of war.

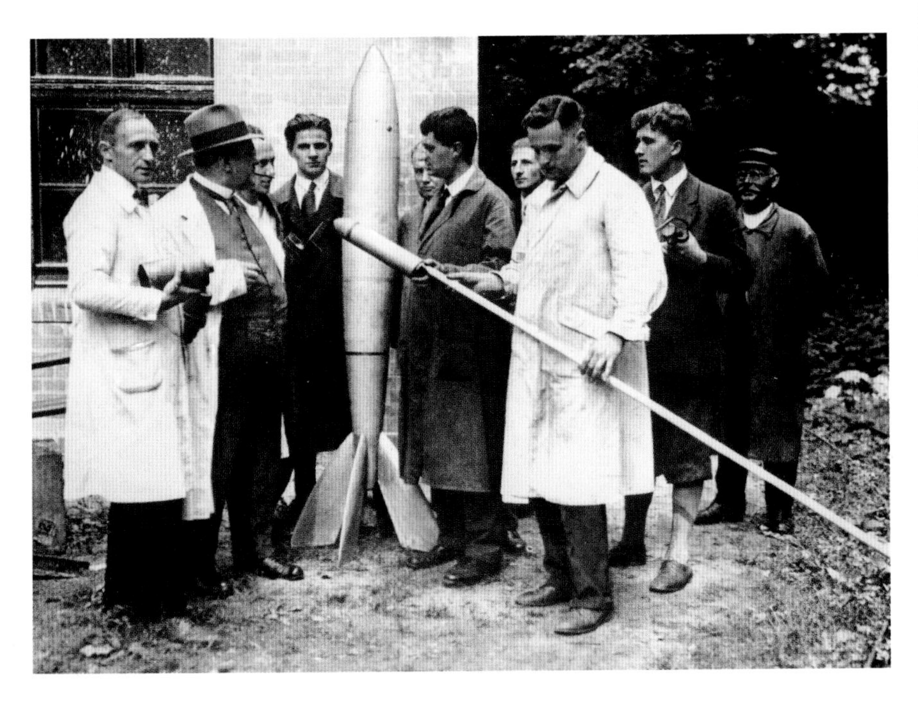

The Society for Spaceship Travel at work in 1930. Left to right: Rudolf Nebel, Franz Ritter, Mr. Baermueller, Kurt Heinisch, Hermann Oberth, Klaus Riedel, Wernher von Braun, and an unidentified person. Klaus Riedel holds an early version of their Mirak rocket.

Selling Space Exploration

During the 1920s, a new craze for all things extraterrestrial seemed to grasp the global popular imagination. Science fiction and science fact increasingly appeared to be reinforcing each other to make the dream of spaceflight seem closer and more achievable than ever before.

Filmmakers following in the footsteps of the French cinema pioneer Georges Méliès (1861–1938) took space exploration to millions around the globe. Méliès's 13-minute-long 1902 film *A Trip to the Moon* depicts a group of travelers climbing inside a cannon shell–like capsule that is literally fired at the Moon. Upon arrival, the travelers are swiftly captured by lunar inhabitants, escape, and return to Earth by pulling the capsule off a cliff. Both the director and the star of the movie, Méliès employed a dramatic array of visual effects to bring his story to life. His creation was instantly hailed as a masterpiece on its release but later fell into obscurity before being rediscovered in the 1930s.

In the Soviet Union, the 1924 silent movie *Aelita*, directed by Yakov Protazanov (1881–1945), was one of the earliest feature-length science fiction films to depict an actual rocket. In the story, a man named Los and some fellow Soviet citizens travel to Mars. The visitors soon learn that the planet is ruled by a brutal hierarchy composed of the empress Aelita and her ruling elite, the Elders. Eventually the Soviets inspire the workers to revolt and rise up against their oppressors. In the ensuing conflict, Aelita is killed. But in a shocking twist ending, the film's final scenes reveal the whole Martian escapade was merely our hero's daydream.

Another widely influential sci-fi film of the silent era was the 1929 German feature *Woman in the Moon*, directed by Fritz Lang (1890–1976). Lang wanted the space travel depicted in his movie to be as accurate as possible and so turned to the German rocket expert Hermann Oberth (p. 30) to be his main technical advisor. Oberth and science writer Willy Ley (1906–69; p. 82) helped Lang with his sets and built a spacecraft that looked realistic. The film depicted a two-stage rocket long before such technology existed and even featured a countdown to increase dramatic tension ahead of the rocket flight sequence.

BELOW The iconic, and slightly gory, image of the Man in the Moon from the 1902 film *A Trip to the Moon*.

OPPOSITE LEFT The movie poster from 1924 silent Soviet sci-fi film *Aelita*.

OPPOSITE RIGHT This scene from the 1929 film *Woman in the Moon* depicts a rocket resting on the Moon's surface. The film's attention to the details of rocketry was superb.

The *Buck Rogers* Phenomenon

1928

An American pop-culture icon whose name became a byword for tales of space travel and adventure.

Created by Philip Francis Nowlan (1888–1940) in 1928 as the hero of the novella *Armageddon—2419*, published in the science fiction magazine *Amazing Stories*, Anthony "Buck" Rogers is a young engineer who becomes trapped in a Pennsylvania mine in the year 1927. Knocked out by "radioactive gas," he remains unconscious for nearly 500 years, awakening in 2419 to discover the America he knew in ruins and under attack.

Starting on January 7 the following year, Nowlan's story was adapted into a syndicated comic strip drawn by Dick Calkins (1894–1962) under the title *Buck Rogers*. It was an immediate success and was followed by the *Buck Rogers* radio program in 1932, which aired four times a week for the next 15 years. Next came a movie, *Buck Rogers in the 25th Century: An Interplanetary Battle with the Tiger Men of Mars*, and then a 12-part serial in 1939 with Buster Crabbe (1908–83) in the title role. There were TV series in the 1950s and 1980s and countless toys, models, video games, and memorabilia, all of which helped *Buck Rogers* retain its prominent position in popular culture to this day.

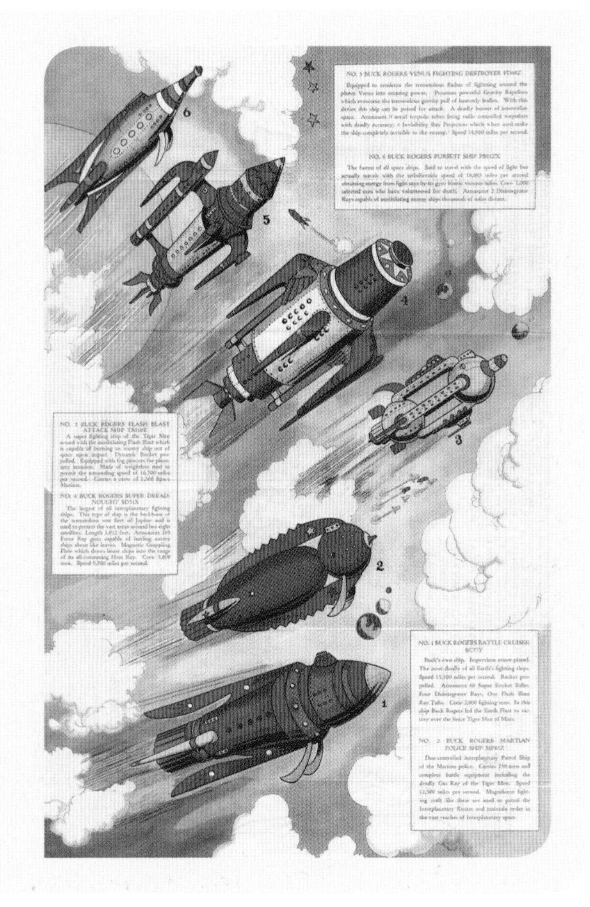

This advertisement depicts the six toy spacecraft of the Buck Rogers Battle Fleet, which could be ordered from the Buck Rogers Company of Chicago.

Rocket Societies around the World

During the 1930s, worldwide rocket groups and societies began to spring up across the world. Inspired by both an interest in science fiction and the practical experiments of the Society for Spaceship Travel in Germany (p. 30), rocketeers in the United States formed the American Interplanetary Society (AIS), later renamed the American Rocket Society (ARS), in New York in 1930. On November 12, 1932, the group undertook its first static test of a liquid oxygen-gasoline-fueled rocket, and on May 14, 1933, they successfully launched their first rocket. While this initial attempt attained an altitude of only 250 feet (7.5 m), the group's next launch, on September 9, 1934, flew up over 1,300 feet (400 m). Thereafter, due to the great cost and risks involved, the group confined itself to further static engine tests, publishing its results in its newsletter.

Meanwhile, on America's West Coast, a group of students at the California Institute of Technology (Caltech), led by Frank Malina (1912–81), began their own experiments with rockets. Called the Suicide Squad because of their reputation for dangerous activities, the Caltech group began to develop so-called sounding rockets for high-altitude atmospheric research. During the early 1930s, they performed static tests for various rocket engines with mixed results. It was not until November 28, 1936, that one of their rocket motors actually ran at all, and even then it was only for 15 seconds. It took the group more than another year to gather enough results to put together Caltech's first scholarly paper on rocketry, "Flight Analysis of the Sounding Rocket," in the *Journal of the Aeronautical Sciences* in 1938. Their results indicated that with proper fuels and motor efficiency, a rocket could be capable of ascending to a height of 1,000 miles (1,600 km).

The British Interplanetary Society (BIS) came into being on October 13, 1933. More oriented toward theoretical studies than practical rocket experimentation, throughout the 1930s the BIS became a haven for writers and other intellectuals interested in space exploration. Its periodical, the *Journal of the British Interplanetary Society*, began publication in January 1934 and quickly became a powerful proponent of space exploration.

BELOW LEFT Diagram of the AIS-2 rocket engine that successfully launched on May 14, 1933.

BELOW RIGHT Rocket Motor No. 4 was used in the American Rocket Society's test flight on September 9, 1934.

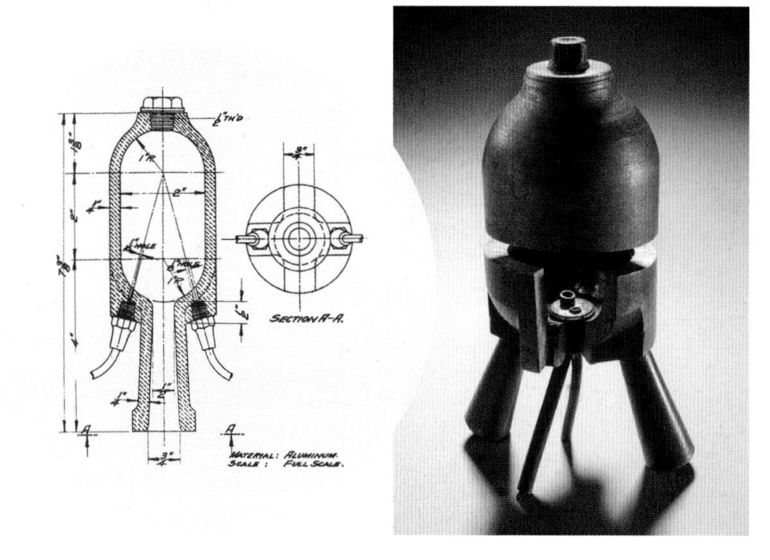

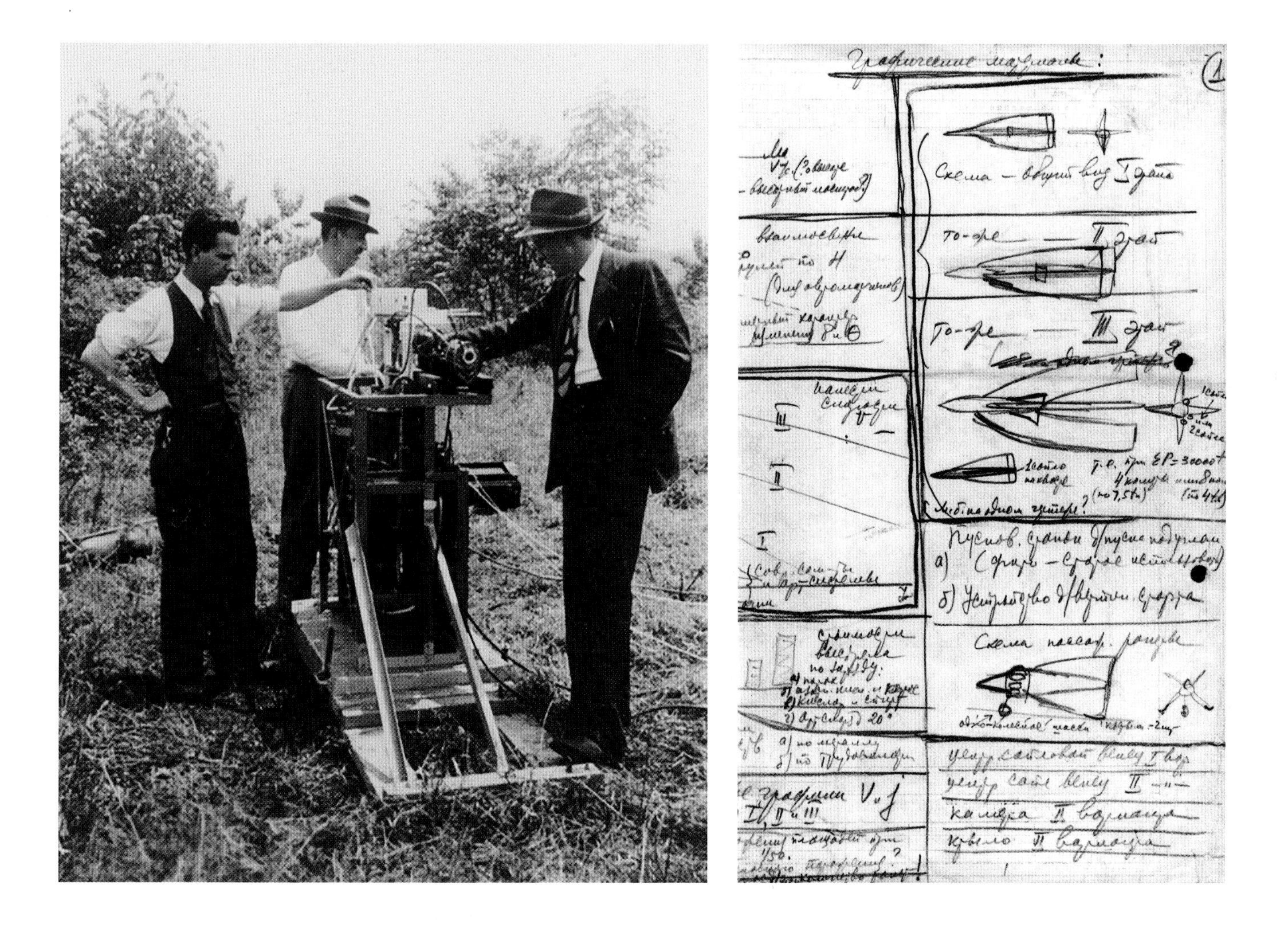

In France, instead of a society of enthusiastic theorists or experimenters, it was one of the country's top aircraft designers, Robert Esnault-Pelterie (1881–1957), who became the leading exponent of space exploration. Independent of Tsiolkovsky (p. 22), Esnault-Pelterie formulated his own equation for calculating the energies required to reach the Moon and nearby planets. He went on to publish a collection of his theories under the title *The Astronautics* in 1930 and also helped establish the world's first prize for astronautics, awarded from 1929 to 1939.

Finally, in the Soviet Union, also taking a cue from the German rocket societies, Sergei Pavlovich Korolev (1907–66) helped found the Moscow rocketry organization known as the Group for Investigation of Reactive Motion, or GIRD, in 1930. GIRD lasted only two years before the Soviet army, recognizing the military potential of rockets, brought the group under its control, renaming it the RNII (Reaction Propulsion Scientific Research Institute). The RNII developed a series of rocket-propelled missiles and gliders during the 1930s, but Korolev was denounced and imprisoned in 1937 during one of Stalin's purges. Despite this, he would later go on to become a key figure in Soviet rocketry (p. 260).

ABOVE LEFT M15-G1 rocket engine being tested on the American Rocket Society's Test Stand No. 2 in June 1942. Left to right: Hugh Pierce, John Shesta, and Lovell Lawrence, who would go on to become three of the founders of Reaction Motors, Inc.

ABOVE RIGHT Some of Sergei Korolev's hand-drawn rocket designs and notes, created long before he became the design mastermind of the Soviet space program.

2

World War II Paves the Way for Space Exploration

World War II changed everything in the field of rocket development. Prior to the conflict, progress in rocketry had been erratic, but the looming war served to highlight the military potential of the technology. Most of those with rocket development experience were swiftly identified, contracted, or recruited, given clear objectives, and supplied with funds and resources to aid their work. The experiments of the various rocket societies around the world largely went on hiatus as many of their leading members, such as Wernher von Braun in Germany (p. 53), Sergei Korolev in the Soviet Union (p. 70), and Robert Goddard in the United States (p. 28), all became entangled in the war efforts of their respective nations.

By the early 1940s, virtually every country involved in World War II had developed some type of rocket technology. Many looked to rockets to provide an effective form of artillery, just as Sir William Congreve's rockets had in the early nineteenth century (pp. 19–21). Some developed smaller one-man rocket launchers to help infantry soldiers destroy enemy tanks and fortifications. Germany and Japan even produced rocket-powered aircraft.

But it was undoubtedly Germany that had the greatest success in rocket development over the course of the conflict. Wernher von Braun and his close-knit team of engineers built the world's first ballistic missile at the Peenemünde research facility on Germany's Baltic coast in 1942. Officially designated the A-4, it is now better known around the world as the V-2 rocket (p. 48). Outside all of the work on rocket development itself, the war also prompted influential research into the effects of high-altitude, high-speed, and long-range flight on the human body, all of which would prove invaluable to later space programs.

OPPOSITE A captured and restored V-2 rocket on display at the National Air and Space Museum.

Using Rockets as Artillery

Katyusha was the nickname Red Army troops gave the multibarreled rocket launchers Soviet forces used during World War II, several of which are seen here unleashing a rocket salvo on the enemy in 1945.

World War II saw a massive expansion in research and development in rocketry as various combatant nations sought to harness this emerging technology against their foes. As in previous centuries (p. 18), artillery proved a popular rocket application, with virtually every nation involved in the conflict fielding some type of rocket-based artillery, or barrage rockets, to rain down explosives on their enemies from afar.

One of the most prominent examples of this kind of artillery rocket system was the Soviet Katyusha. Developed by the Reaction Propulsion Scientific Research Institute (RNII) in the late 1930s (p. 35), each launcher was capable of holding 10 or more tubes filled with 6-foot- (1.8-m-) long solid-fueled rockets, each carrying almost 50 pounds (20 kg) of explosives in its warhead. A highly mobile weapon, the Katyusha could be fired from the ground or from a mounting on the back of a truck, enabling the Soviet Red Army to bring their extra firepower to bear over a wide range of terrain with greater ease.

First tested in the late 1930s, the weapon proved extraordinarily effective when massed and used in saturation bombardment. A battery of four truck launchers could deliver 4.35 tons (4 t) of high explosives on a single sector in around 10 seconds, with little or no warning. Such salvos used in combination with Red Army infantry advances made the Katyusha a tactical weapon of the first order.

The United States Army also fielded mobile rocket launchers in combat, and equipped their allies with this technology. The T34 multiple rocket launcher, which could be mounted on an M4 Sherman tank, fired a barrage of 4.5-inch (114-mm) rockets from 60 launch tubes mounted on the tank's turret. Developed in 1943, soldiers quickly nicknamed it the Calliope because it looked like a calliope steam organ. While the T34 saw limited frontline service, its big brother, the T34E2, which fired 7.2-inch (183-mm) rockets, saw more action. The greater firepower the T34E2 provided proved exceptionally useful for attacking fortifications and enemy tanks.

Some nations took the whole notion of mobile artillery platforms a stage further by mounting rockets on aircraft for ground-attack or air-to-air defense purposes. For example, the British developed the RP-3 (Rocket Projectile 3-Inch) for an air-to-ground attack role. British aircraft fired RP-3s against tanks, trains, transports, and fortifications. The German military used the R4M rocket and the Ruhrstahl X-4 air-to-air and air-to-ground missiles to attack Allied bomber forces over Europe, and occasionally to attack objectives on the ground, too. As a further defense against the threat of Allied bombers, Germany pursued the Rheintochter R1—named in reference to the mythical water nymphs from Richard Wagner's opera cycle *Der Ring des Nibelungen*. Built by the Rheinmetall-Borsig company for the Luftwaffe, the Rheintochter was one of the largest solid-fuel rockets of the war. However, its development program proceeded slowly, which led to its cancelation during the waning months of the war.

The British had more success with developing surface-to-air missiles that could bring down enemy aircraft. They developed a 3-inch (7.6-cm) rocket-propelled

The Benefits of Rocket Artillery versus Conventional Artillery

- Rocket launchers produce little or no recoil when firing, while conventional gun artillery systems produce significant recoil.
- Rocket launchers are capable of delivering mass strikes, launching multiple rockets from the same launch system simultaneously. Conventional artillery typically fires one shell at a time.
- Rocket launchers are less vulnerable to enemy counterattack because of their tendency to leave clear smoke trails.
- Rocket launchers tend to be much lighter and more mobile than conventional artillery systems.

March 22, 1945: Two crew members of an American Third Army tank load rockets into a T34 Calliope 4.5 in a multiple rocket launcher mounted on the top of their armored vehicle.

How the Bazooka Got Its Name

The 2.36-inch- (60-mm-) caliber Rocket Launcher M1 got its nickname, the bazooka, from its similarity in shape to the homemade trombone created by American radio comedian Bob Burns (1890–1956), which the comic referred to as a bazooka.

Rocket Launcher M1 was originally designed by United States Army officer and rocket enthusiast Colonel Leslie Skinner (1900–78). His superiors were extremely impressed by the weapon's potential and swiftly ordered it into production. By May 1942, General Electric had built 5,000 ready for use in combat.

aerial-defense system that could be deployed in two different ways. The more conventional, and usually more successful, deployment launched a rocket with a warhead that exploded at a predetermined altitude, usually the same as that favored by German bombers. The second approach had defenders fire a rocket that was attached to an extremely long length of cable. As the rocket rose, the long cable would unwind at high speed from a bobbin attached to the ground. As the rocket reached the apogee of its trajectory, it would deploy a parachute, slowing its descent. The trailing cable would then be perfectly positioned to shear off the wings or propellers of any enemy aircraft that crossed its path.

Various naval forces also adopted rocket technology during the conflict for use in both anti-aircraft defense and surface-to-surface bombardment ahead of amphibious assaults. In the Pacific, the United States Navy relied heavily on rocket artillery from their amphibious Landing Ship Medium (LSM) transports during their assaults on various islands during the Pacific campaign. Japan also used naval rockets, deploying their Type 4 8-inch (20-cm) rocket launcher against American troops at Iwo Jima and the Battle of Luzon in 1945. While they did not inflict much damage, the rockets registered a significant psychological effect on the soldiers who encountered them, who called them "Screaming Mimis" because of the loud and upsetting sound they made as they neared impact.

Collectively, the rockets deployed during World War II demonstrated the technology's awesome potential for war, but, perhaps more importantly, they also helped a generation of scientists and engineers around the world build a practical knowledge base about how to construct efficient and effective rocket engines.

The United States Navy firing a rocket barrage at the shores of Pokishi Shima, near Okinawa, five days before the U.S. invasion, May 21, 1945.

Shoulder-Mounted Rocket Launchers

Invaluable anti-armor weaponry for the common infantry soldier.

The need for powerful yet portable weapons that could stop a tank became painfully obvious to Allied commanders in the early years of World War II, when Panzers roared through the Netherlands, Belgium, and France with comparative ease. Small shoulder-mounted rocket-propelled explosives gave the average infantryman in the field the necessary firepower to halt, or at the very least slow down, these heavily armored vehicles. Accordingly, every Allied nation developed some form of tube-launched solid-propellant rocket with a high-explosive anti-tank (HEAT) warhead. Such weapons came in various designs; the Americans led the way with what they called the bazooka, but German and Italian forces captured examples of these weapons early in the war and reverse-engineered them to build their own versions. Engineers in the United Kingdom developed their own rocket launchers for infantry, called PIATs. They fired anti-tank rockets and were famously used against the German Panzers defending the bridge across the Rhine River at the Battle of Arnhem in 1944.

ABOVE LEFT An American soldier in 1943 taking aim with his bazooka, a highly durable and light anti-tank weapon.

ABOVE RIGHT The small rockets used in bazookas came with explosive warheads specifically designed to breach the armor of enemy vehicles.

Rocket Aircraft in World War II

Postwar concept art showing the He 176, the first rocket-powered aircraft, in flight in 1939.

With aircraft playing an increasingly important role in warfare during World War II, many nations involved in the conflict devoted significant resources to developing ever more advanced forms of airplane design.

Among the Allied powers, scientists and engineers in both the United States and the United Kingdom studied the concept of rocket-powered aircraft but built no prototypes. The idea was far more thoroughly investigated by the Axis powers of Germany and Japan, where rocket-powered aircraft were viewed as a potentially war-winning technological advance. Rocket planes offered the possibility of faster aircraft capable of attaining a higher altitude, and consequently a greater range, than the propeller-driven aircraft of the time. Both countries invested heavily in building rocket planes for use in combat, but ultimately found, to their own cost, their limited effectiveness in this role.

Germany flight-tested the world's first rocket-powered aircraft, the Heinkel He 176, on June 20, 1939. It had a liquid-fueled rocket engine that was far more efficient than those developed earlier in the decade by the various rocket societies around the world (p. 34). On its first flight, Luftwaffe test pilot Erich Warsitz demonstrated the aircraft's impressive speed, reported to have been in excess of 500 miles per hour (800 km/h), for the watching German leadership. But in other respects, the He 176 did not perform so well; its control system was not a good fit for the aerodynamics of a rocket-powered aircraft, and it carried only enough fuel to fire its engine for less than five minutes. After its fuel ran out, the aircraft was effectively a glider, making it worse than useless in a dogfight. Development on the plane came to an end soon afterward.

The Germans had more success with the Messerschmitt Me 163 Komet. Built as an interceptor to defend against the growing Allied bomber offensive over Germany in 1944, the Me 163 was designed by legendary aeronautical engineer Alexander Lippisch (1894–1976). It was the only rocket-powered fighter aircraft ever to become operational during the war and was also the first to exceed 700 miles per hour (1,125 km/h) in level flight. But despite the fact that over 300 of these aircraft had been built by the war's end, the Me 163 did not make a decisive impact on the Allied air forces, registering fewer than 20 enemies shot down in combat. The Luftwaffe's assumption had been that the Me 163's superior speed would give it an advantage against American B-17s, and even their fighter escorts, but the increased speed came at the cost of incredibly poor fuel efficiency that severely limited the Me 163's range and prevented it from remaining over the battlefield for more than a few minutes. Worse, the aircraft's hydrazine hydrate and methanol fuel was highly corrosive both to the Me 163 and to its pilots, and risked dissolving both when its fuel tank failed.

Toward the end of World War II, the Imperial Japanese Army Air Service engaged in the desperate tactic of kamikaze attacks, in which pilots deliberately crashed their explosive-laden planes directly into their enemies. The Japanese

Eugen Sänger and Irene Bredt-Sänger

German aircraft designer Eugen Sänger (1905–64) and wife Irene Bredt-Sänger (1911–83). Here they are shown discussing the Silverbird, the dynamic, soaring space plane they developed in 1936. Irene Bredt-Sänger was an internationally respected engineer, mathematician, and physicist in her own right and became one of the founding members of the International Academy of Astronautics in 1960.

A Messerschmitt Me 163 Komet on the runway.

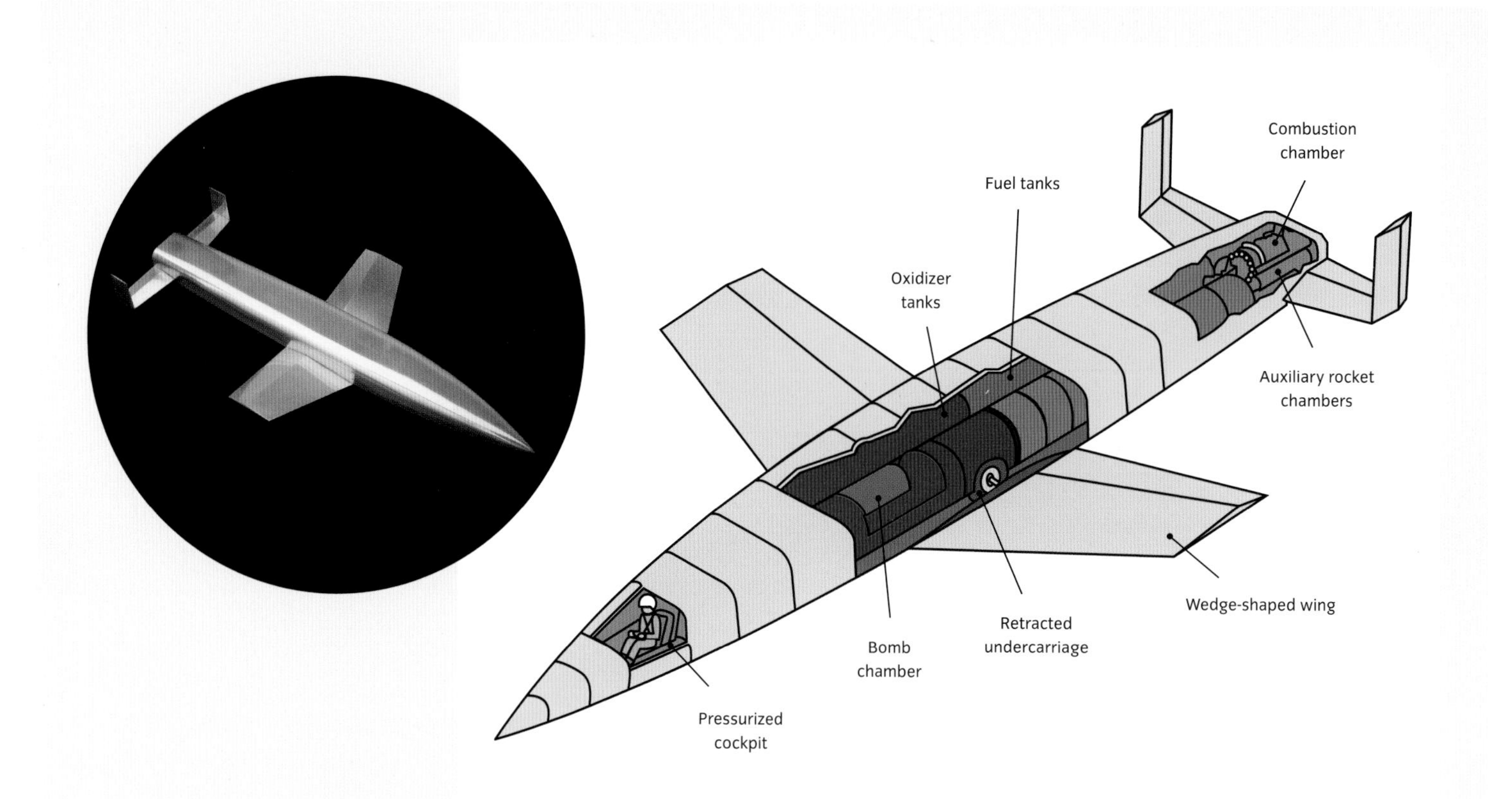

The Silverbird

1929–44

The influential long-range rocket plane that never quite made it into the sky.

Perhaps the most influential rocket plane to be designed but never built before the end of World War II was the proposed intercontinental space plane the Silverbird. Conceived by the Austrian aerospace engineer Eugen Sänger while he was still a doctoral candidate at the Viennese Polytechnic Institute in 1929, the Silverbird was conceptualized as a reusable, rocket-powered plane that could cover more than 5,000 miles (8,000 km) in a single flight. An extremely long range meant the Silverbird would be capable of reaching across continents, making intercontinental bombing raids a tantalizing possibility. This alone made Sänger's space plane a milestone in the early studies of suborbital winged spaceflight. Sänger privately published a detailed description of the Silverbird in 1933in a volume entitled *Techniques of Rocket Flight*, and he would go on to refine his design in collaboration with his wife, the mathematician Irene Bredt-Sänger, over the next 30 years.

Powered by a liquid-fueled rocket engine using kerosene and liquid-oxygen propellants, the Silverbird would have been capable of reaching 6,600 miles per hour (10,620 km/h) at an altitude of 100 miles (160 km). Sänger also predicted that his aircraft would be able to greatly increase its range and efficiency using a technique he termed dynamic soaring: literally skipping the vehicle off the atmosphere at extremely high altitudes. As the vehicle followed an arcing ballistic trajectory in the upper atmosphere, each skip would slightly reduce the forward momentum. Consequently, each skip would be a little lower and shorter until it could no longer be maintained. Despite the obvious appeal of the Silverbird's long range, the demonstrable success of the V-2 garnered more support from Germany's wartime leadership, and the Silverbird's development was sidelined as a result.

After the war, Sänger and his wife eventually moved to America, where the pair continued their rocket-powered aircraft research.

ABOVE LEFT An aerodynamic model for the Silverbird from 1940.

ABOVE RIGHT A cutaway for a possible Silverbird bomber.

mostly used a variety of piston-driven propeller aircraft in these attacks, but also developed a rocket plane, the Yokosuka MXY7 Ohka, for rocket-powered suicide missions. Approximately 850 of these aircraft had been built by the end of the war, but although they represented a deadly danger to their targets, the kamikaze attacks failed to alter the course of any battle in which they were deployed.

While the rocket planes that were designed, tested, and flown during World War II generally proved to be of limited use in combat, their legacy is far from one of failure. The development of rocket-plane technology greatly increased the knowledge base around rocket-powered aviation, both among the nations that had originally pursued the research during the war and among the nations that gained access to it in victory. It was only after the end of the conflict, when the immediate wartime pressures over limited resources abated, that a new group of scientists and engineers were able to explore the noncombat potential of the technology.

Rocket-Powered Aircraft, 1928–45

Type	Country	Role	Date	Comments
Lippisch Ente	Germany	Research	1928	First rocket-powered aircraft
Opel RAK.1	Germany	Research	1929	First purpose-built rocket-powered aircraft
Cattaneo Magni RR	Italy	Research	1931	Test aircraft
Cheranovsky RP-1	Soviet Union	Research	1932	Liquid-fuel rocket-powered glider; tests ended in 1933
Korolyov RP-318	Soviet Union	Research	1936–40	Designed by Sergei Korolev, with further testing and development work conducted in 1938. Flown in 1940
Heinkel He 176	Germany	Research	1939	Liquid-fuel, rocket-powered test aircraft
DFS 194	Germany	Research	1940	Rocket-powered glider
Bereznyak-Isayev BI-1	Soviet Union	Fighter	1942	Short-range interceptor
Messerschmitt Me 163	Germany	Fighter	1944	Tailless rocket-powered interceptor used in World War II
Messerschmitt Me 263	Germany	Fighter	1944	Developed from Me 163
He P.1077 Julia	Germany	Fighter	1944	Proposed rocket-powered interceptor; not built
Focke-Wulf Volksjäger	Germany	Fighter	1944	Three units under construction at the end of the war
Ju EF.127 Walli	Germany	Fighter	1944	Proposed rocket-powered interceptor; not built
Northrop XP-79	United States	Fighter	1944	Flying wing. Converted to jet power for first and only flight
Mitsubishi J8M	Japan	Fighter	1945	Reverse-engineered version of the Messerschmitt Me 163
Rikugun Ki-202	Japan	Fighter	1945	Improved version of the Mitsubishi J8M with an elongated fuselage for longer flight endurance
Mizuno Shinryu II	Japan	Interceptor	1945	End of the war halted development

The First Pressure Suits

BELOW Diving suits were the inspiration for early flight pressure suits, but they were heavy and bulky for flight and had to be redesigned. Here is a 1930s side view of a diver in a pressure suit descending into water from the side of a boat.

OPPOSITE Test pilot Joe Walker in his pressure suit with the X-1E. Such pressure suits were the starting point for the subsequent development of spacesuits.

Soon after they first began to regularly attain altitudes in excess of 12,000 feet (35,000 m) in the nineteenth century, early balloonists discovered that they required plenty of warm clothing to protect against the extreme cold found at high altitudes. On November 4, 1927, the balloonist and United States Army officer Hawthorne Charles Gray (1889–1927) lost consciousness and died in a balloon flight near St. Louis, Missouri, after reaching an altitude of 42,470 feet (12,800 m). His demise from asphyxiation highlighted the low oxygen levels found in the air at high altitude.

Eventually, scientists identified the area between sea level and around 10,000 feet (3,000 m), in which most humans can function normally, as the "physiologically efficient zone." Above that, and especially at altitudes of over 12,000 feet (3,600 m) above sea level, the human body often struggles to draw sufficient oxygen from the air. This can swiftly lead to hypoxia, a dangerous lack of oxygen in the body. Prolonged hypoxia frequently leads to lasting damage to the affected tissue and organs and can even prove fatal.

In the early 1930s, the American air balloonist Mark Ridge worked with Scottish physiologist John Scott Haldane to build a fabric-based full pressure suit. Despite some promising early tests in ground-based pressure chambers, Ridge's suit was never deployed in flight. The first practical pressure suit design came from engineer Yevgeny Chertovsky in the Soviet Union. Tested in Leningrad in 1931, Chertovsky's initial design lacked joints and was subsequently difficult to move in once pressurized. Shortly afterward, the high-altitude American aviator Wiley Post (1898–1935) worked closely with engineers at the rubber goods manufacturer BFGoodrich to design a jointed pressure suit consisting of a helmet, a pressurized rubber bladder, and an outer layer made of rubberized parachute fabric. This primitive pressure suit enabled Post to fly a plane above 30,000 feet (9,000 m), as he demonstrated in his 1934 flight over Chicago.

Two years later, on September 28, 1936, Royal Air Force Squadron Leader Francis Ronald Downs Swain (1903–89) set a new world record for high-altitude ballooning at 49,967 feet (15,000 m) while wearing a different kind of suit that maintained a near-constant atmospheric pressure and made provision to help keep its occupant sufficiently warm at altitude. The Italian aviator Mario Pezzi (1898–1968) wore a similar pressure suit when he reached an altitude of 56,047 feet (17,083 m) in a propeller-driven Caproni Ca.161

biplane in 1938. This remains the all-time world record for the highest altitude achieved by a biplane to this day.

The advent of high-performance monoplane aircraft in the late 1930s also brought with it a new problem for pilots; the apparent weight-altering effects extreme acceleration has on the body as aircraft turn sharply, dive steeply, or climb quickly in the Earth's atmosphere. This pressure on the body is commonly known as g-force. At sea level, humans typically experience 1 g, but by the late 1930s, planes were capable of producing upward of five times that effect on pilots. And those numbers went up considerably during World War II in the new aircraft built for aerial combat. Under such conditions, blood can be forced away from a pilot's brain, causing them to black out and lose control of the aircraft. Throughout World War II, various air forces attempted to devise flight suits that would help reduce the effects of high g-forces produced by extreme maneuvers, and by the war's end, the United States Army Air Force had developed an effective lightweight high-altitude pressure suit. Through lend-lease programs, this suit made its way to most of America's allies in World War II, and it became the forerunner for postwar pressure suits used around the globe. Made of nylon, the USAAF suit was cooled by water fed in through rubber hoses and featured inflatable bladders on each thigh and calf and on both sides of abdomen. This design later also would go on to serve as the basis for most of the first spacesuits.

Early Pressure Suit Chronology

1933 First full pressure suit is manufactured for American balloonist Mark Ridge.

1934 BFGoodrich manufactures a full pressure suit of double-ply rubberized parachute fabric with pigskin gloves, rubber boots, and aluminum helmet for pilot Wiley Post.

1934 Engineers in the Soviet Union begin developing their own pressure-suit designs.

1935 English pilot Francis Ronald Downs Swain breaks two world records with the Mark Ridge Suit.

1935 French engineers M. Rosensteil and Paul Garsaux design a full pressure suit with the backing of the Potex Airplane Company.

1935–45 German engineers design a range of full pressure suits using laminated silk and rubber with a reinforced net of silk cord. The suits maintain pilot mobility but are very heavy.

1937 An Italian pressure suit is flown at 51,000 feet (15,500 m).

1943–46 Engineers at the University of Southern California design the capstan partial pressure suit and test it to altitudes of 80,000 feet (24,000 m). They make three different models that become the standard for high-altitude pressure suits thereafter.

1948–53 The first operational capstan partial pressure suits are produced in custom sizes for early rocket-powered X-Plane test pilots by the David Clark Company.

Rockets as Ballistic Missiles

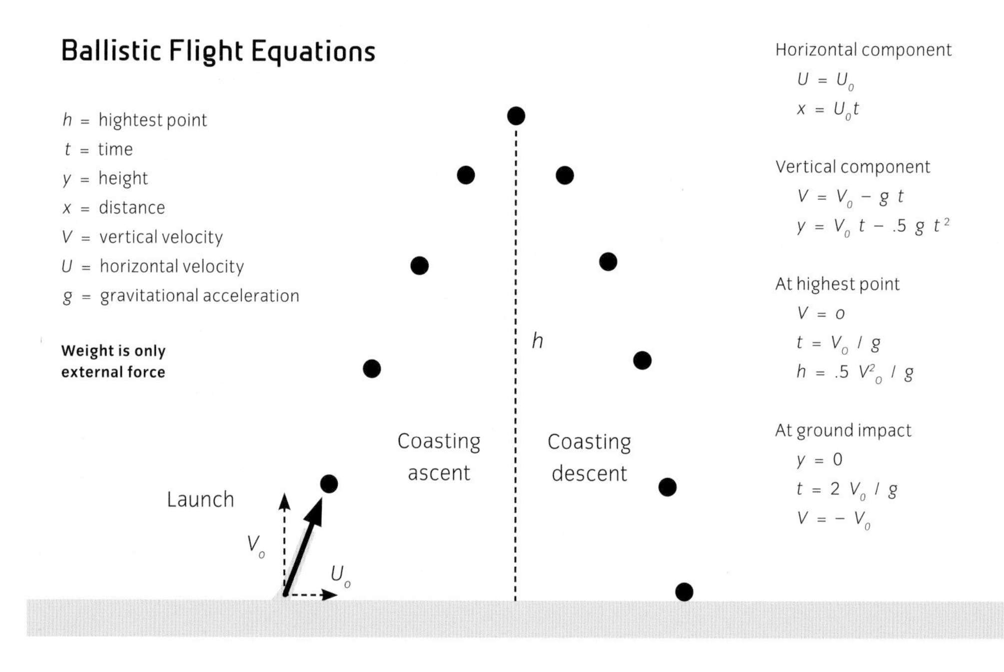

RIGHT The basic calculations involved in assessing ballistic missile trajectories, excluding the factors of drag and thrust.

BELOW The first training and testing battery of V-2 (A-4) rockets, showing three rockets in launch position, probably in Poland, 1943.

The term *rocket* can be used to describe any engine from which propulsive force is obtained by the release of exhaust gases from the combustion of fuel escaping through a nozzle. Rocket-powered devices of various kinds are often also called rockets. They can serve as missiles, projectile weapons hurled through the air at a specific target over relatively short ranges, but the term *ballistic missile* is reserved for projectiles that follow a high arc-like, or ballistic, trajectory to reach their target, often traveling long distances to do so.

The world's first rocket-based ballistic missiles were built in Germany during World War II, largely due to the work and influence of one man: Wernher von Braun. The German military initially learned of von Braun's abilities from his colleagues in the Society for Spaceship Travel (p. 30). In 1932, they persuaded von Braun to work for them by promising to fund his Ph.D. studies and by giving him the chance to build a large-scale research and development program to make rockets. Once von Braun had been recruited, he was placed under the supervision of artillery captain Walter Dornberger (1895–1980). Dornberger immediately saw the potential of rockets to deliver explosives to a target more effectively than conventional artillery and bombers.

Dornberger swiftly assisted von Braun in the initial recruitment of a team of 80 engineers, including some members from the Society for Spaceship Travel, and secured funding from various branches of the German military to start building rockets in the Berlin suburb of Kummersdorf. By late 1934, von Braun's team had designed the first two rocket-powered missiles in the Aggregat series: the A-1 and A-2.

A Sample of V-2 Test Launches

Rocket number	Date	Burn time (s)	Range (km)	Launch pad	Comments
V-1	March 18, 1942	—	0	P-VII	The first A-4 flight test was witnessed by German Armaments Minister Albert Speer, but the rocket fell and was damaged prior to launch. The engine subsequently failed during a test firing (again witnessed by Speer) and was junked without any further launch attempt.
V-2	June 13, 1942	36	1.3	P-VII	Photographed by a British Spitfire on May 15, 1942, the second flight test model was damaged during its fourth firing test on May 20, 1942. A gyroscope malfunctioned immediately after its eventual launch, with the propellant system failing soon afterward. Telemetry ended at 54 seconds while the rocket was at an altitude of 15,000 feet (4,500 m), and the missile crashed into the Baltic and exploded.
V-3	August 16, 1942	45	8.7	P-VII	Nose broke off.
V-4	October 3, 1942	58	190	P-VII	First successful test launch. The rocket reached an altitude of around 56 miles (90 km).
V-5	October 21, 1942	84	147	P-VII	Steam generator failed.
V-6	November 9, 1942	54	14	P-VII	Reached an altitude of 40 miles (67 km).
V-26	May 26, 1943	66.5	265	P-VII	After a series of 20 failed launches, the German government's Commission for Long-Range Firing/Weapons/Bombardment viewed two successful launches.
V-49	October 6, 1943	68	—	P-VII	Successful launch that resulted in a 272-second flight. It was also the first launch after the British bombing raid on the Peenemünde site on August 17, 1943.
MW 17003	January 27, 1944	3	—	P-VII	First test of a V-2 rocket built at the Mittelwerk facility. It detonated three seconds after ignition without liftoff.
V-259	October 30, 1944	—	—	P-VII	Commander-in-chief of the German air force Hermann Göring attended the rocket launch and was so impressed that he declared it "terrific."
MW 21400	February 20, 1945	—	350	Karlshagen	Last V-2 launch at Baltic test site.

General Characteristics of the V-2

1944–52

The low-down on the rocket that influenced all those that followed it.

The historic first successful launch of a prototype V-2 rocket on October 3, 1942, at Peenemünde, Germany.

The V-2 was given its name, short for *Vergeltungswaffe Zwei* (Vengeance Weapon Two), by the Nazi Propaganda Ministry when its existence was first announced to the world in November 1944. As well as being the first ballistic missile, the V-2 rocket was also the first large-scale liquid-propellant rocket vehicle, and as such, it serves as the ancestor of all later rocket-powered space exploration vehicles.

The most important technological breakthrough in the V-2 was its rocket engine, an entirely novel design that proved remarkably versatile. At the time of launch, the rocket's propellants were mixed in the rocket motor and ignited to a temperature of almost 5,000 degrees Fahrenheit (2,760°C). Unignited fuel was forced through a jacket surrounding the rocket combustion chamber, helping to keep it cool, before being pumped back into the main ignition chamber through more than 1,200 pinholes. Regeneratively cooled rocket engine designs subsequently became the standard for all future liquid-fueled rockets.

V-2 Specs

Engine	8,380 pounds (3,810 kg) ethanol/water fuel and 10,800 pounds (4,910 kg) liquid oxygen oxidizer
Launch mass	27,576 pounds (12,500 kg)
Length	46 feet (14 m)
Diameter	5 feet, 5 inches (1.65 m)
Wingspan	11 feet, 8 inches (3.56 m)
Speed	Maximum: roughly 3,580 mph (5,760 km/h)
Range	200 miles (322 km)
Flying altitude	55 miles (88 km) maximum
Warhead	2,150 pounds (977 kg)
Guidance	Gyroscopes for attitude control
Launch platform	Mobile

After the successful launch of an A-2 rocket, von Braun's team received significant political support and resources from various figures and groups both from within the German military and across the broader hierarchy of the National Socialist Party, also known as the Nazis.

In 1937, von Braun's rocket team was moved to a secret laboratory complex at Peenemünde, in the far northeast of Germany's Baltic Sea coast. At its height, the Peenemünde research and development facility was home to more than 5,000 engineers and technicians working on various advanced weapons programs, including von Braun's rockets. By 1941, his team had designed its third missile, dubbed the A-3. The following year, after much demanding work and numerous failures, they designed and built the first true ballistic missile. Designated internally as the A-4, it is better known today as the V-2 rocket. Whereas the A-1 and A-2 rockets had been comparatively modest in stature, each standing at around 5 feet (1.5 m) tall, the V-2 was nearly three times this height, standing at an impressive 46 feet (14 m).

On October 3, 1942, a V-2 rocket roared into the skies over Peenemünde, broke through the sound barrier, and continued up to an altitude of 56 miles (90 km) before arcing downward to crash 118 miles (190 km) away. This was the first successful launch of a true ballistic missile.

But while the V-2's engine was a great success, its guidance systems were never particularly accurate. For navigation, the V-2 used four external tail fins with rudders and four graphite vanes at the exit of the rocket nozzle. These, along with two gyroscopes, provided stabilization and enough general directional control to ensure that the rocket was traveling in the right direction, but little more. This crucial weakness rendered the V-2 more a weapon of terror than one of strategic targeting.

V-2 Combustion Chamber

This V-2 engine shows the rocket's combustion chamber and nozzle. The engine's 18 fuel injectors are visible at the top of the combustion chamber.

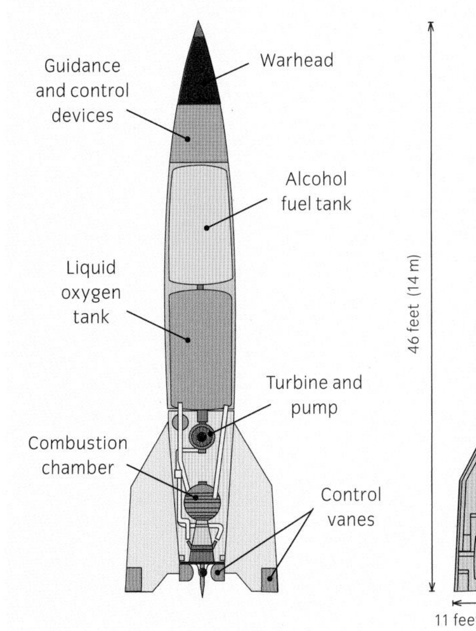

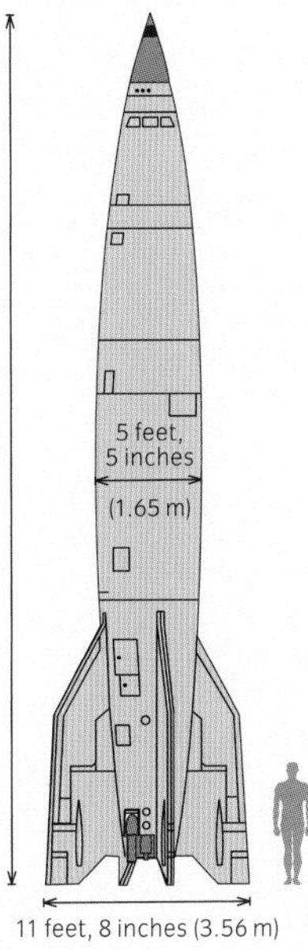

LEFT This cutaway drawing of the V-2 illustrates the dimensions and internal workings of the rocket.

ABOVE German Rocket Society veteran Kurt Heinisch in the control room at Test Stand No. 1, at Peenemünde.

The Human Cost of the V-2

A love of space travel might have driven the young Wernher von Braun to first pursue rocketry, but the V-2 rocket he and his rocket team eventually built was designed as a weapon. While many of the core technologies, including its engine design, fuel, aerodynamics, stabilization, and control systems, could be, and later were, easily adapted to help build a rocket-powered space exploration vehicle, the heavy human cost of the V-2 must be acknowledged.

Once successfully tested, German officials swiftly ordered the V-2 into mass production in 1943. Initially, production work took place at the research and development site at Peenemünde, but following an air raid targeting the facility in 1943, the decision was made to move a substantial amount of the V-2 construction work to the Mittelwerk factory in central Germany. The factory itself was located in a series of underground tunnels close to the Mittelbau-Dora concentration camp. Over 1944 and 1945, the factory built about 6,000 V-2 rockets using slave labor from the concentration camp. Around 10,000 workers from the camp died while building V-2s. Meanwhile, starting from September 8, 1944, and running until the end of the war, a total of 1,155 V-2s were fired toward the United Kingdom, with another 1,675 launched against Antwerp and other continental targets. It has been estimated that by the end of the war, the rocket had killed more than 9,000 people in missile strikes.

Among those more sympathetic to von Braun's technological achievements, many find solace in the anecdote that in March 1944 he was arrested and placed in a Gestapo prison for a few days. The arrest was apparently sparked by some loose talk about what the rocket team wanted to accomplish in space exploration. After a few days in jail, von Braun was released and put back to work. While this incident doesn't negate the engineer's position within the Nazi hierarchy, or exculpate him from the use of slave labor in the construction of V-2s at the Mittelwerk facility, it does at least suggest that he and his associates were more committed to the goal of successfully developing rocket technology than they were to the ideals of the Nazi regime.

The German military prepares a V-2 for a test launch.

V-2 Rocket Engine Firsts

- First rocket to use a turbopump to ensure fuel supply to combustion chamber.
- First rocket to burn water/alcohol fuel using a liquid-oxygen oxidizer.
- First rocket to use a cluster of 18 small propellant injectors rather than a single large injector.
- First rocket to produce 56,000 pounds. (249,100 N) of thrust.

General Erich Fellgiebel (left) congratulates members of the von Braun rocket team from Peenemünde for their successful October 3, 1942, V-2 flight. While there are no clear images of Wernher von Braun in military uniform, he was a major in the Allgemeine SS.

Wernher von Braun's Double Legacy

1912–77

A legendary rocketry genius who spearheaded the development of a deadly missile with a design that would influence almost every space rocket that followed.

Wernher von Braun was one of history's most important rocket developers and champions of space exploration. His career spanned from the beginnings of rocketry in the late 1920s through the Moon landings and on to the end of the Apollo program in 1972. Born in Wirsitz, Germany, on March 23, 1912, the son of German noble Magnus Maximilian von Braun, the young Wernher was a spaceflight enthusiast from an early age, inspired by the science fiction of Jules Verne and H. G. Wells (p. 16) and the scientific writings of Hermann Oberth (p. 30). Von Braun maintained his interest in rockets as a teenager, becoming involved in the amateur rocket society scene in 1929. In 1932, he went to work for the German military to build rockets, and continued this work even when Hitler came to power in 1933. He went on to receive his PhD in aerospace and engineering, eventually joining the Nazi party himself in 1937. By 1941, his team had developed designs for the ballistic missile that eventually became the V-2. Von Braun's legacy is mixed. He created both a weapon responsible for the deaths of literally thousands of people, and a launch vehicle that would give rise to the rockets that would eventually take humanity into space.

ABOVE Childhood picture of Wernher von Braun (right) with his brothers, Magnus and Sigismund.

America's Wartime Missiles

RIGHT In Arroyo Seco, Los Angeles, in 1940, Theodore von Kármán (black coat) sketches out a plan on the wing of an airplane as his jet-assisted takeoff (JATO) engineering team looks on. Left to right: Clark Millikan, Martin Summerfield, Theodore von Kármán, Frank Malina, and pilot Captain Homer Boushey.

BELOW Project director Frank Malina poses with the fifth WAC Corporal at the White Sands Missile Range, October 11, 1945.

While the German government was enthusiastic in its support of Wernher von Braun and his team's pursuit of rocket technology from the early 1930s onward, backing for similar research in the United States was much less fulsome and comprehensive over the same time frame. The leading American institution to get involved in rocket research at this time was the Guggenheim Aeronautical Laboratory (GALCIT), a branch of the California Institute of Technology (Caltech), under the leadership of mathematician and aerodynamicist Theodore von Kármán (1881–1963). In 1936, the gifted aeronautics graduate student Frank Malina set up a rocket research program at GALCIT. Although their early experiments and engine tests met with limited success, they impressed General Henry Harley "Hap" Arnold (1886–1950), soon to be chief of United States Army Air Corps, when he visited the facility in 1938.

In 1939, the international political climate began to darken in the lead-up to World War II. Despite his misgivings about working on weaponry, Malina and his fellow GALCIT researchers began to explore the potential of rocketry research for military purposes. This led to the National Academy of Sciences funding Malina to continue his research with the specific goal of developing jet-assisted takeoff (JATO) rockets to assist heavy aircraft to take off from short runways. The United States Army knew that many of the airfields from which they had to fly had short runways.

Rockets attached under an aircraft's wings could rapidly accelerate any airplane up to the necessary flight speed, even on a relatively short runway.

By late 1941, the rocket pioneer Robert Goddard (p. 28) had joined the United States Navy's Bureau of Aeronautics (BuAer), bringing his considerable rocket knowledge, technical facilities, and test data to the JATO project, which eventually reached fruition in November 1943. Around this time, the U.S. military started to receive intelligence relating to the existence of the V-2 rocket (p. 48). The success of the V-2 prompted the U.S. military to increase its own rocket research, with the United States Army Ordnance Corps signing a long-term contract for further research on large missiles with GALCIT. Soon afterward, GALCIT changed its name to the Jet Propulsion Laboratory (JPL) and started to become more independent of Caltech, serving more as a military facility.

On January 16, 1945, Malina sent the Army Ordnance Corps a proposal for a liquid-fueled, high-altitude atmospheric research rocket of the kind he had been working toward before the war. This rocket, dubbed the WAC Corporal and considered the little sister of the larger Corporal missile system then under development, would be the first true ballistic missile built in the United States, capable of launching a 25-pound (11-kg) payload of scientific equipment to an altitude of 100,000 feet (30,000 m). The rocket would also go on to play a significant role in America's early space exploration efforts (p. 62).

U.S. Army Air Corps Key Missile Achievements

- Developed America's first operational guided missile, the Republic-Ford JB-2.
- Created first weapon system to put a kiloton nuclear warhead in the hands of a field army.
- First missile launched in 1947; first missile operational in 1954; taken out of service in 1966.
- Total weapon system weight about 100 tons (101 t). It required a 250-man battalion and 15 vehicles to support it.
- Launch time about six to eight hours after site selection.

Robert Goddard Goes to War

1941–45

Giving the American war effort a rocket-powered lift.

Notoriously reclusive, Robert Goddard was not a natural team player. His reticence toward collaborative work changed only when America was drawn into World War II. Upon joining the United States Navy's Bureau of Aeronautics (BuAer) in 1941, he aided the successful completion of the JATO project, and by 1943 he was working on pumps and turbines for nitric acid-fueled rockets at the Naval Engineering Experiment Station in Annapolis, Maryland. His work also went into the development of the XLR-11 rocket engine that propelled the first piloted aircraft to fly faster than the speed of sound, the Bell X-1 in 1947 (p. 76).

Goddard learned of the success of von Braun's group and their V-2 rocket only in 1944. He was apparently stunned by their achievements, perhaps missing how much collaborative effort had also factored into the V-2's creation.

Robert Goddard examines the engine of a V-2 rocket in Annapolis, Maryland, April 14, 1945.

Capturing Nazi Research, Personnel, and Technology

As the defeat of the Axis powers loomed toward the end of World War II, various groups within the Allies began to focus their attention on the postwar world. They had observed the technological advances made by many working within the Nazi regime and now sought to secure the underlying research and the scientists and technicians who had made these advances at the end of the conflict.

By late 1944, Wernher von Braun, the technical director of the V-2 rocket team, had recognized the impending German defeat. He quickly determined that he did not want to be captured by the Red Army, which by January 1945 was rapidly closing in on the secret missile development facility of Peenemünde in northeastern Germany. Ahead of the Red Army's advance, von Braun and his most trusted engineers departed the Peenemünde site with their families. They loaded trucks with technical documentation and, traveling mostly in small groups, went to southern Germany. Since von Braun's brother Magnus spoke the best English, he

ABOVE LEFT Members of the United States Army Air Force's Scientific Advisory Group on May 14, 1945, at the Kaiser Wilhelm Institute, Göttingen University. Left to right: Hugh Dryden, National Advisory Committee for Aeronautics; Ludwig Prandl, pioneering German aerodynamist; Theodore von Kármán, Caltech; Hsue-Chu Tsien, JPL.

ABOVE RIGHT A V-2 rocket arriving in Trafalgar Square, London, to take part in the London National Savings Week campaign during 1945.

was sent out to search for members of the United States Army advancing from the west. He eventually located an American soldier and facilitated the surrender of his entire group. Wernher von Braun and about 120 of his colleagues and their families were soon spirited away to the United States, along with all their technical papers.

By the spring of 1945, the United States and the Soviet Union were effectively engaged in a race to retrieve as many V-2 rockets, parts, technical information, and personnel as possible from Peenemünde and other sites around Germany. The American military succeeded in capturing more than 300 railcar loads of V-2 parts, which were shipped off to America. The Soviet Red Army captured Peenemünde itself, and with it numerous V-2 rocket parts and technicians.

By the summer of 1945, military intelligence officials in the United States started work on an effort that would later become known as Operation Paperclip: a plan to recruit and expatriate a wide range of key scientific and technical personnel, including von Braun's group, from Germany to America. Many of those secured would eventually go on to prominent roles in the American space program. Among them was Hubertus Strughold (1898–1986), a physician well known for his investigations into the strain extreme environments place on the human body. While many of his colleagues faced war crimes trials for their experiments using human test subjects from the Dachau concentration camp, Strughold soon found himself at the School of Aviation Medicine at Randolph Field (later Randolph Air Force Base), San Antonio, Texas. There he moved from aviation to aerospace biomedicine, helping to found the field of science that is dedicated to keeping astronauts and cosmonauts alive in space.

"If someday rockets should open the gates to space, then we must be oriented as to its environmental conditions and their possible biological effects. To study this, and to find means of protection, is one of the tasks of space medicine."

HUBERTUS STRUGHOLD,
GERMAN-AMERICAN
SPACE PHYSICIAN

Photograph taken shortly after Wernher von Braun and his rocket team surrendered to American forces in early May 1945. Von Braun's arm is in a cast due to an injury he had recently sustained in a car accident.

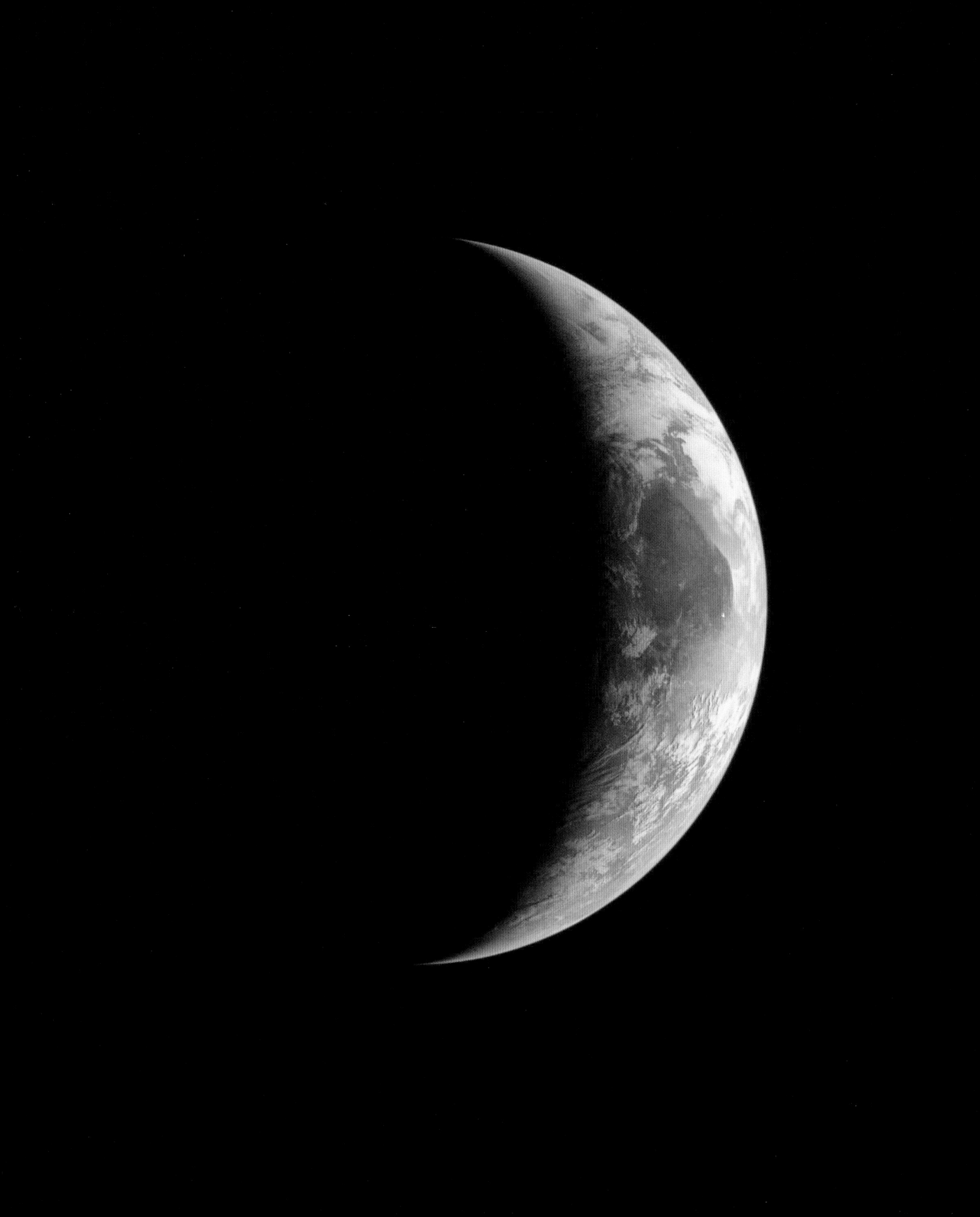

3

Making Space Exploration Real

The dozen or so years that immediately followed the end of World War II witnessed the first glimmerings of the dawning of the space age. They were also years that saw tension, distrust, and conflict rapidly escalating between the Soviet Union and its former wartime allies. Both the United States and the Soviet Union had emerged from the war technologically and politically stronger, and hostility rapidly grew between these emerging superpowers, leading to the bitter rivalry known as the Cold War.

The Cold War pushed both superpowers into increasing funding for weapons research and development. Both the Americans and the Soviets rushed to unravel the secrets of the V-2 rockets they had each recovered from Germany. The German rocket mastermind Wernher von Braun greatly aided the Americans (p. 62) as he attempted to gain their full confidence and support. Von Braun also helped build a number of new rocket designs and worked to popularize the idea of manned space flight among the American public (p. 82). In the Soviet Union, Sergei Korolev and his teams of engineers made great strides in adapting von Braun's work themselves.

At the same time, nations around the world continued their wartime work in developing new designs of aircraft to break the sound barrier (p. 51). The first to actually succeed in this goal was the American Bell X-1, a rocket-powered aircraft that kicked off an entire lineage of extraordinary, experimental X-planes (p. 76).

The fast pace of developing technology soon helped to make the notion of space exploration seem a closer and more realistic proposition than ever before, and a new wave of movies featuring space travel promised audiences that an age of human space exploration was just around the corner.

OPPOSITE Photograph of crescent Earth taken from Apollo 11 during the return trip to Earth, July 21, 1969.

Where Does Space Begin?

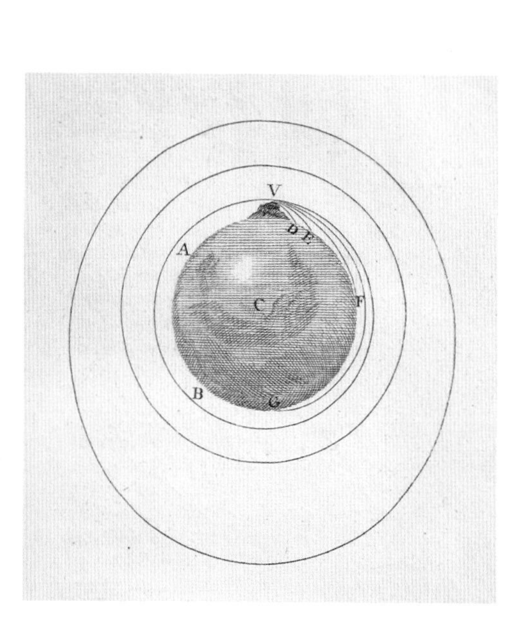

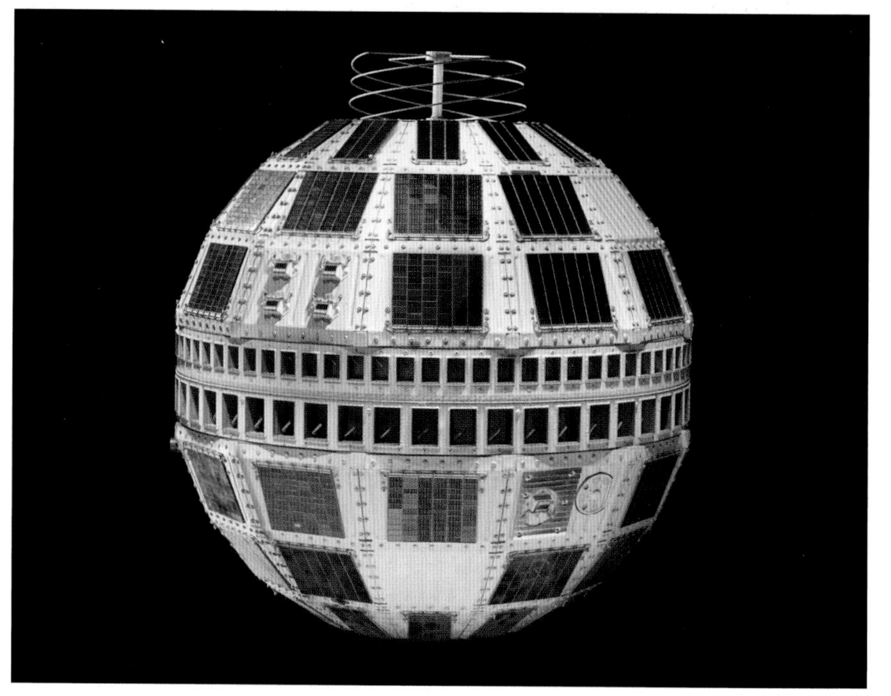

ABOVE LEFT This illustration from Isaac Newton's *The System of the World* (1728) demonstrates that Earth orbits are a consequence of gravitational forces. The velocity of an object in orbit serves to cancel the effect of gravity pulling that object back toward the larger body.

ABOVE RIGHT Launched on July 10, 1962, Telstar 1 was the world's first commercial communication satellite. It enabled the first transatlantic television transmission between the United States and France, but its electronics were soon compromised by radiation damage caused in part by the Van Allen radiation belts, resulting in the satellite's deactivation in February 1963.

OPPOSITE LEFT Artificial satellites may be placed in low Earth orbit, medium Earth orbit, high Earth orbit, and geosynchronous orbits. They may also be placed in polar orbits.

It is difficult to determine with precision the exact point where the Earth's upper atmosphere truly ends and the vacuum of space takes over, but the definition most widely recognized today has space beginning at a distance of 62 miles or around 330,000 feet (100 km) above Earth. This altitude is also known as the Kármán line after the founding director of the Jet Propulsion Laboratory (JPL), Theodore von Kármán. Kármán calculated that the atmosphere at such heights was too thin to support any vehicle in aeronautical flight that was not also traveling at sufficient speed to reach orbit.

Even at the Kármán line, the Earth's atmosphere is still dense enough to swiftly drag any spacecraft back to the ground. For a stable orbit, an object or vehicle must be boosted to at least 225 miles (362 km) above the Earth's surface. This altitude and up to around 1,200 miles (2,000 km) is known as low Earth orbit, or LEO. It is the region of space in which all manned space stations and spacecraft, with the exception of the Apollo Command/Service Module and Lunar Module, are designed to operate.

LEO is also the region of space in which most satellites operate, particularly those monitoring the weather or offering two-way voice telecommunications. Objects in a LEO orbit complete an orbit around every two hours. Satellites in medium Earth orbit (MEO) operate at about 12,550 miles (20,200 km) above the Earth. These orbits are especially useful for navigation satellites, such as those that compose the American Global Positioning System (GPS) and other, similar services. Satellites in MEO orbit the Earth between every 2 and 12 hours. Geosynchronous satellites are so named because they orbit in harmony with the rotation of the Earth, remaining above the same spot on the Earth's surface at an altitude of roughly 22,235 miles (35,786 km). As the name implies, it takes a full 24-hour day to complete an orbit. This is the prime location for communications satellites specializing in data transfer, particularly those used to maintain the Internet. However, the latency of any signal traveling to this altitude and back creates delays that are too long for telephone services.

It is important to note that for most satellites, their orbital plane—the route by which they travel around the Earth—seldom aligns with the Earth's equator. Orbital planes are usually inclined at an angle that will give the satellite coverage of a specific geographic area linked to the needs of a spacecraft's mission. Highly inclined orbits make it possible to observe much of the surface of the Earth as the planet rotates under the orbiting satellite. Most satellites designed for such purposes are placed into near-polar orbits that pass over the Earth's North and South Poles.

For a spacecraft to reach and stay in orbit, it must travel very fast. In his posthumously published 1728 work *The System of the World*, Isaac Newton explained how a cannonball fired from atop a mountain could orbit the Earth if it traveled fast enough. In Newton's example, the larger the charge used to fire the cannonball, the faster and farther it would travel before Earth's gravity pulled it to the ground. If it achieved enough speed, roughly 17,500 mph (25,000 km/h), the downward curve of its trajectory would match the curve of the Earth and the cannonball would then be in orbit: continually falling toward the Earth but also traveling fast enough to keep missing it.

It takes a lot of energy to attain the velocity to reach any Earth orbit, but once a spacecraft has achieved orbit, it takes far less energy to de-orbit back to Earth, transfer to another orbit, or even leave Earth orbit entirely. As the old space exploration saying goes, "Once you're in low Earth orbit, you're halfway to anywhere" you might want to visit next.

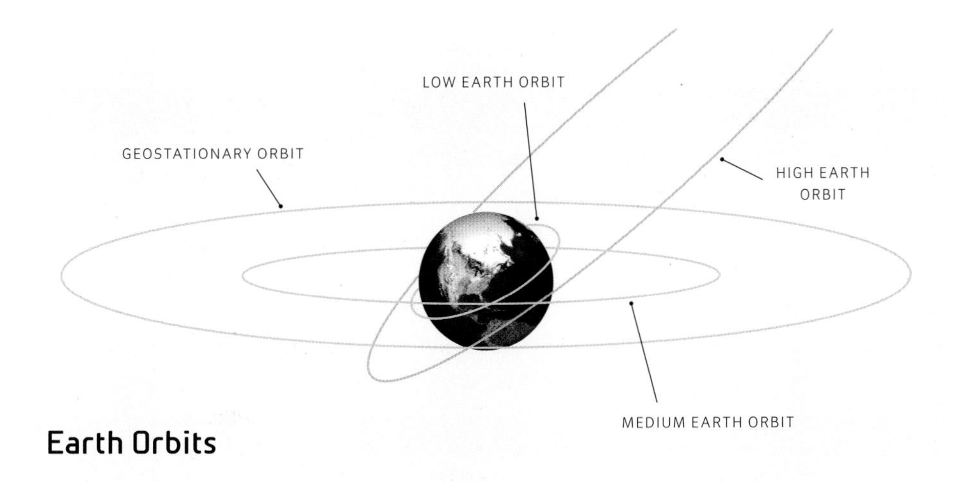

Earth Orbits

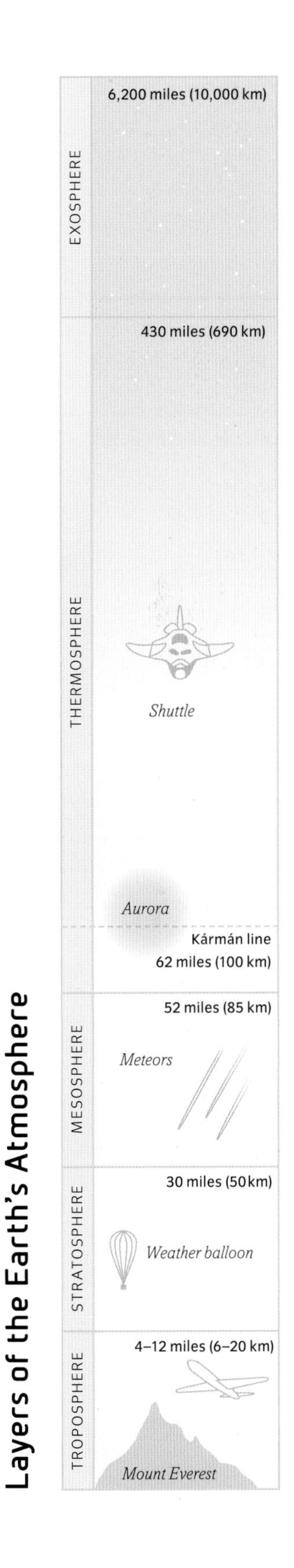

Layers of the Earth's Atmosphere

The V-2 and the Pursuit of Postwar Missile Technology in America

ABOVE Wernher von Braun holding a model of his Jupiter missile in the 1950s.

OPPOSITE ABOVE LEFT A V-2 rocket is hoisted into a static test facility at White Sands, New Mexico in 1947. Many of the German engineers and scientists who developed the V-2 came to the United States at the end of World War II and continued testing the rocket under the direction of the United States Army. More than 60 V-2s were launched in the United States.

OPPOSITE ABOVE RIGHT Bumper 8, shown here during launch on July 24, 1950. The WAC Corporal rocket is mounted on the nose of the larger V-2.

OPPOSITE BELOW The German rocket design team shortly after their arrival at the Redstone Arsenal, Alabama, in 1950.

In September 1945, Wernher von Braun and about 120 of his rocket team were taken to Fort Bliss in El Paso, Texas. There, they were contracted to oversee the building of launch facilities for a new V-2 test program at the nearby White Sands Proving Ground in New Mexico. Called Project Hermes, the program saw the Americans successfully launch the first of their captured V-2s at the White Sands facility on April 16, 1946. Between 1946 and 1951, a total of 67 V-2s were reassembled or adapted for test launches in nonorbital flights at the site, allowing the American scientists and engineers present to significantly expand their knowledge and expertise in rocket design.

Through the rest of the 1940s and into the early 1950s, von Braun's rocketeers took an active role in conducting ever more demanding tests showcasing their rocket's key capabilities, and also its value as a vehicle for gathering scientific data (p. 65). Among the most important tests of the V-2 was Project Bumper. This saw some of the captured V-2s adapted into a first-stage booster rocket to propel the smaller WAC Corporal rocket to a higher altitude than it would have been able to reach on its own. The WAC Corporal had been designed by Frank Malina and his team at the recently established Jet Propulsion Laboratory (p. 54) to obtain data on conditions in the upper reaches of the Earth's atmosphere, but Project Bumper also facilitated the gathering of data from even greater heights as well as enabling the engineers at White Sands to investigate the principles of two-stage rocket design. The first fully successful V-2/WAC Corporal launch took place on February 24, 1949, with the second-stage WAC Corporal rocket reaching an altitude of 244 miles (360 km) and a velocity of 5,150 miles per hour (8,288 km/h). The last two tests in the project took place on July 24, 1950, at the Long Range Proving Ground in Florida, a site that would later become part of the Cape Canaveral launch facility, the primary focus for space exploration launches in the United States to this day.

With the American military having gleaned all they could from the V-2, von Braun and his team were given the opportunity to relocate to the Redstone Arsenal, a United States Army facility in Huntsville, Alabama, in the summer of 1950. Von Braun and a number of his team would go on to become American citizens not long afterward. While at the Redstone Arsenal, von Braun and his team worked on the development of a variety of more advanced rocket designs. These included the short-range Redstone ballistic missile, which served largely as a test project for the longer-range missiles that would follow it. The Redstone was later redeveloped into the Jupiter-C interregional ballistic missile, which would become the launch vehicle used to place the first American satellite into orbit in 1958 (p. 98).

Turning a Weapon into a Research Aid

In late 1945, Wernher von Braun and his team arrived in the United States to help demonstrate their V-2 rockets in action to their new colleagues in the American military (p. 62). The United States Army's main aim for these tests, dubbed Project Hermes, was for their engineers to learn how to build, operate, and improve upon such rockets themselves. But it wasn't just the engineers who benefited from these launches. Thanks to some far-sighted military officers, especially Major Generals Holger Toftoy and John Medaris, the rockets also contained a variety of scientific instruments that provided invaluable space science data.

The addition of payloads of scientific equipment turned the V-2s into bigger and more powerful versions of the WAC Corporal sounding rocket Frank Malina and his JPL team had been working on (p. 54). The selection and organization of these payloads fell to Homer Newell (1915–83) from the United States Naval Research Laboratory (NRL) and the newly established V-2 Upper Atmosphere Panel. As the minutes of its first meeting in February 1946 record, the panel was assembled to "develop a scientific program, assign priorities for experiments to fly on the V-2s, and to advise the Army Ordnance Department on matters essential to the success of the program." Although its exact name and the scope of its responsibilities changed repeatedly during the next several years, the panel continued to coordinate activities with the army's various rocket programs until the birth of the National Aeronautics and Space Administration (NASA) in 1958 (p. 102). One of the panel's first decisions was to prioritize the use of these V-2 launches to study solar and other stellar ultraviolet radiation, the aurora borealis, and the composition of the upper atmosphere. This provided a wealth of new data, including measurements of the ionosphere, solar radiation, and cosmic radiation and the discovery of evidence of micrometeorites in the upper atmosphere. It also represented the very beginnings of the field of space science.

Following the panel's lead, scientists across the United States, and then across the world, soon recognized the massive benefits of sending instrument packages into the upper atmosphere, and later into orbit. Using rocket flights, space experts were soon expanding scientific understanding of the Earth's exact size, geology, and composition, as well as its physical properties. Prior to this, Earth scientists of all disciplines often needed to undertake extensive fieldwork to gather their data and develop their ideas. Scientific sounding rockets made it possible to take a literally global view of their subject from the comfort of their own laboratories. It was a truly revolutionary new capability.

OPPOSITE Preparing a V-2 for launch in 1946 at White Sands Proving Ground, New Mexico. The nose contained a small instrument to measure cosmic rays at the upper reaches of Earth's atmosphere.

BELOW This rugged solar spectrograph was one of the many scientific instruments specially designed and manufactured to be flown on a V-2 rocket.

Peter Hurd

This original 1951 painting by American artist Peter Hurd (1904–84) depicts a Project Hermes launch of a modified V-2 missile at White Sands Proving Ground in New Mexico. The General Electric Company, the main contractor for Project Hermes, commissioned this painting and presented it to those working at the facility, which was renamed the White Sands Missile Range in 1958. The artwork hangs in the range's command suite to this day.

Of the 67 captured V-2s test-launched in the United States between 1946 and 1951, most had some kind of scientific payload aboard. Once the supply of captured German V-2s was expended, American scientists continued to hitch rides for their instruments in WAC Corporal rockets and in the Aerobee and Viking rockets that followed. Other countries later conducted similar rocket-based upper-atmosphere research (p. 74). The importance of this work was probably best expressed by the American geochemist Harrison Brown (1917–86) in 1961, when he said: "It seems likely that in the years ahead we will learn more about the Earth by leaving it than by remaining on it."

ABOVE This metal wind tunnel model of an adapted V-2 missile from the National Air and Space Museum was likely used in preflight testing during Project Hermes.

LEFT Two scientists from the Naval Research Laboratory prepare a cosmic-ray experiment for launch into the upper atmosphere.

The First Images from Space

1946–50

During the early V-2 rocket launches at White Sands Proving Ground, the NRL's John Mengel (1918–2003) led the way in placing 35mm cameras on the rockets to capture photos of the curvature of the Earth. These grainy, black-and-white images represented the first true expressions of the Earth as seen from the high reaches of the upper atmosphere. Sadly, perhaps because of their poor image quality, these early photographs did not immediately capture the imagination of the American public.

Not that the public was the core audience for the images. Recalling one trip to recover the film from a crashed V-2, Army draftee Fred Rulli reported that the scientists with him "were ecstatic, they were jumping up and down like kids." He added that when the film was developed and projected for all to see, "the scientists just went nuts."

Scientists acquired more than 1,000 photos of the Earth from various V-2 flights from altitudes as high as 100 miles (160 km) between 1946 and 1950. The photos received particular attention from meteorologists and geographers who studied the landscape and climate in the expansive southwest United States. Many were also published in newspapers and magazines for the public's appreciation. In 1950, Clyde Holliday, who developed the camera for the missions, wrote in *National Geographic*: "Results of these tests now are pointing to a time when cameras may be mounted on guided missiles for scouting enemy territory in war, mapping inaccessible regions of the Earth in peacetime, and even photographing cloud formations, storm fronts, and overcast areas over an entire continent in a few hours." This was an understatement if ever there was one; photographic images and data from the Earth captured by satellites in orbit are now part of our everyday lives, from weather data to Google Earth images.

Aerial mosaic of Earth taken over the United States and Mexico, March 1947.

Robert Gilruth and Postwar Rocket Research in America

RIGHT The first launch at NACA's Wallops Island facility was a small rocket fired on June 27, 1945. Seven identical rockets were also fired later that day.

BELOW Engineer Robert Gilruth (left) enjoys a moment of levity with astronauts Alan Shepard Jr., John Glenn, and Charles "Pete" Conrad in Houston's Mission Control Center during the Gemini XII mission of 1966.

OPPOSITE BOTTOM This image from 1952 depicts the tabulation of flight test results at the Langley Memorial Aeronautical Laboratory. It was laborious work that was given largely to women.

By the latter part of World War II, the leadership of the American National Advisory Committee for Aeronautics (NACA), the federal agency responsible for aeronautical research, had become interested in the possibilities of rockets for both space science and human space exploration. Working more broadly than Wernher von Braun and his rocket team had done (p. 62), NACA created the Pilotless Aircraft Research Division (PARD) to test various rocket designs alongside other flight technologies. Under the leadership of a promising young engineer named Robert Gilruth (1913–2000), who would later go on to take a leading role in the development of human space exploration in the United States (p. 105), PARD developed significant knowledge about the workings of numerous rocket propulsion systems, the aerodynamics of flying to space, and the guidance and control systems necessary to enable a rocket-powered vehicle to reach Earth orbit.

Primarily based at NACA's Langley Memorial Aeronautical Laboratory in Hampton, Virginia, Gilruth established a rocket test site for his research division on Wallops Island on Virginia's Atlantic Coast. The first launch from the site took place on June 27, 1945, and between then and 1950 Gilruth's team went on to fly a total of 386 rockets of various types from the facility, ranging from simple single-stage rockets to various multistage, hypersonic, and solid- and liquid-fueled launch vehicles.

Like many other organizations in the pre-electronic computation age, Langley employed "human computers," a group of women who would carry out the often laborious and time-consuming mathematical calculation work required in analyzing the data from the various launch tests. Every rocket test was meticulously documented, with measurements of stress on every component of the vehicle at every speed, altitude, and circumstance. This data had to be assessed, analyzed, and placed into a format that could be accessed by other engineers working on follow-up projects. The human computers at Langley were responsible for completing this process, and the result was trailblazing technical reports that often became the principal resource for anyone pursuing subsequent rocket development projects in the United States. The first human computer to start work at the Langley laboratory was Virginia Tucker (1911–85) in 1935, and by 1942 the facility employed 75 women in the role. That number rose further as World War II progressed, with African American women, who had never previously been allowed to work at the laboratory in such a role, being accepted from 1943 onward. The work of these women would expand over time, contributing greatly to the success of America's early human space exploration program.

The progress Gilruth oversaw in developing various rocket technologies and expanding the field of flight aerodynamics made PARD a critical hub of the American quest for space in the 1950s, with Gilruth himself becoming a vocal champion of the idea of human space exploration.

Rocket Tests for Jet Engines

Not all rocket tests conducted at Wallops Island were focused on rocket technology itself. This F-23 research rocket was used between 1950 and 1954 to test the air-breathing ramjet engines mounted on either side of the rocket's body under actual flight conditions.

Postwar Rocket Developments in the Soviet Union

By the end of May 1945, Soviet forces had secured much of eastern Germany, including both the Peenemünde rocket test site and the Mittelwerk factory used by Wernher von Braun and his team. Soon afterward, Sergei Korolev, who was by now serving as a colonel in the Red Army, traveled to Soviet-occupied Germany along with a team of his fellow rocket experts to see what they could learn about German rocket development. Korolev interviewed dozens of German V-2 engineers and technicians who had not been captured by other Allied powers, and Soviet troops recovered enough hardware to rebuild 12 V-2s.

Under Korolev's direction, the Red Army rounded up a group of about 200 German employees of the Mittelwerk V-2 factory on the night of October 22, 1946, and sent them to relatively comfortable living quarters at Lake Seleger, located between Moscow and Leningrad (now St. Petersburg). Led by Helmut Gröttrup (1916–81), a technician with expertise in the V-2's guidance and control systems, these Germans had little direct contact with their Soviet counterparts. Aside from assisting in the launch of a few V-2s, they mainly just answered the Soviet engineers' written questions. The first Soviet V-2 launch took place on October 30, 1947, from the newly established rocket launch site of Kapustin Yar, located some 75 miles (120 km) east of Stalingrad (now Volgograd). More followed. Gradually, as the Soviet engineers grew more familiar with the rocket's design and function, the German technicians were returned to Soviet-controlled territory in eastern Germany.

A rare photo showing Colonels Georgiy Tyulin (left) and Sergei Korolev, the father of the Soviet space program, in Germany in 1946 during their V-2 missile recovery operations. Alongside Korolev, Tyulin would become one of the most important managers of the Soviet space program in the 1960s.

As the Soviet Union attempted to tighten its grip on its wartime territorial gains, its relations with the United States and other former allies worsened. In the face of this, and spurred on by the success of the V-2 rocket tests, Soviet leader Joseph Stalin defiantly signed a decree ordering the development of a series of long-range ballistic missiles for the Soviet Union. His minister of armaments, Dmitri Ustinov, was placed in charge of the project. In turn, Ustinov personally appointed Korolev, whose engineering skills and organizational abilities he admired, to be the chief constructor responsible for delivering the missiles. By August 1946, Korolev was in charge of the newly established Scientific Research Institute NII-88, and his work toward designing the missiles began. First, the V-2's proven design was slavishly copied with all-Soviet components as the R-1. The R-1 was then quickly redeveloped into progressively more powerful missiles capable of carrying heavier payloads over greater distances. The need for such missiles became even more pressing after the Soviet Union conducted its first nuclear test in 1949. Korolev's missiles needed to serve as a delivery system for the Soviet Union's nuclear weapons. By April 1, 1953, as he was preparing for the first launch of his R-11 rocket, Korolev received approval from the Soviet Council of Ministers to further develop his earlier R-7 rocket engine into what would become the world's first intercontinental ballistic missile (ICBM)—a weapon with a range so great that it could literally cross continents to reach its target.

To concentrate on the development of the R-7, Korolev's other projects were reassigned to a new design bureau in Dnepropetrovsk (now Dnipro), a city some 240 miles (390 km) southeast of the Ukrainian capital, Kiev, and headed by Korolev's assistant, Mikhail Yangel. This was the first of several design bureaus that would be spun off and established once Korolev had perfected an innovative technology. Despite being designed as a weapon, Korolev's R-7 rocket was not well suited to its stated role. It required enormous launch pads and complex assembly and launch procedures. Furthermore, the nuclear warhead it had to carry weighed

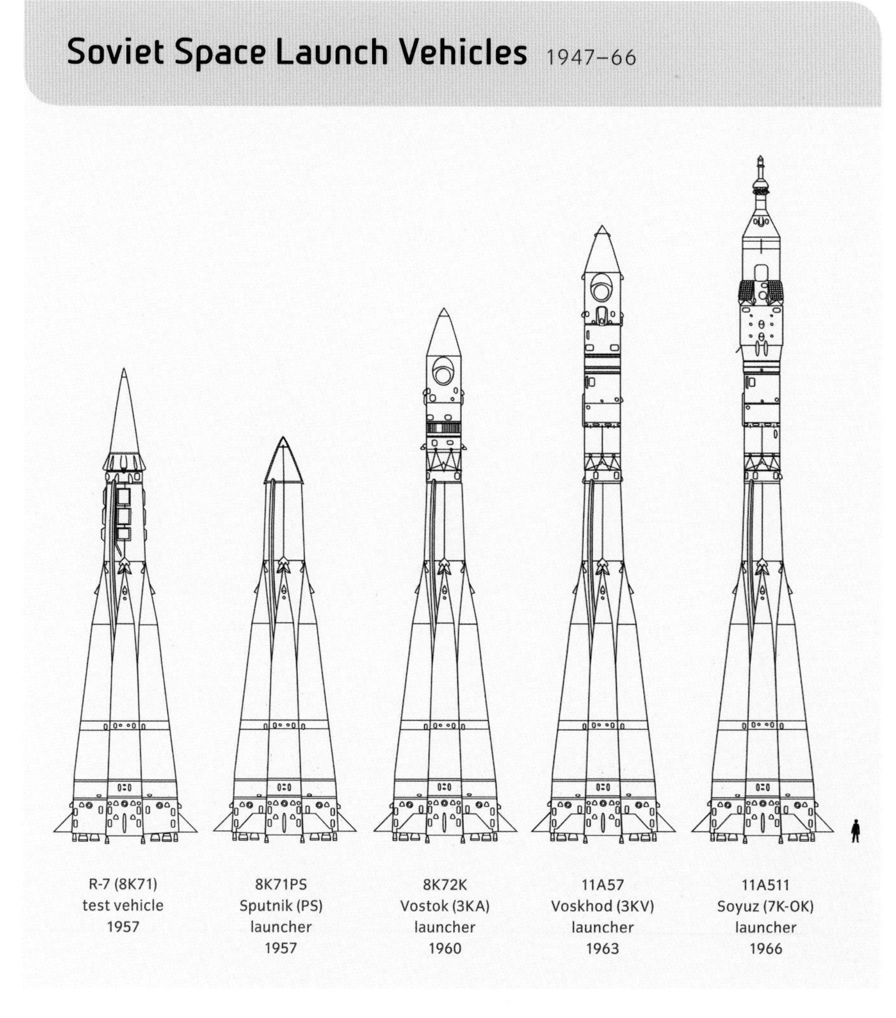

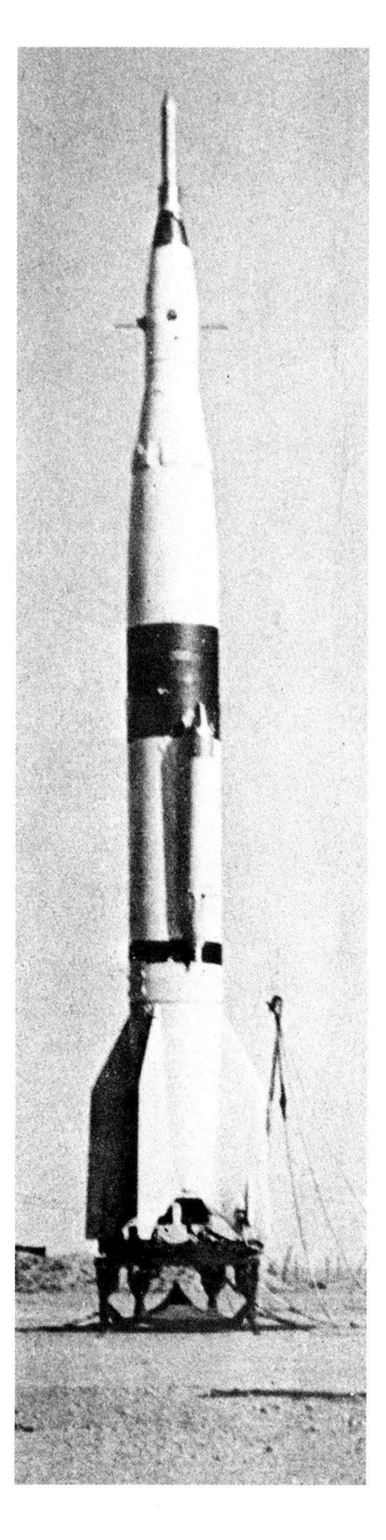

One of the R-2A missiles built by Korolev's engineers. Thirteen of them were launched between 1957 and 1960.

Sergei Pavlovich Korolev

1907–66

The chief designer of the Soviet space program, who masterminded its early successes and the development of its core technology.

Soviet chief designer Sergei Korolev stands at the Kapustin Yar firing range in 1953, the same year that he joined the Communist Party and was elected a corresponding member of the Soviet Academy of Sciences.

Sergei Pavlovich Korolev was an early advocate of space exploration in the Soviet Union. He had experimented with rockets in the 1930s and was eventually recruited to work in what would become the state-funded Reaction Propulsion Scientific Research Institute.

It was at this institute that Korolev developed a series of rocket-propelled missiles and gliders in the 1930s. During the paranoia of the great purges that gripped the Soviet Union between 1936 and 1939, Korolev was accused of deliberately slowing down the work of the institute. A number of his fellow rocket enthusiasts denounced him, and in 1937 Korolev was arrested and tortured until he confessed.

He was then convicted and sentenced to hard labor at a gulag in Siberia, where he lost most of his teeth due to malnutrition. He later received a retrial that reduced his sentence and was moved to a *sharashka*, a prison camp reserved for scientists and engineers. There Korolev designed weapons and even some rockets for the Red Army.

He was finally released in 1944, by which time the Soviet Union was at war with Germany. He joined the army and was sent to recover whatever he could from the German V-2 rocket program at the end of the war.

After designing the Soviet Union's first ICBM, he went on to become a leading figure in his country's space program (p. 90), where his success was enthusiastically embraced and supported by Soviet premier Nikita Khrushchev until his removal from office in 1964. After Khrushchev's ousting, Korolev was still expected to deliver success for the Soviet space program, but he had far less financial backing.

The stress of this reduced funding was a strain on Korolev's health, which had never really recovered from his experiences in Siberia. He died on January 14, 1966, reportedly from a botched hemorrhoid operation, although some sources suggest he was having surgery for cancer. Throughout his life and career, Korolev's identity was closely guarded as a state secret, but in death he received the public recognition he was never allowed in his lifetime. He received a hero's funeral in Moscow and is buried in the Kremlin Wall.

more than 5,600 pounds (2,540 kg). The immense weight of this payload would limit the missile's range to about 3,500 miles (5,600 km), barely enough to reach the northern United States from the Soviet Union, even by the shortest route, over the North Pole. As a result, the R-7 ICBM would be deployed as a weapon at only eight launch pads in northern Russia and would be superseded as a weapon by other rocket designs in the early 1960s. However, with some minor adjustments, the R-7 was far more suited to serve as a launch vehicle for space exploration, as the world would soon discover (p. 90).

Affectionately known as Semyorka (number seven) in Russian, this example of the R-7 is on display at the Exhibition of Achievements of National Economy, Ostankinsky, Moscow.

R-7 Rocket

Specifications

Weight
308 tons (280 t)

Length
112 feet (34 m)

Diameter
9 feet 11 inches (3.02 m)

In service
1957–68

Variants

R-7 Semyorka
First launch
May 15, 1957

Last launch
February 27, 1961

Launch attempts
27, 18 of which were successful

R-7A Semyorka
First launch
December 23, 1959

Last launch
July 27, 1967

Launch attempts
21, 18 of which were successful

Postwar Rocket Research in Europe

French Diamant Rocket

A French Diamant B taking off from the Kourou launch facility in French Guiana. It was the first satellite launch vehicle built by a nation other than the United States and the Soviet Union, and it was used to place five satellites into orbit in the early 1970s. France then abandoned its own rocket program in favor of the development of a pan-European launch vehicle.

Outside of the superpowers of the United States and the Soviet Union, other nations also pursued rocket research in the immediate aftermath of World War II. In March 1945, French authorities established the Groupe Opérationnel des Projectiles Autopropulsés (GOPA) to study V-2 technology captured in Germany. The French army brought about 90 German rocket engine experts, led by the propulsion engineer Karl-Heinz Bringer (1908–99), to the town of Vernon in Normandy, France, to collaborate with French scientists on developing liquid-propellant rockets at the Laboratoire de Recherches Balistiques et Aérodynamiques (LRBA).

This research led to the development of the first version of the Véronique sounding rocket in 1952. Similar to the German anti-aircraft Wasserfall missile, which had been under development at the end of World War II, the Véronique was about 21 feet (6.4 m) long and used liquid propellants to carry a 132-pound (60-kg) payload to an altitude of up to 40 miles (65 km). Regularly launched from the French military base at Colomb-Béchar/Hammaguir, Algeria, Véronique was the first European-built liquid-fueled rocket used for space research, but by 1958 it had largely been superseded by the Diamant rocket, the first rocket-powered launch vehicle in Europe capable of placing a satellite in orbit.

The British also obtained V-2 technology from Germany. Since the United Kingdom had suffered most from the V-2 bombings of World War II, the British military made recovering V-2 hardware a priority, and had significant success at a V-2 launch site located in Cuxhaven on Germany's North Sea coast. As part of Operation Backfire, the British took more than 400 German rocket engineers into custody, including Peenemünde V-2 launch director chief Kurt Debus (1908–83). As part of the operation, in October 1945, Debus demonstrated the launch of three V-2 rockets to Allied personnel. Based on these tests, the British moved propulsion engineer Walter Riedel (1902–68) and several other V-2 veterans to the United Kingdom, where they participated in the development of the Blue Streak missile system and the subsequent Black Arrow launch vehicle (p. 187).

The latter half of the 1950s saw an increasing range of nations become involved in upper atmosphere research using sounding rockets. Most of these rockets were derived from military missiles and, in many instances, were primarily military flight tests conducted with scientific instruments riding piggyback. These rockets ranged in capability from the small Canadian Black Brant 4A, which could lift a small payload of 8 pounds (4 kg) to an altitude of 114 miles (185 km), to the French Véronique 61M, which could take a 440-pound (200-kg) instrument package to an altitude of 136 miles (220 km), but more capable rockets would soon follow.

LEFT Blue Streak was a British medium-range ballistic missile of the 1950s and early 1960s. Although it swiftly became obsolete as a weapon, it was later proposed as the first stage of the Europa satellite launch vehicle.

BELOW A Véronique sounding rocket being prepped for launch in 1953 in Algeria (at the time a French colony).

X-Planes and the Postwar Aircraft Boom

> *"If you can walk away from a landing, it's a good landing. If you use the airplane the next day, it's an outstanding landing."*
>
> CHUCK YEAGER, AMERICAN TEST PILOT

The decade after the end of World War II brought remarkable transformations in aerospace technology across the world as the massive wartime research and development projects embarked upon by various countries during the conflict reached fruition, producing new planes that could fly faster, farther, and higher than any that had come before. Aside from revolutionizing the aerospace industry, this aircraft boom also provided for a significant amount of research that would feed into the expertise of many of the scientists and engineers who would later work on early spaceflight missions.

A key milestone in this journey toward spaceflight was building an aircraft capable of surpassing the speed of sound, or Mach 1. By 1946, British aircraft designers at the de Havilland Aircraft Company had developed the de Havilland DH 108, a new tailless, swept-wing plane designed to reach speeds of up to 1,000 miles per hour (1,600 km/h). But a test flight on September 27, 1946, resulted in the plane's catastrophic structural failure and the death of its pilot, Geoffrey de Havilland Jr. (1910–46). It was not until November 14, 1947, that American pilot Charles "Chuck" Yeager (1923–) became the first person to fly faster than the speed of sound in level flight in the Bell X-1 rocket plane. Built by the Bell Aircraft Company, the X-1 was the first plane to be commissioned by the American National Advisory Committee for

Aeronautics (NACA) for its X-plane program of experimental aircraft designs. The aircraft was flown chiefly by pilots from the newly formed United States Air Force, and the X-1 would go on to make more than 80 test flights over the course of its relatively short service history before being retired in 1951. While NACA itself has since been superseded by the National Aeronautics and Space Administration (NASA), the X-plane moniker is regularly applied to various American experimental aircraft developments to this day.

The X-1 represented a landmark achievement in piloted rocket-powered aviation research and an important step toward developing aircraft capable of reaching beyond the atmosphere, but it was soon surpassed by other rocket-powered aircraft reaching speeds in excess of Mach 3. Thereafter, attention within the X-plane program shifted toward the next great flight milestone: building an aircraft capable of reaching and surviving speeds in excess of Mach 5 (p. 122). At such velocities, aircraft experience significantly increased heating rates, a problem that engineers would have to overcome before reliable space exploration could become a reality.

Outside of the X-1, the first production fighter aircraft to fly faster than the speed of sound was the American F-86 Sabre jet on September 15, 1948. It was followed in 1952 by Sweden's Saab 32 Lansen. The Soviet Union, eager to ensure its aircraft would be competitive in any potential conflict with the United States, developed its MiG-19 to fly at supersonic speeds in level flight in 1953.

NACA's High-Speed Research Aircraft

This photo from 1953 depicts a range of aircraft from NACA's high-speed research program. They are, clockwise from bottom left, a Bell X-1A, a Douglas D-558-1 Skystreak, a Convair XF-92A, a Bell X-5, a Douglas D-558-2 Skyrocket, a Northrop X-4 Bantam, and a Douglas X-3 Stiletto in the center.

OPPOSITE TOP The Bell X-1 rocket plane, nicknamed "Glamorous Glennis," on the ground in 1947.

OPPOSITE BOTTOM Chuck Yeager poses in the cockpit of his Bell X-1 at Muroc Air Force Base, California, in May 1948.

LEFT "Glamorous Glennis" being carried aloft by its mothership, a Boeing B-29. Bell X-1 rocket planes were not capable of takeoff under their own power and were instead mated to larger aircraft to be launched while in flight.

The Beginnings of the UFO Phenomenon around the World

This controversial image of a supposed UFO near McMinnville, Oregon, on May 11, 1950, was reprinted in magazines and in newspapers across the United States and is considered one of the most famous UFO images captured on film.

For centuries people around the globe have reported seeing things in the sky that they could not explain. In the past, these were described variously as evidence of angels, demons, and deities, but soon after the end of World War II, another explanation arose: alien visitation via unidentified flying objects, or UFOs. Between 1947 and 1960, there was a total of 6,523 reports of UFO sightings in the United States alone. Thorough records were not necessarily kept for other nations, but anecdotal stories suggest that such sightings occurred worldwide.

A key catalyst for many of these sightings appears to have been the experience of private pilot Kenneth Arnold (1915–84), who claimed that he saw an object he described as a "flying saucer" near Seattle, Washington, on June 24, 1947.

This sparked a wave of similar sightings around the globe. In the latter half of the 1940s, numerous UFO sightings were reported over Scandinavia, causing the Swedish Defense Staff to begin investigations into these "Ghost Rockets." About 2,000 sightings were logged in Swedish airspace between May and December 1946, and of those, more than 200 were verified with radar returns. On October 15, 1948, a radio signal at Kyushu Island, Japan, prompted a pilot to chase what he claimed to be a stubby, cigar-shaped flying object. The object accelerated beyond his ability to pursue it. On July 30, 1952, Flight Sergeant Roland Hughes of the British Royal Air Force reported sighting a "silver metallic disk" near Oldenberg, West Germany. A further wave of UFO sightings swept Europe in September and October 1954, as people in France, Italy, and other countries reported seeing strange objects in the skies; this large-scale phenomenon had some military officials at the North Atlantic Treaty Organization (NATO) concerned that the Soviet Union had developed some new technical capabilities that could threaten Western Europe. National security officials in the United States soon began investigating UFO reports, although they were more concerned about the security ramifications of unexplained intrusions into American airspace than the idea of alien visitors.

A photograph of a supposed UFO taken in the Soviet Union in 1961. Countless similar claims of UFOs were made across the Soviet Union during the Cold War.

Alien Autopsy?

In 1995, the British musician and film producer Ray Santilli first released 17 minutes of black-and-white footage purporting to show the autopsy of a body of an extraterrestrial recovered from the wreckage of a "flying disk" discovered in Roswell, New Mexico in 1947. Santilli initially claimed he received the footage from a retired military cameraman who wished to remain anonymous. While the footage did not convince any scientists and was viewed with suspicion by special effects experts, it nevertheless captured the public imagination. It even inspired a comedic feature film entitled *Alien Autopsy* in 2006.

Certainly, some extraterrestrial anxiety can be seen in science fiction films of the time. *The Day the Earth Stood Still* (1951) and *Earth vs. the Flying Saucers* (1956) both feature powerful, potentially hostile, UFO-flying aliens. In the latter film especially, the fear of the alien turns out to be entirely justified, with the flying saucers eventually attacking buildings and people across the U.S. capital.

Eventually, in 1957, the United States Air Force issued a report that rejected the notion that UFO sightings were evidence of either a hostile alien threat or any kind of Soviet weapon system. As the report put it: "First, there is no evidence that the 'unknowns' were inimical or hostile; second, there is no evidence that these 'unknowns' were interplanetary space ships; third, there is no evidence that these unknowns represented technological developments or principles outside the range of our present day scientific knowledge; fourth, there is no evidence that these 'unknowns' were a threat to the security of the country; and finally there was no physical evidence or material evidence, not even a minute fragment, of a so-called 'flying saucer.'" Despite the continued popularity of UFOs in popular culture and further revelations of various other government-funded reports over the years, to date there has not been one scintilla of physical evidence to support the idea of contact between humanity and any alien civilization.

RIGHT Many magazines featured UFO visitation. *Fate Magazine* in the United States specialized in this type of sensationalism.

FAR RIGHT *The Day the Earth Stood Still* (1951) features a humanoid alien visitor named Klaatu, who comes to Earth with an 8-foot- (2.4-m-) tall robot called Gort.

Major Jesse Marcel holding debris recovered from a ranch near Roswell, New Mexico, in 1947.

The Roswell Incident

1947

The iconic UFO sighting that sparked countless conspiracy theories and an alleged alien autopsy video.

In late June 1947, rancher William Brazel spotted some unusual debris on the land he was working some 30 miles (48 km) north of the town of Roswell, New Mexico. It was not until a few days later, after he heard reports on the radio of UFO sightings, that he first considered that the debris might be of extraterrestrial origin. Brazel immediately went out and collected some of the debris he had seen and took it to the local sheriff's office. He also reported that there was a "large area of bright wreckage made up of rubber strips, tinfoil, a rather tough paper, and sticks." The sheriff contacted the nearby Roswell Army Air Field (now Walker Air Force Base) and turned the material over to it. Major Jesse Marcel, an intelligence officer at the base, received the material and subsequently went with a detail of enlisted men to gather the rest of the debris on the ranch.

From that point, reports of the case begin to get somewhat cloudy. The Air Force issued a statement that the debris consisted of parts of a weather balloon, but rumors began to circulate that they instead belonged to a UFO. These rumors were then reported in the local newspaper. Journalists from farther afield began to descend on the town.

Upon learning of the story, General Roger Ramey ordered that all recovered materials from Roswell be flown to the Fort Worth Army Air Field, now the Naval Air Station Joint Reserve Base, Fort Worth. At the base, officials stated that the debris was indeed the remnants of a downed weather balloon and its instruments. But this explanation did not satisfy everyone.

In the 1990s, there was a fresh investigation into the incident, which concluded that the debris came from a crashed high-altitude electronic sensor that had been part of the then-secret Project Mogul, a military project to monitor the upper atmosphere for signs of Soviet atomic tests.

The sensor was designed to detect sound waves in the ionosphere from a possible Soviet nuclear test and was essentially a microphone attached to a high-altitude balloon, which kept the device at a constant altitude. On June 5, 1947, Mogul flight 4 launched from the Alamogordo Army Air Field in New Mexico (now Holloman Air Force Base). It is now presumed that this balloon lost altitude and crashed back to Earth as the debris Brazel found. Meanwhile, the city of Roswell has gone on to become a tourist destination for UFO aficionados worldwide.

Space Visionaries Fire the Public Imagination

"We have the scientists and the engineers.
We enjoy industrial superiority.
We have the inventive genius. Why, therefore,
have we not embarked on a major space program
equivalent to that which was undertaken
in developing the atomic bomb?
The issue is virtually the same."

"WHAT ARE WE WAITING FOR?"
EDITORIAL; *COLLIER'S*, MARCH 22, 1952

In 1952, while he was still working for the United States Army in Huntsville, Alabama, Wernher von Braun burst into the lives of the American public with a series of articles in *Collier's* magazine about the near reality of spaceflight. These articles were largely inspired by the work of the science fact and fiction writer Willy Ley, a former member of the German rocket society of which von Braun had once been a member (p. 30). Ley, who had fled Germany in 1935 to settle in the United States, organized the Space Travel Symposium at the Hayden Planetarium in New York City in 1951. He was a skilled promoter of space exploration and asked participants to emphasize that "the time is now ripe to make the public realize that the problem of space travel is to be regarded as a serious branch of science and technology." By chance, two writers for *Collier's*, a periodical that had the fourth-highest circulation in the United States during the early 1950s, attended this meeting. Their interest led to a series of important articles accompanied by vivid artwork by the American illustrator Chesley Bonestell and others that appeared in the magazine between 1952 and 1954. These articles swiftly captured the imaginations of millions around the world.

Von Braun achieved celebrity status in the United States largely because of his involvement in Ley's symposium and the articles in *Collier's* that followed from it.

Von Braun's first article appeared in the magazine on March 22, 1952. It gave the rocket enthusiast an opportunity to present his own unfiltered vision of the future of space travel. He advocated for the orbiting of artificial satellites followed by orbital flights by humans, the development of a reusable spacecraft for travel to and from Earth orbit, the building of a permanently inhabited space station, and human exploration of the Moon and Mars by spacecraft launched from that space station. Ley and various others also elaborated on further aspects of space exploration. These articles framed the exploration of space in the context of the Cold War rivalry with the Soviet Union. One edition of the magazine contained a particularly strident editorial that stated: "*Collier's* believes that the time has come for Washington to give priority of attention to the matter of space superiority. The rearmament gap between the East and West has been steadily closing. And nothing, in our opinion, should be left undone that might guarantee the peace of the world. It's as simple as that."

The *Collier's* series catapulted von Braun into the public spotlight in a manner that his previous scientific and engineering work had never managed to achieve. Over three million copies of the magazine were produced each week, and its publishers further claimed that four or five people read each copy sold. A significant number of the magazine's readers thus were opened up to the possibilities of spaceflight.

One particularly notable *Collier's* reader was the American animator and entertainment entrepreneur Walt Disney, who subsequently became fascinated with space exploration. His enthusiasm for the subject led him to produce three classic documentaries about the prospects for flight beyond Earth, including one episode each on the Moon and Mars. Moreover, when Disneyland opened in Anaheim, California, in 1955, it featured a section called "Tomorrowland," with a "Moonliner rocketship" and a "Rocket to the Moon" experience in a futuristic-style theater.

Earth-Orbiting Satellite

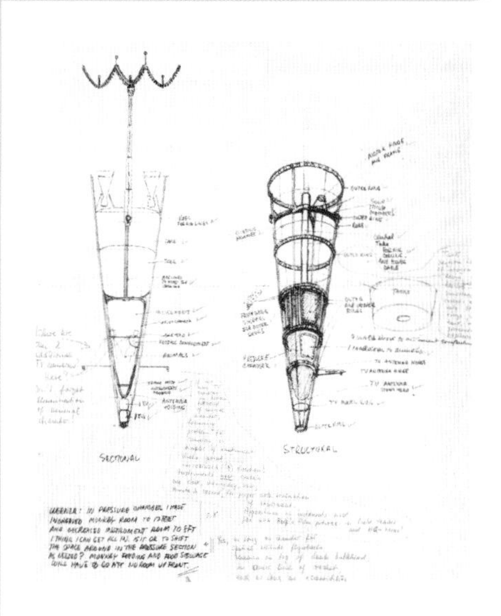

Chesley Bonestell and Wernher von Braun corresponded in detail for their collaborations in *Collier's*. This page of sketches of a proposed Earth-orbiting satellite includes handwritten notes from both men.

LEFT This depiction of the exploration of Mars by Chesley Bonestell captured the imaginations of *Collier's* readers in the years before the dawn of the space age.

OPPOSITE The issue of *Collier's*, dated September 22, 1952, energized public excitement for space exploration, promising that "man will conquer space *soon*."

Transforming Human Expectations

In December 1949, Gallup pollsters in the United States found that only 15 percent of Americans believed humans would reach the Moon within the next 50 years, while 15 percent had no opinion and a whopping 70 percent believed that it would not happen within that time frame, if at all. By October 1957, less than a decade later, only 25 percent of Americans believed that it would take more than 25 years for humanity to reach the Moon, while 41 percent firmly believed that it would happen within 25 years and 34 percent were not sure. An important shift in perceptions had taken place between the two polls, one that reflected the well-known advances being made in rocket technology, the increasing prevalence of space travel in popular culture around the globe, and the increasing popular support for human space exploration that had been actively fostered by a variety of public intellectuals, including Wernher von Braun. Around this time, the German-American writer Willy Ley, a former member of the prominent German Society for Spaceship Travel rocket club (p. 30) alongside von Braun, was producing a steady stream of fiction and nonfiction books about space that fired the imaginations of millions. His 1949 book, *The Conquest of Space*, stunningly illustrated by the artist Chesley Bonestell, presented space exploration as humanity's destiny. Even more successful was Ley's *Rockets, Missiles, and Space Travel*, which went through 21 printings in multiple languages between 1944 and 1969. This was undoubtedly one of the most significant works in English on the prospects of space exploration available to a mass audience in the mid-twentieth century.

Amazing Stories was a major outlet for science fiction and fantasy from the 1920s through the 1950s, during which time it published work from a variety of world-renowned authors.

The constant drumbeat of excitement for space exploration was also served by pulp magazines such as *Amazing Stories* and various children's comic books, such as *The Eagle* in the United Kingdom. Such titles were filled with exciting interplanetary tales of contact with alien civilizations and space exploration. It is easy to forget that many of these visions of space travel were originally published before the first satellites reached orbit. Collectively, they prepared people around the globe for the excitement of the golden age of space exploration that was soon to begin in earnest.

Filming Space Exploration

1950–57

The postwar years of the 1950s saw a rebirth in the popularity of science fiction tales both at the cinema and on TV.

In addition to writings about space exploration, many people in the immediate postwar era were introduced to the possibilities of space exploration by a steady stream of science fiction films that found enormous audiences, and high praise, around the globe. Crafted by producer-director George Pal (1908–80), *Destination Moon* (1950) told the story of a privately financed expedition to reach the Moon, and was written by science fiction master Robert Heinlein (1907–88). The production team paid special attention to conveying the reality of spaceflight by including elements such as weightlessness, spacesuits, and finite fuel.

On British TV screens, viewers thrilled to the adventures of the first human mission into space in *The Quatermass Experiment* in 1953. Another popular movie was the 1956 feature *Forbidden Planet*. Directed by Fred McLeod Wilcox, the film was a science fiction retelling of William Shakespeare's play *The Tempest* and introduced to the world Robby the Robot, an early cinematic hint that artificial intelligence might become a reality.

Don't tell the Screen Actors' Guild, but Robby the Robot got top billing alongside his human costars in this poster for the 1956 cinematic smash hit *Forbidden Planet*.

Conquering Space

On this book jacket, Chesley Bonestell's art depicts astronauts disembarking from a modified V-2-type rocket during the first human expedition to the Moon. The painting appeared on the dust jacket of a 1949 book by Bonestell and popular science author Willy Ley that laid out a realistic program of space exploration for the following 50 years.

4

The Space Age Dawns

The space age truly dawned on October 4, 1957, when the Soviet satellite Sputnik 1 (p. 90) became the first human-made object to be placed into orbit above the Earth. It was an achievement that marked humanity's first tentative steps on the journey toward leaving our home planet and exploring the wider universe.

But Sputnik 1's launch also sounded the starting pistol in the new arena of space exploration, an area of competition that would come to be seen by many as a space race between the superpowers of the Soviet Union and the United States.

The Soviet Union pursued an aggressive program that prioritized achieving as many new space "firsts" as quickly as possible, while the Americans came to take a longer view, establishing the civilian National Aeronautics and Space Administration (NASA) in 1958 and setting a series of space exploration goals (p. 102). The next dozen or so years following Sputnik 1's launch saw many twists and turns, plenty of eye-catching, headline-grabbing feats, and plenty of hard work that brought an unparalleled level of global public engagement to the enterprise of space science.

Beginning in 1961, when Yuri Gagarin became the first human to orbit the Earth (p. 114), the Soviets and the Americans embarked upon what is often now seen as the heroic era of human space exploration. With Project Mercury (p. 105), the Americans were both attempting to catch up and overtake their rivals. Almost every mission pushed back the frontiers of the unknown and increased human knowledge, establishing the basics of how best to survive and explore in the hostile environment of space. The stakes were huge, but the rewards were equally great.

OPPOSITE The Russian American artist Boris Chaliapin created this artwork for the cover of *Time* magazine to celebrate Yuri Gagarin's historic trip into space on April 12, 1961.

The International Geophysical Year and Peaceful Uses of Outer Space

RIGHT Sidney Chapman (center) addresses an International Geophysical Year conference with a group of his fellow scientists, among them Marcel Nicolet (second from right).

BELOW The logo of the International Geophysical Year, the scientific grand challenge that provided much of the impetus for the first artificial satellite launches.

The age of space exploration truly began with the International Geophysical Year (IGY), a series of international scientific experiments and projects conducted between July 1, 1957, and December 31, 1958. The idea for organizing this scientific grand project, intended as an opportunity to gather more data about the geophysical properties of the Earth, had started in a most mundane manner several years earlier. On April 5, 1950, while on a visit to the United States, British scientist Sidney Chapman (1888–1970) had dinner with a group of his fellow experts at the home of astrophysicist James Van Allen (1914–2006) in Silver Spring, Maryland. Over dinner, Chapman discussed an idea he had for setting up a series of events to gather various scientific data about the Earth on a transnational basis, as had been done for the Arctic in the 1880s and for both poles in the 1930s.

Everyone present agreed such an initiative would have real value, and the group soon convinced the Belgian aeronomist Marcel Nicolet (1912–96), a scientist who specialized in upper atmosphere research, to propose the idea at the International Council of Scientific Unions (ICSU). The ICSU took up Nicolet's proposal and soon began organizing the endeavor, timing it to coincide with the peak of the Sun's 11-year solar cycle.

In October 1954, at a meeting in Rome, the ICSU expanded the scope of the scientific research effort to encourage the creation of artificial satellites as part of the data-gathering effort. This made space exploration a specific goal for the IGY. Both the Soviet Union and the United States accepted the challenge to design and launch such a satellite. The Soviets placed Sergei Korolev (p. 70) in charge of their effort, while, after a brief but fiercely contested competitive process, the United States Naval Research Laboratory, which was already developing its own Viking scientific rocket program, was given the mandate to design and launch the American IGY satellite.

In stark contrast to later space exploration missions, neither nation showed much initial urgency in its IGY satellite efforts, with progress quickly getting bogged down in bureaucratic or engineering difficulties. Korolev was engrossed in the development of his R-7 intercontinental ballistic missile (ICBM) (p. 73). The satellite program was, at best, of secondary importance. By contrast, the team working on Project Vanguard, the American satellite effort, had a far brighter public spotlight on them. However, they were confident in the superiority of their technology.

Although today the launch of the first satellite, and the rush toward further space exploration that followed it, has somewhat overshadowed the rest of the IGY, it is important to note that by the end of 1958, scientists and governments from nearly 70 countries had participated in its data-gathering success, making it the largest and most ambitious scientific collaboration the world had ever seen to that point.

BELOW LEFT A team of American scientists in Antarctica awaiting the delivery of supplies to erect a scientific base in support of the International Geophysical Year on March 28, 1957.

BELOW RIGHT A test launch of the Viking rocket in 1952, the launcher that was to be used for Project Vanguard, America's artificial satellite designed to gather data during the International Geophysical Year.

The Sputnik Satellites in Earth Orbit

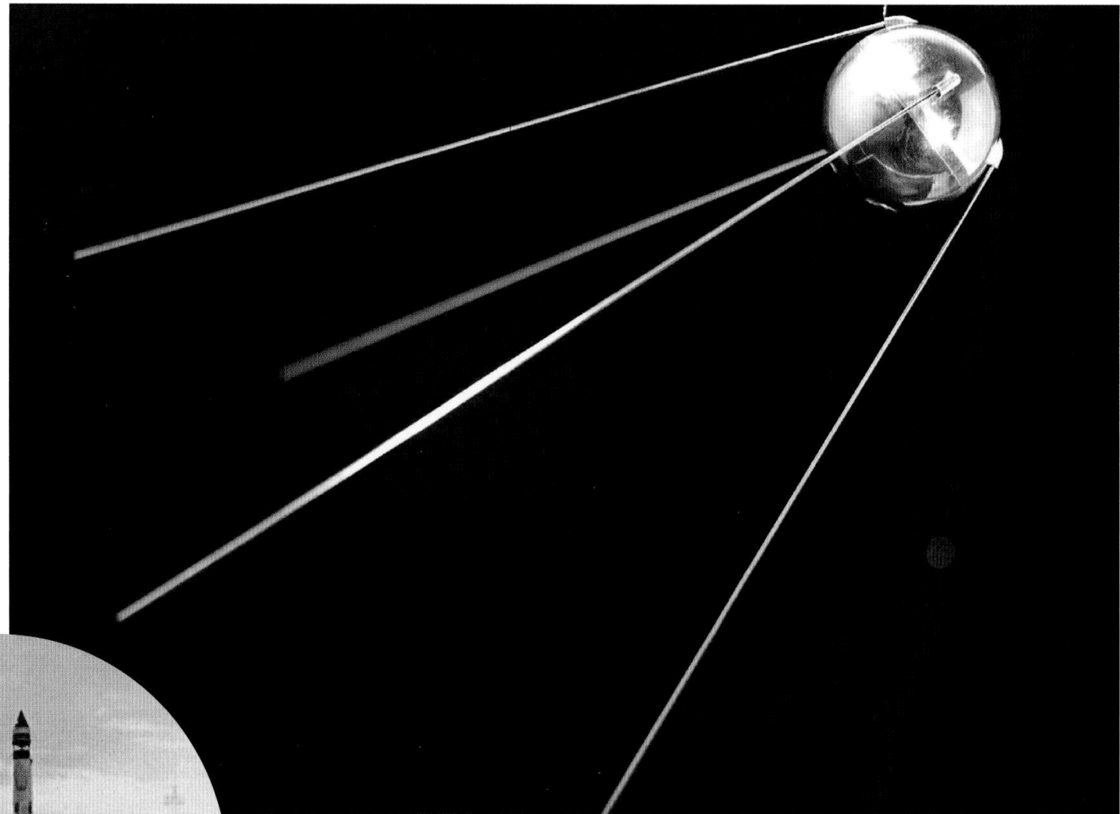

In the fall of 1957, the Soviet rocket engineer Sergei Korolev drove his team hard preparing for the launch of Sputnik 1, the world's first artificial satellite. Technical difficulties had forced them to abandon more complex satellite designs in favor of a simple metal sphere about the size of a beach ball. It weighed about 183 pounds (83 kg) and carried a radio transponder that would enable people to track its movement as it orbited the Earth. Soviet scientists announced Sputnik 1's planned orbital details and the radio frequencies it would be using in advance so that it could be tracked by anyone with access to long-wave radio equipment, whether they were scientists involved with the International Geophysical Year (IGY) or ham radio enthusiasts. The radio signal, a set of beeps broadcast at 20 and 40 MHz, ensured that no one could claim that the satellite launch was a hoax.

On October 4, 1957, Korolev and his team used a modified version of the R-7 rocket, the world's first intercontinental ballistic missile (p. 73), to launch Sputnik 1 into orbit from a newly established rocket testing facility best known today as the Baikonur Cosmodrome, in the desert near Tyuratam in the Kazakh Republic (modern-day Kazakhstan). Everyone involved breathed a sigh of relief when the

ABOVE LEFT The R-7 rocket lifts Sputnik 1, the world's first artificial satellite, into orbit on October 4, 1957, the first orbital flight from Tyuratam (now Baikonur Cosmodrome), Kazakhstan.

ABOVE RIGHT This back-up version of Sputnik 1 is on display at the Memorial Museum of Cosmonautics, Moscow.

two-stage rocket successfully deployed its tiny satellite cargo and Sputnik 1 began transmitting. This was the proof that the satellite, and the upper stage of the R-7 that had carried it, had reached orbit.

The elliptical orbit Sputnik 1 achieved took it around the Earth every 90 minutes while its radio beacon's regular beeps revealed its location. The satellite transited the United States twice before anyone in the military even knew to search for it. Once the military was made aware of the satellite, they were greatly concerned. Some worried, without any evidence, that the satellite's signal might reveal targeting information on American ballistic missile deployments.

In the United States' capital, Washington, D.C., American scientist Lloyd Berkner (1905–67) was attending a reception at the Soviet embassy. It was the closing event of a long week of meetings relating to the IGY. Berkner was the senior American delegate. He had heard some worrying hints from Soviet scientists that they were "on the eve of the first artificial Earth satellite," months ahead of its scheduled launch date in 1958. An American reporter at the event soon confirmed Berkner's worst fears: the Soviets had succeeded and beaten the beleaguered American satellite program into orbit. With grace in defeat, Berkner clapped his hands for attention and said: "I wish to make an announcement. I've just been informed by the *New York Times* that a Russian satellite is in orbit at an elevation of 900 km. I wish to congratulate our Soviet colleagues on their achievement."

Cocktails in hand, the scientists and other delegates present then adjourned to the roof of the embassy to see if they could spot Sputnik 1 in orbit. The satellite itself was too small to be seen, but if they looked hard enough, some of those present might have made out the upper stage of the R-7 rocket that had carried it into orbit glistening in the sunlight.

Key Technological Challenges to Reaching Space

- Minimizing drag and heating of the launch vehicle.
- Finding materials that can withstand extreme stress and high temperatures.
- Creating guidance systems that can guide vehicles into orbit.
- Developing reliable propulsion units with adequate power.

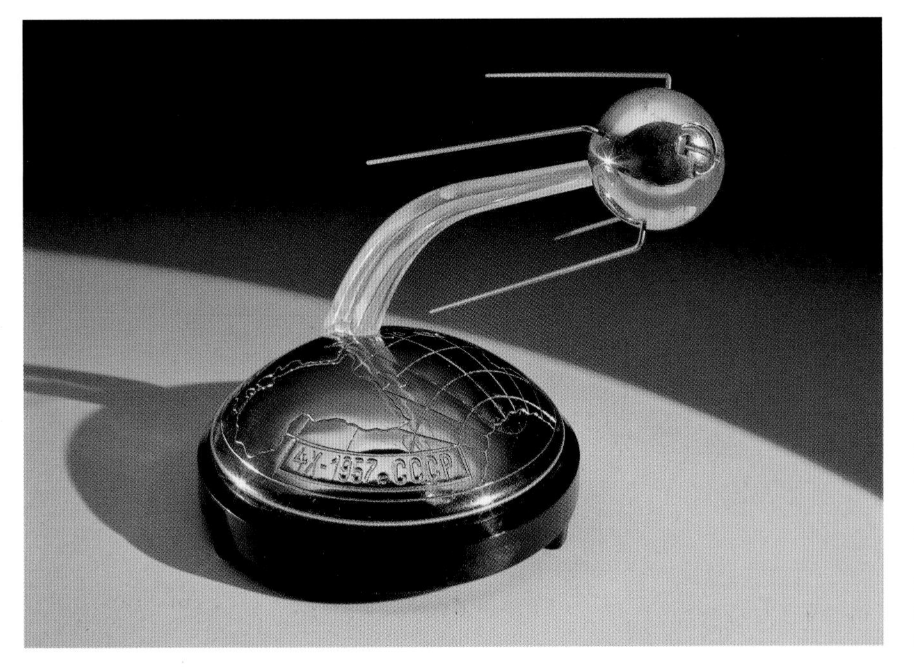

The news of Sputnik 1's success sparked a wave of national pride across the Soviet Union. Various pieces of Sputnik 1 memorabilia followed, including this commemorative music box purchased in 1964. The box is inscribed with the date of the satellite's launch and plays a short tune, followed by the characteristic beeps of the satellite's signal.

Rocket Propellant Applications

Rocket engine technology went through two revolutions after World War II. First, engineers in the United States, the Soviet Union, France, and Britain created large liquid-fuel engines based on the German V-2. American firms were also commissioned to design and build large rockets using new, high-energy solid propellants.

In the United States, the most important line of liquid propellant rocket engines evolved from the Air Force's Navaho cruise missile project. North American Aviation's propulsion division, later called Rocketdyne, began by copying the V-2 motor for Navaho, then created ever-larger engines for that project and for the Redstone, Jupiter, Thor, and Atlas missiles, Saturn launch vehicles, and the Space Shuttle.

Solid propellant ballistic missiles became increasingly important in the 1950s because such propellants could be stored safely in the rockets over long periods, while liquid propellants could not. This gave solid propellant rockets an advantage in military applications, where constant readiness was a key requirement. Solid propellant rockets were later used to provide additional thrust for liquid propellant space launch vehicles.

Back in the Soviet Union, it seemed few people initially grasped the immense propaganda value of the satellite's launch, least of all Soviet premier Nikita Khrushchev. The first announcement about Sputnik 1 from the Soviet news agency, TASS, was restrained. It was not until Khrushchev saw the three-column headlines in the *New York Times* and other western newspapers the next day that he realized the worldwide acclaim this success had achieved. As a result, his public support for his nation's space program increased almost immediately. This was followed by massive political and financial support. Sputnik 1 ushered in a new age in the story of human progress and a new chapter in the narrative of the Cold War. The Soviet Union had pulled off a tremendous propaganda coup, assuming apparent leadership in a major new technological field overnight, and with it an immense level of prestige. The question now was how this achievement would be followed.

The cover of *Life* magazine, October 21, 1957, showing Smithsonian Observatory scientists working to calculate Sputnik 1's orbit.

Sputnik 2 and Laika

1957

The pioneering pup that went where no dog had gone before.

The success of Sputnik 1 swiftly led to an even more audacious follow-up mission. On November 3, 1957, less than a month after Sputnik 1's launch, Soviet scientists aimed to put the first animal into orbit on board Sputnik 2. The animal was a dog named Laika, one of several strays from the streets of Moscow. She was female, weighed 13 pounds (5.9 kg), and apparently earned many nicknames during her training, including Kudryavka, the Russian diminutive for "curly," but Laika—literally "barker"—was the one that stuck.

The modified satellite that carried Laika into orbit was significantly heavier than its predecessor, weighing 1,120 pounds (508 kg) in total. It also contained far more equipment, including a telemetry system, a guidance and control unit, a temperature control system for the cabin, and two spectrophotometers for measuring solar radiation and cosmic rays. Laika herself was placed in a separate small, sealed, pressurized cabin, which provided just enough room for her to lie down or stand. Laika's biological signs were monitored throughout the mission, with the data transmitted to Earth every 15 minutes along with video footage. Food and water were placed onboard to be dispensed periodically, but there was no provision for Laika to return from her mission. From the start, the scientists in charge knew Laika would die in orbit when the oxygen ran out, but they were nevertheless unprepared for the strong public reaction her death provoked around the world.

Despite this, Sputnik 2 proved to be a massive success; it stayed in orbit for almost 200 days and returned the first scientific data on the behavior of a living organism in orbit. It also helped reinforce the image of the Soviet Union's technological prowess, a situation that provoked the United States to increase its support for its own space endeavors.

TOP RIGHT Laika became the first animal in orbit when she was launched aboard Sputnik 2 on November 3, 1957.

RIGHT A replica of Sputnik 2, the spacecraft in which the dog Laika was hurtled into space, on exhibit in the Russian pavilion at the Brussels Universal and International Exhibition in 1958.

Establishing the Open Skies Doctrine

While the launch of Sputnik 1 came as a surprise to the world, and a particular shock to many in the United States, it also served to establish an important principle in international law. The Open Skies doctrine is the principle of overflight that allows any given country to place a satellite in orbit, where it will travel across international borders without violating the sovereignty or airspace of any country below.

By the 1950s, international custom had established that nations could force down any aircraft violating their territorial airspace. But where did a nation's claim to airspace end? There was no clear consensus or agreement.

On February 14, 1955, United States defense officials raised the question of the international laws governing territorial waters and airspace in a two-volume, 190-page assessment entitled *Meeting the Threat of Surprise Attack*. The response to the report among American politicians was that space should be recognized as international territory. By contrast, their Soviet counterparts argued that territorial limits extended above a nation into infinity.

As discussion of orbiting artificial satellites was raised as part of the efforts surrounding the International Geophysical Year (IGY; p. 88), the notion that nations should have unimpeded freedom of movement in space became an increasingly significant issue. The imposition of territorial prerogatives outside the atmosphere could legally restrict any nation from orbiting satellites without the explicit permission of all nations that might be overflown. Since the United States government viewed itself to be in the strongest position to capitalize on such freedom, it favored Open Skies. Other nations, particularly those without the imminent capability to launch their own satellites, had little interest in allowing America to orbit satellites over their heads, particularly if those satellites could be capable of carrying out spying and reconnaissance.

On July 21, 1955, American President Dwight Eisenhower tried to negotiate a precedent for open skies at an international summit in Geneva, Switzerland, but Soviet leader Nikolai Bulganin rejected the proposal, claiming that it was an obvious American attempt to "accumulate target information." Eisenhower later admitted, "We knew the Soviets wouldn't accept it, but we took a look and thought it was a good move." With the diplomatic route blocked, the Americans subsequently started to search for other ways to pursue their policy goal.

Then, on October 4, 1957, Sputnik 1 reached orbit, overflying the United States and many other nations of the world in the process. It wasn't until four days later, on October 8, 1957, that Eisenhower advisor Donald Quarles identified an unexpected benefit to the turn of events, noting "the Russians have...done us a good turn, unintentionally, in establishing the concept of freedom of international space." Eisenhower immediately seized on Sputnik 1 as the precedent his administration

Corona Reconnaissance Satellite Key Facts

The world's first spy satellite, this U.S. project consisted of a camera and film capsules. The capsules reentered Earth's atmosphere, deployed a parachute, and were recovered in the air by C-119 aircraft.

The first successful air recovery was of a capsule designated Discovery 14 on August 18, 1960.

The last of the program's 145 missions took place on May 25, 1972.

Corona satellites photographed all Soviet missile bases and submarine types, establishing accurate estimates of the number and capacity of Soviet nuclear missiles.

OPPOSITE TOP American President Dwight Eisenhower and Soviet Secretary Nikita Khrushchev sit across from each other at the Geneva Summit in 1955, at which Eisenhower proposed the Open Skies doctrine to allow overflight of each nation by the other.

OPPOSITE BOTTOM United States Deputy Secretary of Defense Donald Quarles (left) was the first to spot the political opportunity represented by Sputnik 1's orbital flight.

LEFT President Eisenhower at a press conference a week after the launch of Sputnik 1. He promised that an American satellite would soon follow.

had been looking for, and gave the go-ahead to pursue the design and launch of an American reconnaissance satellite. The subsequent launches of scientific satellites by both the Soviets and the Americans only served to further cement the precedent set by Sputnik 1, and by the end of 1958, the principle of Open Skies had been firmly established for all practical purposes. At the height of the Cold War, the Soviet space program helped implement a policy position of the U.S. government.

Subsequent treaties and international practice came to follow the Open Skies principle, using the Kármán line, the rough boundary 62 miles (100,000 m) above the Earth's surface, as an effective demarcation of the upper limit to territorial airspace and the start of international space.

Perhaps surprisingly, although Open Skies opened the way for the operation of orbiting reconnaissance satellites, these spacecraft did not end up fanning the flames of Cold War tension. Instead, they served as a stabilizing influence as the Americans and, soon after, the Soviets could see what their rival was actually doing. Both sides benefited from this capability, and the world became a safer place because of it, as each power could independently verify the claims of the other.

Under development in the late 1950s, Project Corona was a major successful reconnaissance satellite program of the United States. The first reconnaissance satellites were essentially still-film cameras that briefly orbited their target territory, took pictures, and then jettisoned a small capsule containing the film back toward the Earth. During its descent, this capsule would be intercepted and scooped out of the air by a cargo aircraft before being taken back to base for processing and analysis. The first Corona satellites flew in the early 1960s, and the program continued well into the 1970s, when it was replaced by more advanced satellite surveillance systems. The Soviets used similar satellite technology in order to spy on the United States.

BELOW A United States Air Force C-119J Fairchild Flying Boxcar aircraft using a trailing grapple to catch and recover a capsule of reconnaissance photographs (lower left), descending by parachute from orbit.

OPPOSITE TOP LEFT Corona satellite image of the first Chinese nuclear weapons test site four days after detonation on October 16, 1964. Such reconnaissance missions were vital for nations seeking to assess each other's capabilities.

OPPOSITE TOP RIGHT The KH-4B camera pictured here was the last and most advanced camera system used in Project Corona, the first photo reconnaissance satellite program. Between August 1960 and May 1972, when the program ended, 145 Corona satellites were launched, and they produced over 800,000 usable images of the Soviet Union and other nations.

"Space photography would be worth ten times what the whole [space] program has cost."

PRESIDENT LYNDON JOHNSON, 1967

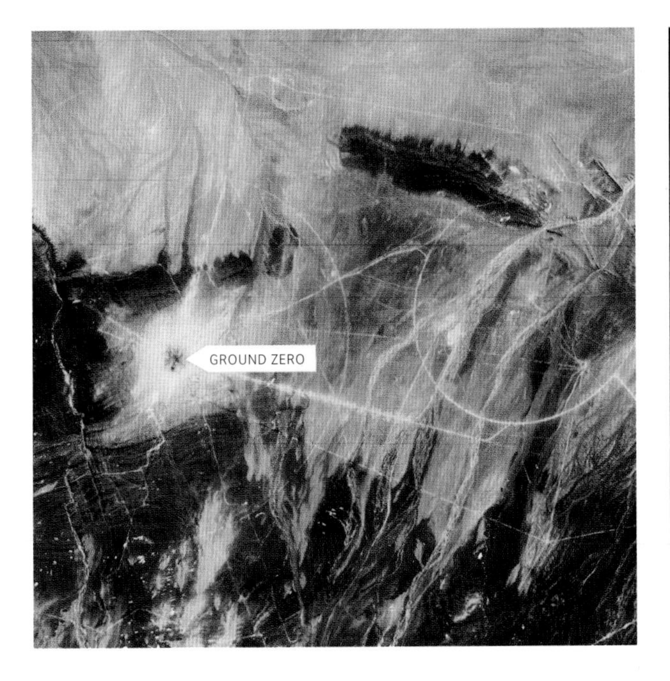

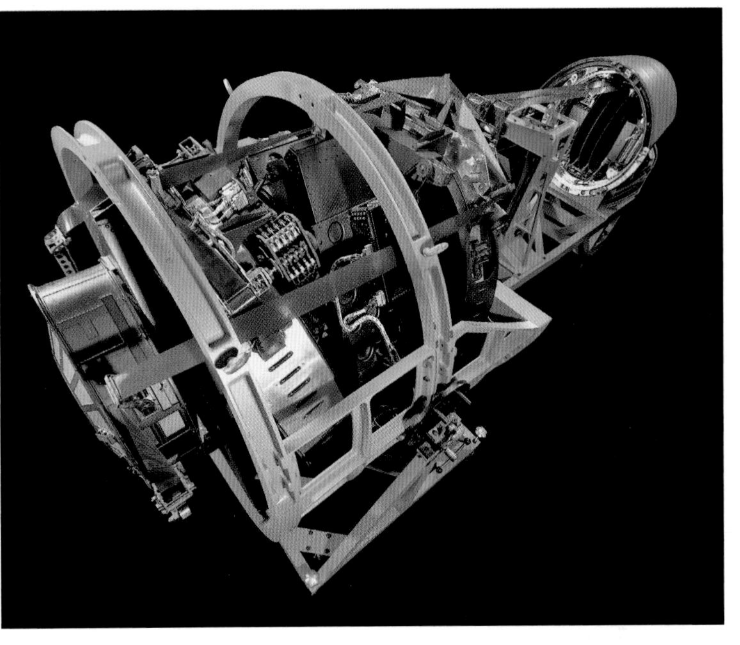

America Responds to Sputnik

1957

The early Soviet space program successes prompted the Americans to reorganize their own space efforts.

Many American politicians called for an aggressive response to the Sputnik launches in 1957, one that would unify the American effort toward its space program. In the year that followed, the American approach to space exploration saw four major changes.

First, the Eisenhower administration took steps to ensure a speedy launch of the American International Geophysical Year (IGY) satellite (p. 88). Second, Eisenhower himself established the President's Science Advisory Committee (PSAC) to focus attention inside the White House on science and technology issues in the government. Third, Congress passed the National Defense Education Act of 1958 to provide greater funding for the training of scientists and engineers. Finally, in 1958, Congress passed the National Aeronautics and Space Act to establish the National Aeronautics and Space Administration, or NASA (p. 102).

The development problems faced by the Vanguard rocket (1957–59) served to underline the need to overhaul America's early space program.

The Explorer Satellites and the Discovery of the Van Allen Radiation Belts

SATELLITE EXTRA

The Huntsville Times

SATELLITE EXTRA

HUNTSVILLE, ALABAMA, SATURDAY, FEB 1, 1958

5c PER COPY

Jupiter-C Puts Up Moon

Eisenhower Officially Announces Huntsville Satellite Circles Globe

Wail Of Sirens Brings In Era On Space Here

Thousands Gather On The Square For Noisy Success Demonstration

9 Labs Here Aided Project Of Launching

Weather Change Sped Launching

AUGUSTA, Ga., Feb. 1 (AP)—President Eisenhower announced early today America's first satellite is in orbit around the earth.

The President's dramatic announcement was issued at his vacation headquarters a few minutes before 1 a.m. EST by White House press secretary James C. Hagerty.

The satellite was launched at Cape Canaveral, Fla., at 10:48 p.m., EST, last night.

Army Reveals Second Moon Is Scheduled

70-Foot Carrier Roars Into Starry Night At Cape Canaveral

This front page, from the newspaper in the city where the Explorer 1 rocket was built, highlights the accomplishment for local readers.

Within weeks of Sputnik 1's launch in October 1957, the Eisenhower administration took steps to catch up to the Soviets' early success. Project Vanguard, the United States Naval Research Laboratory's plan to build a satellite for the International Geophysical Year (IGY) program, was still mired in technical problems. Its attempted launch on December 6, 1957, ended in embarrassing disaster when its modified Vanguard launch vehicle exploded just seconds after takeoff. This prompted the administration to direct the Jet Propulsion Laboratory (JPL) to design, build, and operate a new, simple satellite, ultimately named Explorer 1, that would reuse a cosmic-ray detector and a micrometeorite detector developed for another project. The satellite itself was to be sent into orbit using a Juno 1 rocket, a specially adapted version of the Army's Jupiter-C ballistic missile, produced by the Army Ballistic Missile Agency team headed by Wernher von Braun at the Redstone Arsenal.

The Explorer 1 mission enjoyed a successful launch at Cape Canaveral, Florida, late on January 31, 1958. Tracking sites across the globe followed its ascent to the upper reaches of the atmosphere, but ground controllers had to wait nervously until the satellite had completed its first full orbit before they could be sure that it was functioning properly. Nearly two hours after its launch, the mission's controllers breathed a sigh of relief as they regained contact with the spacecraft, now orbiting safely overhead.

Once the satellite was in orbit, James Van Allen, the principal science investigator responsible for the satellite's instrument package, von Braun, and William Pickering (1910–2004), the director at JPL, presided over a jubilant press conference at the National Academy of Sciences building in Washington, D.C., in the early hours of February 1. They triumphantly held up a model of Explorer 1 for photographers, but the real achievement came from the science data recorded by the satellite's cosmic ray detector, which showed areas of high radiation along Explorer 1's orbital path. These were early evidence of Earth's radiation belts.

The success of Explorer 1 served to cement von Braun's role as the key engineer in the American space program, and catapulted Van Allen to celebrity status. The radiation belts discovered by Van Allen's instruments now bear his name. Van Allen went on to take a lead role in the first four Explorer probes, either serving as the chief scientist or as a collaborator on the science component of the first Pioneer missions (p. 100), several Mariner missions, and the orbiting geophysical observatory satellites.

Explorer 3 and Explorer 4 further charted the Van Allen belts in 1958; they also managed to record the Earth's slight pear shape. It was only with subsequent missions that scientists were able to determine that the belts form two concentric

BELOW LEFT This is an iconic image from the press conference marking the success of the Explorer 1 mission on January 31, 1958. Pictured (left to right) are William Pickering, director of the Jet Propulsion Laboratory; James Van Allen, principal scientific investigator for the mission; and Wernher von Braun, technical director of the Army Ballistic Missile Agency, which built and launched the Redstone rocket that placed Explorer 1 into orbit.

BELOW RIGHT The Van Allen radiation belts are a series of concentric arcs of radiation that ring the Earth, trapped by the planet's magnetic field. This radiation belt serves as a shield for the surface of the Earth from deadly cosmic rays and radiation from the Sun. It is one of the major reasons why life as it exists here is able to survive.

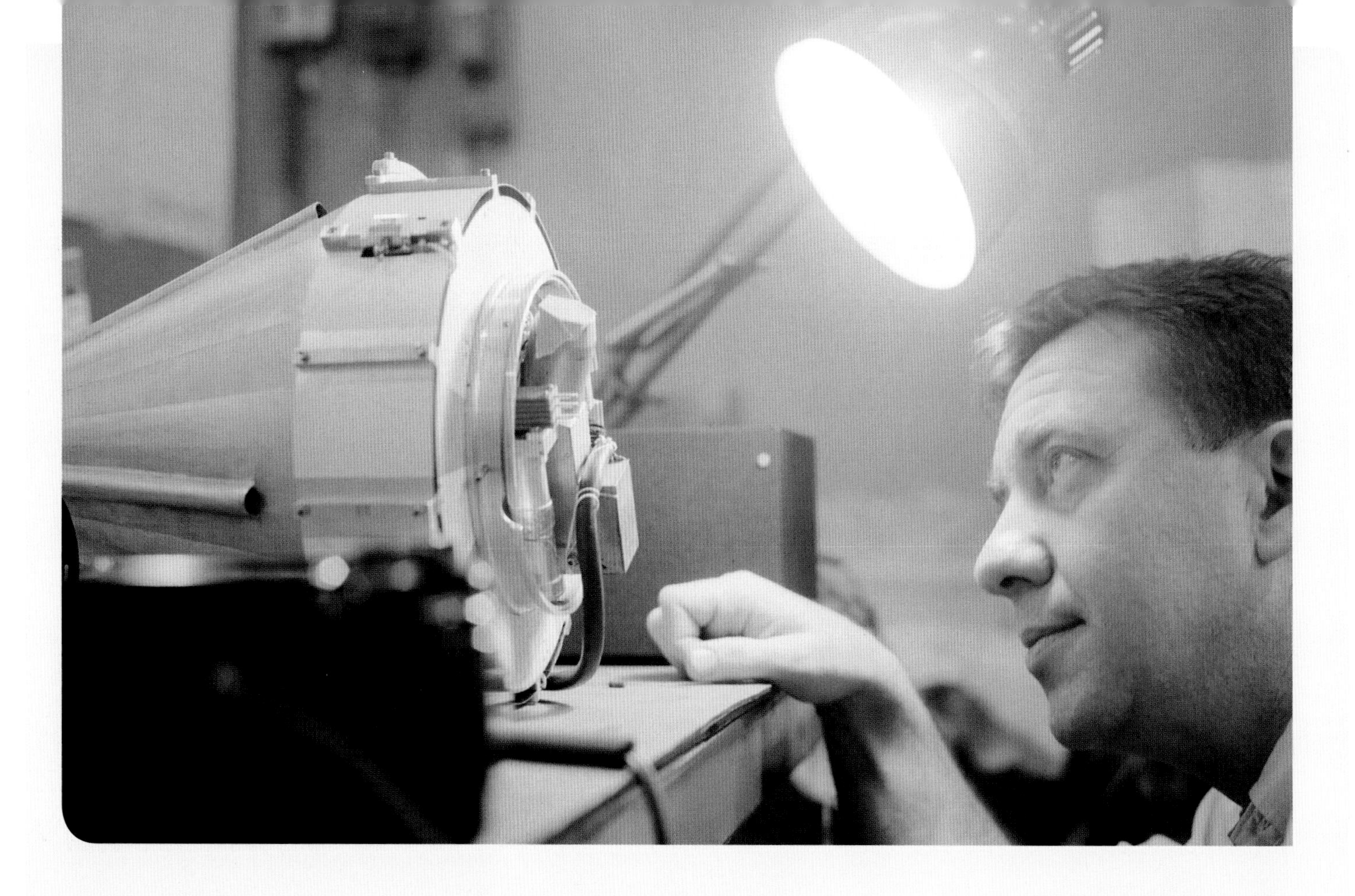

James Van Allen

1914–2006

The iconic academic who dedicated his career to the pursuit of space science.

James Van Allen in his workshop in 1957, working on instruments for the later Pioneer IV lunar probe.

James Alfred Van Allen was born and grew up in the small town of Mount Pleasant, Iowa. From a young age he was fascinated with science, mathematics, and engineering, and he proved particularly skilled in any classes at school that required mechanical skills. After attending college in his home state, Van Allen enrolled at the California Institute of Technology (Caltech) in the fall of 1935. He completed his masters in science the next year before proceeding to his doctorate, defending his dissertation in June 1939.

Thereafter he accepted employment with the Department of Terrestrial Magnetism at the Carnegie Institution of Washington. Under the watchful eye of his department director, the renowned geophysicist Merle Tuve, Van Allen completed a Carnegie Research Fellowship measuring nuclear decay caused by gamma rays. He also became involved in several other research projects at the institution, working on the development of proximity mines for use by the United States Navy.

After World War II, Van Allen returned to civilian life and began working in high-altitude research, first for the Applied Physics Laboratory at Johns Hopkins University and, from 1950, at the University of Iowa. Around this time, his experiments were among those selected to be flown in the upper-atmosphere research experiments conducted in New Mexico using captured V-2 rockets.

Van Allen's career took an important turn in 1955 when he began designing scientific instruments for the Army Ballistic Missile Agency's proposed satellite in support of the International Geophysical Year (IGY; p. 88). But the Navy's Project Vanguard, which proposed a more complex satellite, was chosen over the Army's proposal. In the wake of the success of Sputnik 1's launch, Van Allen was soon recruited to develop scientific instruments for Explorer 1. He remained at the University of Iowa for the rest of his career, serving as Emeritus Carver Professor of Physics from his formal retirement in 1985 up to his death.

bands of radiation that encircle the Earth. They further discovered that their radiation is composed of highly energetic charged particles and is held in place by the planet's magnetic field, its magnetosphere. Together, the belts serve to shield the surface of the Earth from deadly cosmic rays and radiation from the Sun.

The discovery of the Van Allen belts and their relationship to the magnetosphere gave rise to the scientific field of planetary magnetospherics and subsequently helped establish the link between the interactions of charged particles from the Sun, also known as the solar wind, and the polar fringes of the magnetosphere that resulted in the aurorae of the northern and southern lights. Like these cascading light shows, a stream of scientific illumination flowed from Explorer 1. The satellite demonstrated that scientific satellites could provide invaluable insights into the workings of our planet and the wider universe.

The upper part of the Jupiter-C rocket containing Explorer 1 at Cape Canaveral, Florida, in 1958.

Vanguard and Jupiter-C Rocket Specifications

Vanguard Specifications

Length: 66 feet (20 m)

Weight: 64,200 pounds (29,180 kg)

Thrust: 83,000 pounds (370,000 N)

Propellants: Hydrazine and liquid oxygen

Manufacturers: Chrysler (airframe), Rocketdyne (engine)

Jupiter-C/Redstone Specifications

Length: 70 feet, 9 inches (21.6 m)

Weight: 22,600 pounds (10,000 kg)

Thrust: 27,000 pounds (120,000 N)

Propellants: First stage, kerosene and liquid oxygen; second stage, hydrazine and nitric acid; third stage, solid propellant

Manufacturers: Glenn L. Martin Co. (prime), General Electric, Aerojet General, Thiokol (engines)

Establishing NASA

President Dwight Eisenhower commissioning Thomas Keith Glennan (right) as the first administrator of NASA and Hugh Dryden (left) as deputy administrator.

The American National Advisory Committee for Aeronautics (NACA) was founded on March 3, 1915, as a federal agency responsible for aeronautical research in the United States. On October 1, 1958, NACA was succeeded by the National Space and Aeronautics Administration, or NASA. The new organization was created in the wake of the Soviet Union's successful launch of Sputnik 1 to bring a unifying focus to America's space exploration efforts and to disseminate the fruits of the nation's scientific research widely.

NASA's first administrator, Thomas Keith Glennan (1905–95), presided over an organization that had absorbed NACA whole, including its approximately 8,000 employees, while its annual budget of $100 million made up the core of NASA's initial funding. A small headquarters staff in Washington directed operations, but three research laboratories—the Langley Aeronautical Laboratory, established in 1918 (p. 68); the Ames Aeronautical Laboratory near San Francisco, opened in 1940; and the Lewis Flight Propulsion Laboratory built in Cleveland, Ohio, in 1941—provided the bulk of the organization's technical capabilities. There were also two small flight test facilities, one for high-speed flight research located alongside a military base at Muroc Dry Lake, established in the high desert of California in 1943, and one for sounding rockets at Wallops Island, Virginia.

Shortly after NASA was formally established, several other organizations involved in various space exploration projects were also folded into it. One of the most important of these was the space team at the Naval Research Laboratory based in Maryland. Officially becoming a part of NASA on November 16, 1958, this unit became the science-focused Goddard Space Flight Center.

At roughly the same time, NASA gained responsibility for several disparate space satellite programs, two lunar probes, and a research effort to develop a more powerful single-chamber rocket engine. In December 1958, NASA also acquired control of the Jet Propulsion Laboratory (JPL) in Pasadena, California (p. 54).

In May 1960, NASA acquired the Army Ballistic Missile Agency's (ABMA) rocket development group based at the Redstone Arsenal in Huntsville, Alabama (later renamed the General George C. Marshall Space Flight Center). Perhaps most importantly, along with the agency, NASA also acquired the services of its technical director, Wernher von Braun, who was by then one of the most prominent advocates for space exploration in the country. Von Braun's group brought NASA a strong sense of technical competence.

Significantly, while many of its predecessor components had been associated with various branches of the military, NASA was specifically defined from its inception as a nonmilitary organization whose work was to be highly visible to the public. Among the organization's first and most highly anticipated public missions following its establishment was to be Project Mercury, an undertaking to put an American in space (p. 105).

> *"It's human nature to stretch, to go, to see, to understand. Exploration is not a choice, really; it's an imperative."*
>
> MICHAEL COLLINS, AMERICAN GEMINI AND APOLLO ASTRONAUT

BELOW LEFT NASA stationery showing the organization's original logo from 1959.

BELOW RIGHT Wernher von Braun speaks with President Dwight Eisenhower during the dedication of the new General George C. Marshall Space Flight Center in Huntsville, Alabama, on September 8, 1960. Behind them is the engine of the first stage of the Saturn 1B rocket.

NATIONAL AERONAUTICS AND SPACE ADMINISTRATION
NATIONAL AERONAUTICS AND SPACE ADMINISTRATION U.S.A.
NASA HEADQUARTERS
1520 H STREET NORTHWEST
WASHINGTON 25, D. C.
TELEPHONE: EXECUTIVE 3-3260 TWX: WA 785
IN REPLY REFER TO

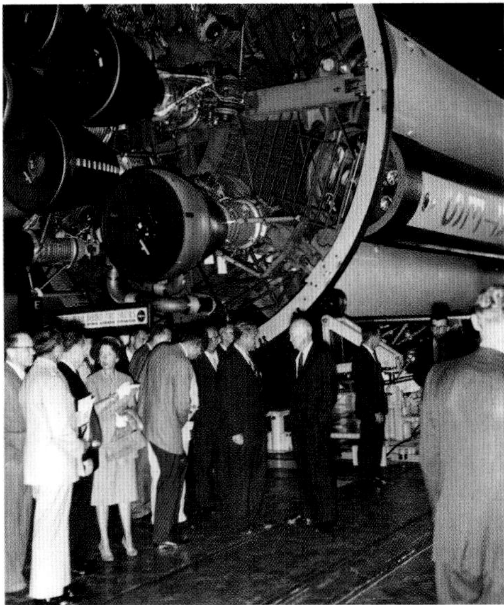

UNITED
STATES

Project Mercury and the Mercury Seven

Less than a week after the establishment of NASA on October 1, 1958, American President Dwight D. Eisenhower (1890–1969) assigned the new space agency the task of placing an American in orbit. For the next five years, NASA worked on Project Mercury: a three-phase project for six spaceflights to put the first Americans in space. The first phase of the project involved selecting and training a group of men for space travel and developing the systems to ensure their safety in space. The second phase involved adapting the Redstone ballistic missile to send humans on suborbital ballistic flights. The third phase involved adapting the more powerful Atlas rocket to launch an occupied capsule into orbit.

To manage Project Mercury, NASA set up the special Space Task Group under Robert Gilruth, who had overseen years of rocket testing and development at the Wallops Island launch facility (p. 102). The group was to oversee the development of all the hardware, project management, contractor support, and ground launch infrastructure. The Space Task Group designed a compact capsule capable of sustaining a single person in orbit and, using a combination of retrorockets and parachutes, returning him safely to the Earth, landing in the sea. The conical

OPPOSITE Technicians work on three Mercury capsules used for tests in various stages of assembly. The Mercury capsule to the right has a launch escape tower that would boost the capsule away from the rocket in the event of a problem during launch.

ABOVE A sequence of images from 1959 showing an abort test of the Mercury capsule with an escape motor, from launch to parachute opening.

Mercury spacecraft was about 11 feet (3.3 m) long and 6 feet (1.8 m) wide at the base, giving its occupant very little room for movement. The Mercury mission also required expanding NASA's spaceflight infrastructure with the establishment of a series of ground tracking stations around the globe, a new mission control center, and an expanded launch complex at Cape Canaveral.

The first "crewed" Mercury test took place at NASA's Wallops Island launch site in Virginia on December 4, 1959, with a suborbital test flight of the Mercury space capsule occupied by a rhesus monkey named Sam. This and many other test flights were conducted using the Little Joe launch vehicle, built from a small cluster of solid-fuel rockets. Much cheaper to produce and fuel than the larger Redstone, Little Joe offered a cost-effective means to test the heat shielding, launch escape systems, and general space-worthiness of the Mercury capsule. More tests followed, and on January 31, 1961, the chimpanzee Ham flew 157 miles (252 km) into space during a 16-minute, 39-second flight.

Away from the capsule design, NASA engineers continued with preparations for the orbital phase of Project Mercury. Putting an astronaut in orbit required a more powerful launch vehicle than the Redstone, and NASA turned to a modified version of the Atlas ballistic missile. After numerous initial difficulties, NASA's technicians eventually achieved their first successful orbital flight with an unoccupied Mercury capsule propelled by an Atlas rocket on September 13, 1961. On November 29, a final test flight took place before NASA was finally ready to send a man into space.

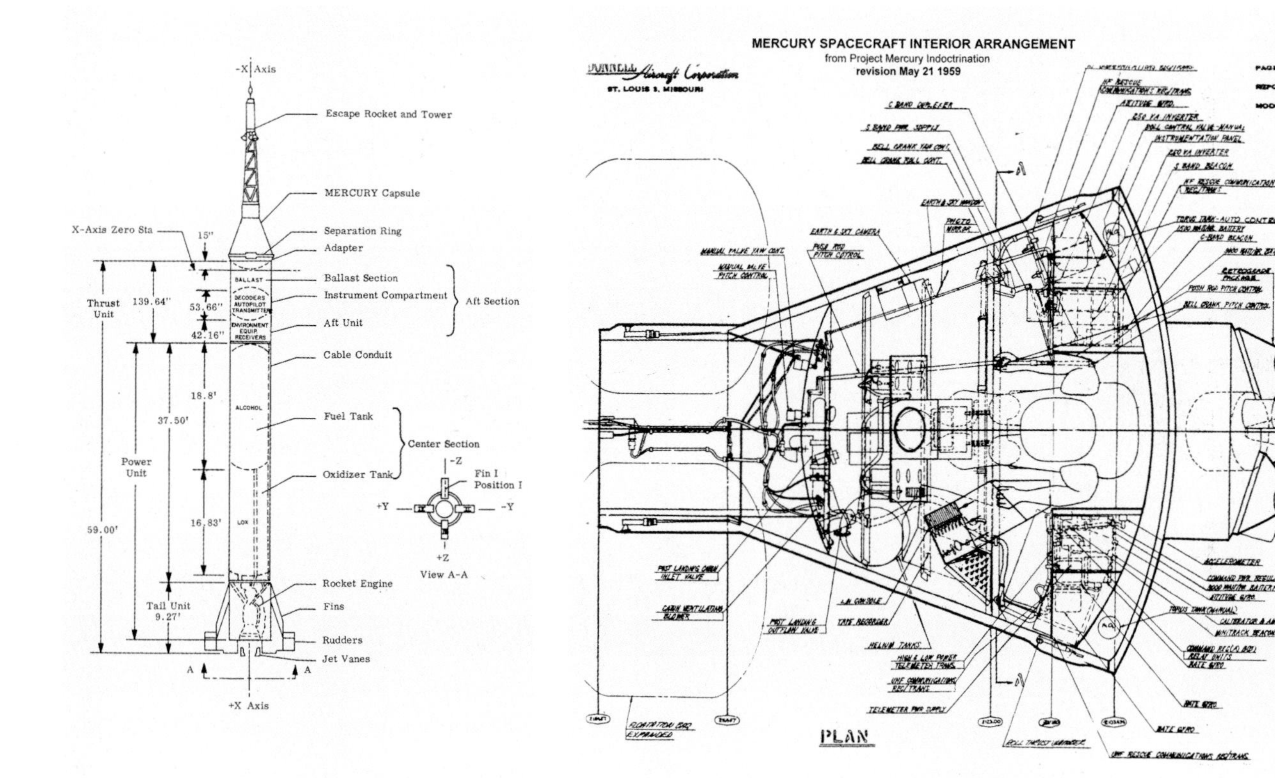

BELOW LEFT Technical drawing of the Mercury-Redstone. The Redstone launch vehicle was not powerful enough to place a Mercury capsule in orbit and was only used in suborbital flights.

BELOW RIGHT The Mercury spacecraft interior arrangement from the Project Mercury indoctrination manual, produced in May 1959.

Choosing the First Astronauts

1959

As Project Mercury began, NASA selected its first batch of spaceflight trainees, or astronauts, as they came to be known. The organization's minimum standards for the mission were straightforward and underlined the need for participants who were not just young and physically fit but were also seasoned pilots and short enough to cope with the cramped conditions of the Mercury capsule. Each man was to be under 40, with a height under 5 feet 11 inches (1.8 m). He also needed to be in excellent physical condition, have a bachelor's degree or equivalent, be a graduate of test pilot school, have amassed 1,500 hours of flying time, and be a qualified jet pilot.

The grueling selection process started with interviews, followed by a battery of written tests, technical interviews, psychiatric interviews, and medical history reviews. By March 1, 1959, the process had reduced the original pool of test pilot volunteers down to just 36. After further interviews and physical exams, the NASA leadership whittled this list down to a final seven in April 1959.

NASA publicly unveiled its astronauts, dubbed the Mercury Seven, in a press conference on April 9, 1959. They were all active serving members of the United States military; Marine Lieutenant Colonel John Glenn Jr. (1921–2016); Navy Lieutenant Commanders Walter "Wally" Schirra Jr. (1923–2007), Alan Shepard (1923–98), and Scott Carpenter (1925–2013); and Air Force Captains Gordon Cooper (1927–2004), Virgil "Gus" Grissom (1926–67), and Donald "Deke" Slayton (1924–93). Almost overnight these seven young men became famous, due in part to a deal NASA struck with *Life* magazine for exclusive rights to their personal stories.

As James Reston of the *New York Times* exulted, "What made them so exciting was not that they said anything new but that they said all the old things with such fierce convictions...They spoke of 'duty' and 'faith' and 'country' like [American poet] Walt Whitman's pioneers...nobody went away from these young men scoffing at their courage and idealism."

ABOVE LEFT These silver boots might look like they belong on a dance floor but were worn by Mercury astronaut Walter "Wally" Schirra. Mercury spacesuit boots were made of a lightweight nylon fabric with tennis shoe–type soles, and were not integrated into the suit. The suit's pressure bladder covered the foot and was then inserted into the boot, which was adjusted separately.

ABOVE RIGHT The Mercury Seven astronauts in their iconic silver spacesuits, 1959. Left to right, back row: Alan Shepard, Virgil "Gus" Grissom, and Gordon Cooper. Front row: Walter "Wally" Schirra, Donald "Deke" Slayton, John Glenn, and Scott Carpenter.

Choosing Cosmonauts

RIGHT Sergei Korolev (lower row, center) with some of the first group of Russian cosmonauts in 1961. Yuri Gagarin (in uniform) is seated to Korolev's right. Standing on the far right is Grigori Nelyubov, who was kicked out of the cosmonaut program for drunk and disorderly behavior. He was airbrushed out of later published versions of this photo.

BELOW Sergei Korolev, chief architect of the Soviet space program in the early 1960s, seen here at the radio during a Vostok mission, played a key role in the cosmonaut selection process.

In contrast to the very public process of choosing American astronauts, the Soviet Union undertook a far more secretive selection process for its space explorers, dubbed cosmonauts. In September 1959, Sergei Korolev established a cosmonaut selection commission under the Scientific Research Institute of the Soviet Air Force. Since American astronauts had just begun their training in the full glare of the world's media, the Soviets, who lacked a focused central organizing body like NASA, were nevertheless able to borrow and adapt much of what they saw of NASA's training methods for their own cosmonaut program.

They too drew from an initial pool of about 3,000 military pilots who had experience flying high-performance aircraft. This group then underwent reviews of their medical and other records, a battery of medical tests, and a range of physical and psychological stress testing. This whittled the group down to 15 pilots, collectively referred to as "Air Force Group One." The group then reported to the newly formed Cosmonaut Training Center, located just outside of Moscow, for assignment in March 1960. The center later evolved into the Star City facility, which is still used for Russia's cosmonaut training today. Of the 15 who were trained, 11 eventually made it into space, completing at least one orbital mission each.

Among those 11 were Yuri Gagarin (1934–68), Anatoli Kartashov (1932–2005), Andrian Nikolayev (1929–2004), Pavel Popovich (1930–2009), and Gherman Titov (1935–65). This group rapidly rose to the top of the candidates in cosmonaut training and soon took on extra responsibilities over their fellow candidates. All of the cosmonaut selectees were aware they were competing for the honor of being the first to fly in space. By early 1961, Gagarin was widely viewed as the frontrunner, both by Korolev and the other cosmonauts. However, the chief designer did not want to give any indication that he had any particular candidate in mind for the mission, fearing that to do so would harm the motivation of the others. Korolev refused to announce, even internally, who would take on the responsibility of becoming the first cosmonaut to fly the Vostok 1 mission almost until the point of its launch (p. 114).

Unfortunately, as Korolev and his team would later learn, while their cosmonaut training process was rigorous, it was not infallible. Cosmonaut Gherman Titov reported experiencing disorientation, extreme fatigue, dizziness, and nausea on his spaceflight. This prompted a reconsideration and overhaul of the initial cosmonaut selection and training processes.

In 1962, Korolev selected a second group of cosmonaut candidates, and this time the physical capabilities and resilience of the group were more rigorously assessed. Moreover, this group included five women, one of whom would go on to become the first woman in space (p. 124).

BELOW LEFT Alexey Leonov photographed in 1963 during his cosmonaut training.

BELOW RIGHT Cosmonaut Gherman Titov trains for the space mission at the Cosmonaut Training Center in Zvezdny Gorodok, also known as Star City.

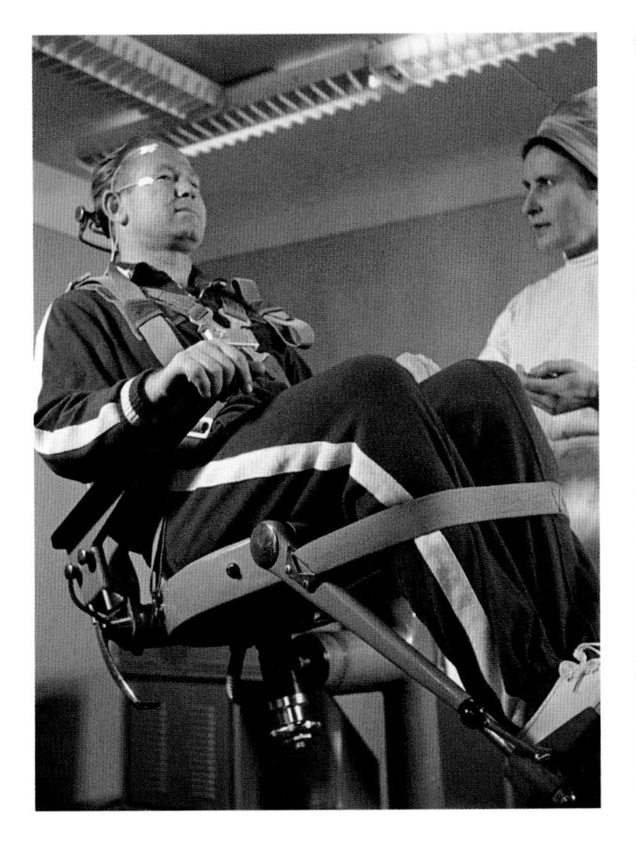

Training the First Space Explorers

The seven original Mercury astronauts participate in survival school at Stead Air Force Base in Nevada to prepare them in the event of an emergency in a remote area. Portions of their clothing have been fashioned from parachute material, and all have grown beards during their time in the wilderness.

Both the Soviet cosmonauts and the American astronauts undertook intense training to prepare for their space exploration missions, with many exotic devices, complex procedures, and physical endurance tests. Throughout their respective programs, astronauts and cosmonauts underwent medical tests before, during, and after their flights. They also underwent physical training, procedure training, pilot familiarization training, and plenty of experiment practice. They would simulate all phases of their missions to assure mission planners that they could accomplish their tasks without error, almost by rote.

Those early space explorers were keenly aware that their simple spacecraft allowed humans to venture into space only for short periods of time. They also knew they had to learn quickly how best to cope with and overcome any technological hurdles they might encounter. Life scientists worked closely with their respective trainee space travelers to ensure both their survival and their missions' success. Engineers worked to help them master their spacecrafts' controls and systems. They even had to undertake Earthly survival training in case they landed in a remote location and needed to sustain themselves while awaiting rescue.

The mission crews also needed to account for the basic human physiological needs, such as a sufficient supply of oxygen, to ensure the comfort and safety of their astronauts and cosmonauts while in space. Scientists, flight surgeons, and the fliers themselves all worked to establish the best protective garments, pressure suits, and spacesuits of various types.

One notable individual who worked hard to test the endurance limits of the human body was United States Air Force Lieutenant Colonel John Stapp (1910–99). Using a rocket-powered sled, Stapp and his colleagues conducted a series of tests simulating the extreme acceleration and deceleration that pilots and astronauts can experience in flight. Astronauts and cosmonauts had their capacity to withstand such pressures assessed in centrifuges that could simulate more than ten times the

BELOW LEFT United States Air Force Lieutenant Colonel John Stapp rides the rocket sled at Edwards Air Force Base, California. Sleds such as this were used to measure human tolerance for high speeds, such as those encountered during spaceflight.

BELOW RIGHT This desalter kit was intended to convert seawater into drinkable water in case an astronaut splashed down in the ocean far away from recovery crews.

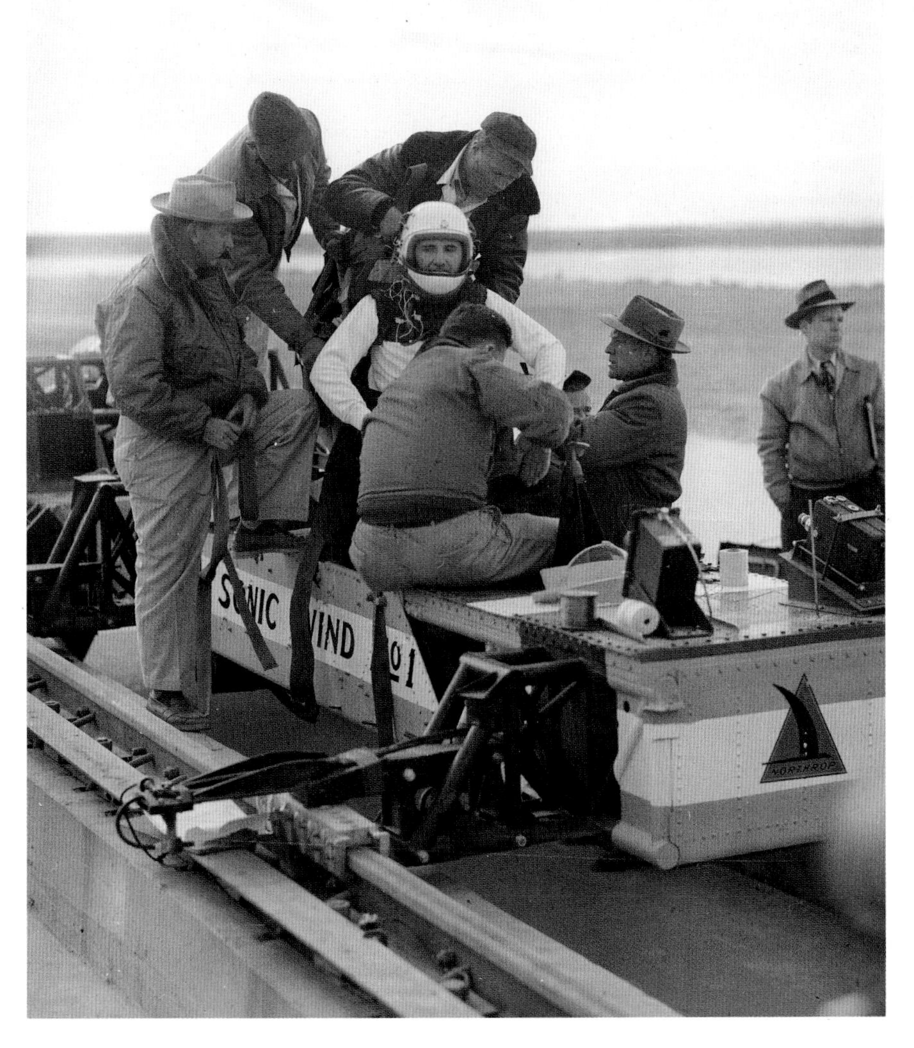

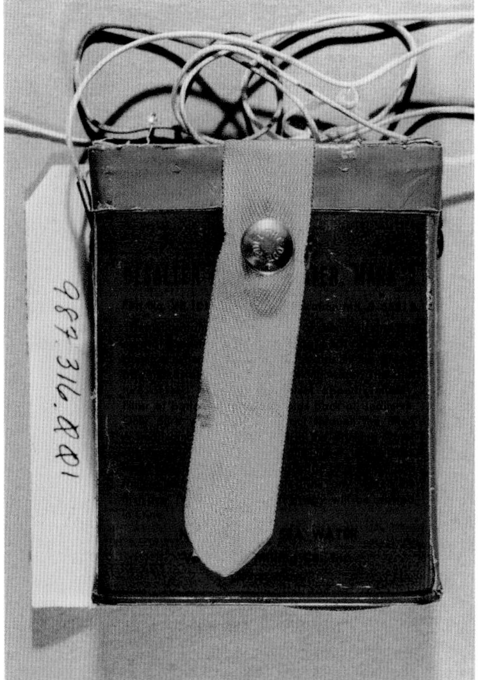

force of gravity as they spun round and round. Low-gravity acclimatization was conducted through flights on aircraft that repeatedly climbed and dived to simulate the constant free fall of orbital space travel.

Typically, astronaut and cosmonaut training could be split into three stages: individual general training; group training; and specific training for missions. Cosmonauts and astronauts learned all the technical and management skills required to understand, operate, and, if necessary, repair their respective spacecraft. In both the Soviet Union and the United States, prospective cosmonauts and astronauts developed their own specializations, and served as liaisons to the technical teams working on building the space vehicles, providing valuable user feedback on their equipment. Finally, as flights neared their launch dates, those training to travel in space would rehearse and rehearse the details of their missions, often in full-size mock-ups of their spacecraft. The goal was that all the space explorers would become so proficient at their various tasks that they could perform them automatically, almost without thinking.

The success of this training could be witnessed throughout the early human space missions. Astronauts and cosmonauts regularly faced glitches and equipment failures, but in each and every case they remained calm and were successful in overcoming the difficulties they faced before going on to complete their missions as planned.

> *"When you explore a frontier, the people who explore bring back images and bring back stories about what these frontiers are like. I feel an obligation to share this experience so that everybody else can at least participate through the eyes of the people who do go into the frontiers."*
>
> DON PETTIT,
> NASA ASTRONAUT

Astronaut Buzz Aldrin practices extravehicular activity in an underwater neutral buoyancy simulator in 1966.

Yuri Gagarin in his orange spacesuit, shortly before the launch of Vostok 1 on April 12, 1961.

The First Spacesuits

1961–65

Built specifically for space travel, spacesuits were designed to ensure safety and comfort.

Survival in space depends on ensuring basic human physiological needs in an intensely hostile environment. Among the essential requirements are: an appropriate atmospheric pressure; breathable oxygen; a moderate temperature; sufficient drinking water and food; the capacity to withstand centrifugal pull on physical systems; and radiation mitigation. There are also other, less immediate, needs that must be considered, especially on longer journeys, such as ensuring adequate provision of toilet facilities.

There are really two ways to ensure space explorers' physiological safety in space: either build a sealed protective spacecraft for them, or clothe them in various layers of protective garments that will do essentially the same job as a spacecraft.

The first clothing for astronauts and cosmonauts was adapted from the pressure suits that were developed during World War II (p. 46). The iconic silver spacesuits used in Project Mercury were transferred virtually intact from flight suits previously used in high-performance X-planes.

Interestingly, their metallic silver surface had no real practical purpose, and was selected mostly for its futuristic look. No NASA spacesuits after Mercury used this fabric. The orange Soviet spacesuit, worn by Yuri Gagarin and other Vostok cosmonauts, was based on fighter pilot gear used in the Soviet Air Force. Its bright coloring was chosen to make it easy to spot downed pilots from far away. Both types of suits offered heating and cooling, air for breathing, and pressure to aid the wearers' comfort and to ensure that they did not pass out in high g-force maneuvers during launch and landing.

It was only after mission planners began to contemplate spacecraft that involved crews of more than one that the question arose of how to sustain life outside a space capsule.

Designers knew that any spacesuit for out-of-capsule use had to be flexible, resistant to extremes of heat and cold, permit its wearer to retain a significant amount of dexterity, and provide a flow of oxygen over the whole body. Gloves, life-support, helmets, and maneuvering equipment all needed to be considered.

The solution first arrived at was an umbilical cord that physically linked the space traveler to the craft as well as supplying oxygen and communications.

Gagarin and the Missile Shot Heard Around the World

RIGHT Yuri Gagarin prepares for the first-ever human orbital mission on April 12, 1961.

BELOW This launch of Gagarin aboard Vostok 1 opened the era of human spaceflight.

April 12, 1961, marked another great milestone in the space race, as Soviet cosmonaut Yuri Gagarin became the first human to reach orbit. He was propelled into the history books by a specially adapted R-7 rocket carrying the 5,400-pound (2,460-kg) spherical Vostok 1 capsule and its 5,000-pound (2270-kg) conical equipment module. Although the craft also possessed manual controls, Gagarin's trip was planned to be entirely controlled from the ground. The Soviets flew their early missions this way for two main reasons. Firstly, there was the concern that, in the largely unknown environment of space, the cosmonauts might become incapacitated and unable to perform any tasks themselves. Secondly, some Soviet officials were worried that if the cosmonauts had complete control of their spacecraft, they might choose to defect to the West, and thereby cause embarrassment to the Soviet Union.

Gagarin regularly reported his status over the course of his 108-minute single-orbit flight, reassuring the flight controllers that he was feeling fine and in good spirits. As his craft entered the final stages of its descent, Gagarin ejected from his capsule, eventually parachuting down to land around 14 miles (22 km) southwest of the city of Engels in eastern Russia. Gagarin's parachute landing was a feature of the Vostok's simple design. Without any reliable method to slow the capsule down sufficiently to make a soft landing possible, ejecting from the falling capsule was viewed as the least dangerous option available to the cosmonaut, especially when the terrain the capsule would come down on could not be guaranteed or even predicted with any degree of accuracy.

Initially, the Soviets attempted to cover up this element of Vostok's design, mainly for fear that it would harm the legitimacy of their accomplishment with the Fédération Aéronautique Internationale (FAI), the official world governing body for air sports and air travel records. At the time, the organization stipulated that pilots had to land in their craft for an aeronautical record to be valid. But the Soviets need not have worried: after Vostok's landing procedure became widely known, the FAI revised its guidelines for spaceflight to allow such landings.

Following Gagarin's triumph, the Soviet leadership held a ceremony to honor him in Moscow's Red Square on April 14, 1961. The great success of the Vostok 1 mission made the gregarious Gagarin a global hero, and he served as an effective exemplar for his nation on his subsequent world tour.

But for all the acclaim and apparent success, Vostok 1 had encountered some significant problems during its short flight. The mission had come close to disaster when the capsule spun dangerously out of control at the beginning of its reentry

"Orbiting Earth in the spaceship, I saw how beautiful our planet is. People, let us preserve and increase this beauty, not destroy it!"

YURI GAGARIN (1934–68), SOVIET COSMONAUT AND FIRST HUMAN IN SPACE

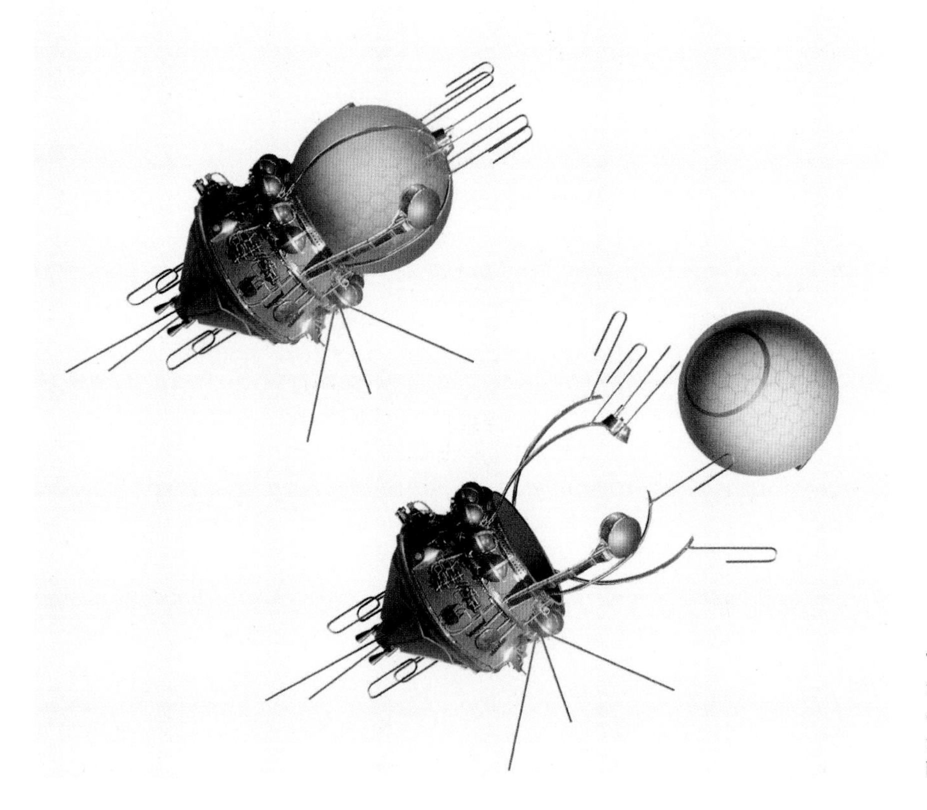

The physical separation of the spherical Vostok 1 descent capsule from its equipment module in preparation for landing back on Earth.

Yuri Aleksevich Gagarin

1934–68

The spaceman who ushered in the age of human space exploration.

After his historic flight, Yuri Gagarin speaks to a crowd of adoring Russians at the Moscow Film Festival in 1961. Among his fans in the front row is Italian actress Gina Lollobrigida (far left).

Cosmonaut Yuri Aleksevich Gagarin was born on March 9, 1934, the son of a collective farmer in the village of Klushino in the Smolensk region of the Soviet Union. Like many Soviet citizens of his generation, his education was interrupted in 1941 by the German invasion of his homeland.

After the war, the Gagarins moved to the town of Gzhatsk, where young Yuri completed his basic education. From there, he went to vocational school in the town of Lyubertsy, near Moscow, specializing in metalworking. He completed his degree in 1951, and later attended the Saratov Industrial Technical School, eventually earning a diploma with honors.

While studying at the technical school, Gagarin joined a local flying club and developed his aviation skills. In 1955 he joined the Orenburg Aviation College, and after completing its course in 1957 he entered the Soviet Air Force and became a fighter pilot. While serving in the military, Gagarin also completed studies at the Zhukov Air Force Engineering Academy.

He was selected for cosmonaut training in 1959 and underwent a series of increasingly rigorous physical and mental exercises to prepare for spaceflight (p. 108). Part of Gagarin's appeal to the Soviet leadership was that he represented his nation's ideal of the worker who rises through the ranks solely on merit. His handsome appearance, thoughtful personality, and boyish charm also made him an attractive figure on the world stage. The importance of these attributes was not lost on Nikita Khrushchev and other senior Soviet leaders.

For Gagarin, the reward for his newfound fame after he completed his space mission was the hard work of being a figurehead for his nation, a duty that took a heavy toll on him.

He soon became viewed as too important to risk losing in an accident and, after serving as a back-up cosmonaut, he was removed from consideration for future space missions. He died just a few years later, in a plane crash while on a training mission for the Soviet Air Force on March 27, 1968.

sequence. As Gagarin reported to officials during his post-flight debriefing: "As soon as the braking rocket shut off, there was a sharp jolt, and the craft began to rotate around its axis at a very high velocity." This spinning was apparently caused when the equipment module failed to separate completely from Gagarin's descent capsule.

After approximately 10 stressful minutes, the final connectors linking the capsule and the equipment module burned through in the heat of reentry; the spacecraft then stabilized, and the rest of the landing proceeded as planned.

The Need to Catch Up

Yuri Gagarin's flight prompted U.S. President John F. Kennedy to congratulate the Soviet Union on its success, and to admit in a press conference held shortly after the announcement of the Vostok mission: "We are behind...the news will be worse before it is better, and it will be some time before we catch up."

The Soviet success prompted Kennedy to seek a space exploration project in which the United States could outdo its Cold War rival. The search for a suitable undertaking had far-reaching consequences for the subsequent course of the Space Race.

Vostok 1 being examined on a stand in 1960.

UNITED

The First Americans in Space

For many Americans, the success of the Soviet Union's first crewed space mission stung. As journalist Hanson Baldwin put it in a *New York Times* article: "Even though the United States is still the strongest military power and leads in many aspects of the space race, the world—impressed by the spectacular Soviet firsts—may believe we lag militarily and technologically." It was not much comfort to wounded pride when around three weeks after Yuri Gagarin completed his historic flight, Alan Shepard (1923–98) became the first American to reach space during a 15-minute suborbital flight on May 5, 1961. Comparisons between the two missions were inevitable, and reflected poorly on the American effort. Gagarin had orbited the Earth; Shepard had barely touched space. Gagarin's Vostok 1 spacecraft had weighed roughly 10,400 pounds (4,730 kg); the Freedom 7 Mercury capsule weighed 2,100 pounds (952 kg). Gagarin had been weightless for 89 minutes; Shepard for only 5 minutes.

Things didn't get much better for American prestige when, during the second suborbital Mercury mission on July 21, 1961, the hatch from the capsule, Liberty Bell 7, blew off prematurely on splashdown. Seawater surged in, and pilot Virgil "Gus" Grissom barely managed to escape. A helicopter rescued him shortly before the capsule sank to the bottom of the Atlantic Ocean. In 1999, a team of undersea archaeologists recovered Grissom's capsule from the seabed. It was subsequently restored and placed on public display at the Kansas Cosmosphere in Hutchinson, Kansas.

It was only on February 20, 1962, that NASA finally sent an astronaut into orbit. John Glenn circled the world three times in his Friendship 7 capsule. His flight was not without incident, however; he had to fly parts of the last two orbits manually because of an autopilot failure. There was also a warning light during his reentry that showed his capsule had a loose heat shield. This prompted NASA experts to advise Glenn to retain throughout his descent the retrorocket pack that he had originally been instructed to jettison on reentry. The retrorocket was affixed to the base of the spacecraft over the heat shield with three strong cables. Under normal circumstances, they would have been severed with explosive bolts, but on this occasion they helped to hold the heat shield in place.

Glenn's flight provided a much-needed boost to American national pride. Quite aside from the technological achievement of his mission, the American public proudly embraced Glenn as a hero. NASA was soon flooded with hundreds of requests for personal appearances, and the

OPPOSITE The launch of Alan Shepard in Freedom 7 on May 5, 1961.

BELOW Astronaut Shepard is recovered after his Freedom 7 capsule splashed down in the Atlantic Ocean.

Mission Control Museum

The mission control facility used for Project Mercury was demolished in 2010, but the mission control facility at the Johnson Space Center that was in active service from 1964 to 1998 has been retained. The public can visit this crucial part of space exploration history on guided tours to this day.

organization quickly became aware of the powerful public relations potential of their astronauts. Glenn addressed a joint session of the United States Congress and participated in several ticker-tape parades around the country.

Three more successful Mercury flights took place during 1962 and 1963. Scott Carpenter made three orbits on May 20, 1962, and on October 3, 1962 Walter "Wally" Schirra flew six orbits. The last Project Mercury flight launched on May 15, 1963, and saw L. Gordon Cooper circle the Earth 22 times in 34 hours. Project Mercury had succeeded in its aim of putting an American in orbit, and it had also given NASA the opportunity to explore aspects of spacecraft tracking and control, as well as the various biomedical issues associated with spaceflight. All this would prove invaluable as the organization readied itself to take on new challenges.

ABOVE The Soviet Mission Control Center in Moscow during the Vostok 2 mission in August 1961. The main screens show both a map of the world and a map of the Soviet Union.

RIGHT As directed by mission control, the Pad 14 gantry at Cape Canaveral moves back during a systems check several days before the last Project Mercury launch on May 15, 1963.

Mission Control at the Johnson Space Center during Project Gemini in 1965.

Mission Control Centers

From almost the very beginning of the period of crewed spaceflight, both American and Soviet programs used Mission Control Centers (MCCs) to manage their operations. Regardless of where it is based, each MCC generally has a flight director, an individual with overall responsibility for the success of each mission and the staff of flight controllers.

For Project Mercury, NASA's MCC was based close to its launch facilities in Cape Canaveral, Florida, but in 1964 it transferred its operations to the Manned Spacecraft Center (later renamed the Johnson Space Center) in Houston, Texas.

The appearance of standard NASA MCCs has varied a little over the years, but in general they have retained a relatively consistent layout. Rows of flight controllers sit at their desks facing a set of maps and large display screens that display basic information about the mission currently underway. The flight director typically occupies the central desk in an upper tier, away from the display screens.

"The Trench" is the popular name for the first row of consoles, which accommodate four control stations. These are for the propulsion systems engineer (BOOSTER), the retrofire officer (RETRO), the flight dynamics officer (FIDO), and the guidance officer (GUIDO). The BOOSTER monitors the propulsion system, while the RETRO, FIDO, and GUIDO handle the direction and trajectory of the spacecraft.

The left side of the second row has the flight surgeon monitoring the health of the astronauts, and the capsule communicator (CAPCOM), who is responsible for voice transmissions between the ground and the spacecraft. CAPCOM is virtually always an astronaut and, in order to avoid any possible confusion, no one else in the MCC is normally permitted to speak directly to the astronauts.

On the right side of the aisle are monitors for specific systems, including: the electrical, environmental, and consumables manager (EECOM); the guidance, navigation, and controls systems engineer (GNC); the propulsion controller; the extra-vehicular activity (EVA) staff; the payload and experiments controllers; and the monitor for the craft's communications and instrumentation (INCO).

On the third row sit the communication officer and the operations and procedures engineer, who is charged with coordinating with other teams and assuring the decision to proceed with the flight. An assistant flight director and the flight director sit in a center console; controllers assigned to flight operations and network issues sit on the right side of the row.

The final row includes representatives charged with communicating with the media; flight operations management; and the overall mission director. During crewed spaceflight missions, this row also includes desks for the Johnson Space Center director, the director of the astronaut office, and NASA headquarters officials.

The First Space Planes

The X-plane program (p. 76) continued long after its initial instigator, the National Advisory Committee for Aeronautics (NACA), was folded into its successor organization, NASA, in 1958. But even after the advent of the space age, the goal of surpassing hypersonic speeds, in excess of five times the speed of sound, eluded American aircraft designers until the development of the X-15 rocket plane.

The three X-15s were built by the North American Aircraft Company. They were designed to study flight at very high speeds and altitudes, pushing the frontiers of atmospheric flight into space, which the United States Air Force considered to start at an altitude of just 264,000 feet (80,000 m) at the time. Like the various X-1 rocket planes, the X-15 was mated to a larger B-52 on the ground, then flown to an altitude of 45,000 feet (13,716 m) for flight launches. The X-15's service life was relatively short, encompassing a total of just 199 flights from 1959 to 1968, but it set many records over this period.

This 1962 photograph of shock waves over a scale model of the X-15 shows the aerodynamic stability of the vehicle.

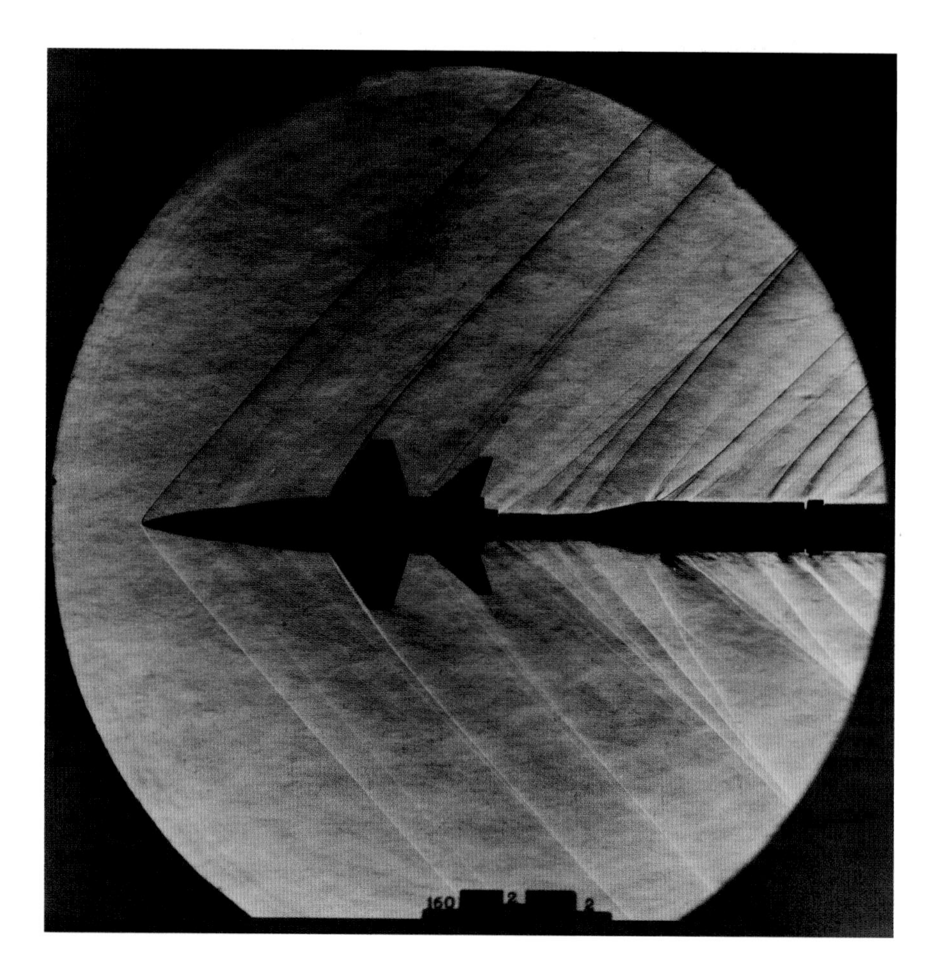

In 1961, Yuri Gagarin became the first human to reach hypersonic speeds as he fell back to Earth at the end of his first spaceflight (p. 114), but the first powered and piloted vehicle to reach the same goal was the X-15 flown by U.S. Air Force Major Robert White (1924–2010). On a test flight on June 23, 1961, White's X-15 reached Mach 5.27 and an altitude of 107,700 feet (32,800 m). In the following year, White flew his X-15 to an altitude of 314,750 feet (96,000 m) above sea level.

The X-15 pilots stretched their aircraft to the limit and beyond. For example, on August 22, 1963, NASA test pilot Joseph Walker (1921–66) flew his X-15 to an altitude of 354,200 feet (108,000 m), well in excess of its design parameters. These tests also highlighted many of the key problems associated with hypersonic flight, including reduced performance and stability; control issues; aerodynamic heating, and heat transfer within an aircraft, although a number of these effects were

not understood immediately. Ultimately, it took many years of study and analysis for NASA's scientists to grasp the full aerodynamic implications of the data obtained from X-15 flights.

As the X-15 program proceeded into the early 1960s, engineers worked to extend the vehicle's speed and altitude capabilities even farther, in order to test the limits of how fast a plane could fly within Earth's atmosphere. The result was a modified X-15 designated the X-15A-2. The X-15A-2 first flew in June 1964, and on October 3, 1967, U.S. Air Force test pilot William John "Pete" Knight (1929–2004) flew the aircraft to the highest speed it ever attained, Mach 6.7, and an altitude of 102,100 feet (31,000 m). This turned out to be the last outing of the X-15A-2 variant. Extensive thermal heating damage sustained during flight made the aircraft unfit for use and not worth the cost of repair. Indeed, the whole X-15 program ended the next year, on October 24, 1968.

As of 2017, a total of 765 research reports have been written using X-15 flight data. This data proved instrumental in establishing the field of hypersonic aerodynamics, an area that has had a profound impact on human spaceflight and particularly the area surrounding atmospheric reentry.

ABOVE LEFT A signed photo of William "Pete" Knight, posing in his flight suit in front of his X-15.

ABOVE RIGHT This view of the forward instrument panel in the cockpit of an X-15 from 1963 shows the panel, center stick, and console with throttle and side stick.

Valentina Tereshkova: The First Woman in Space

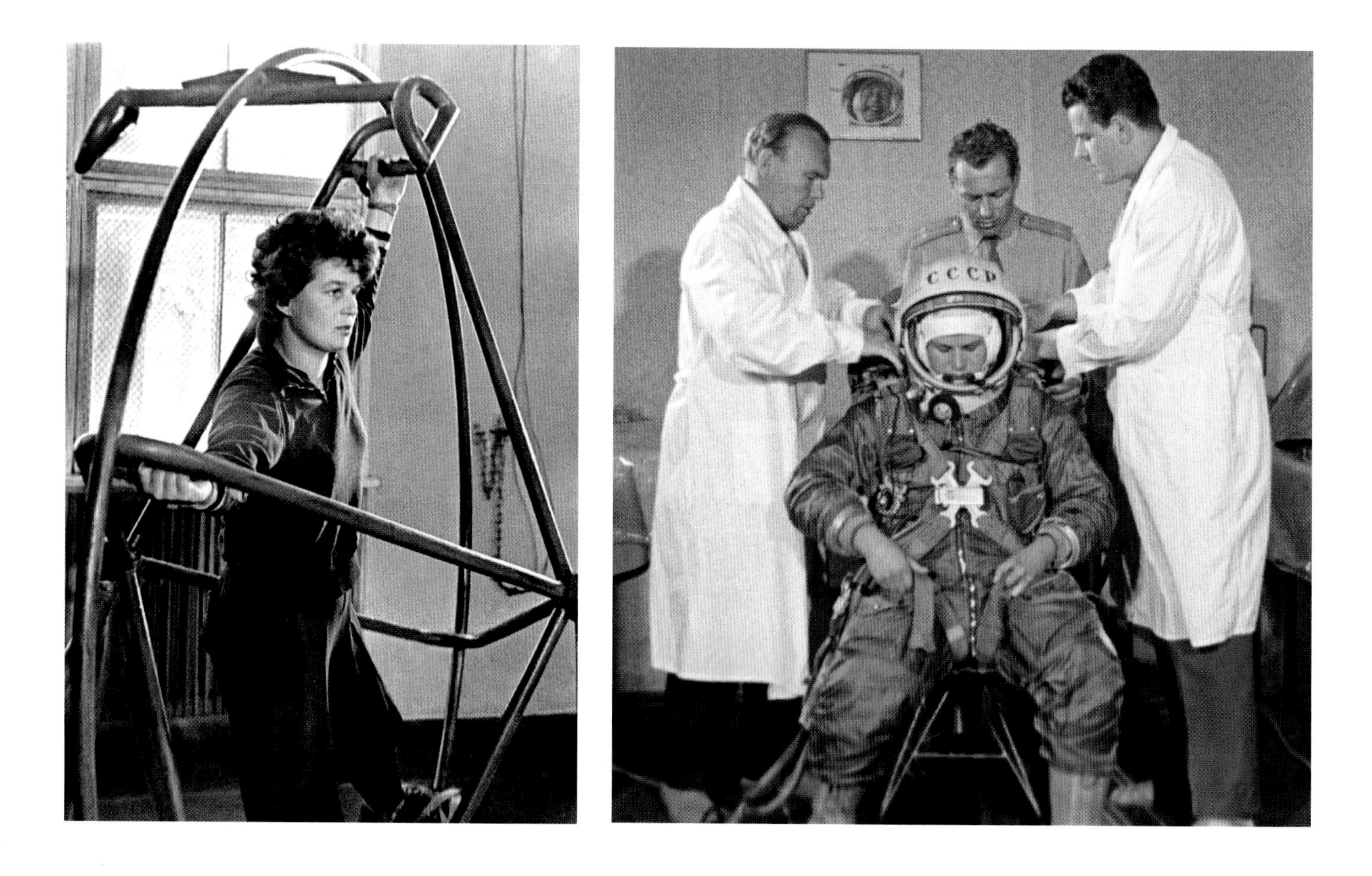

ABOVE LEFT Soviet cosmonaut Valentina Tereshkova, the first woman in space, training in 1963.

ABOVE RIGHT Tereshkova dressed in her spacesuit, preparing for her historic flight.

For the first decade after Sputnik, the Soviet Union's space program seemed to succeed in almost every aspect of space exploration it attempted. It scored another success with the flight of Valentina Tereshkova (1937–), who became the first woman in space on June 16, 1963.

Tereshkova became a cosmonaut on February 16, 1962, after Sergei Korolev persuaded Soviet premier Nikita Khrushchev to approve a plan to put a woman in space. Unlike the male cosmonauts, who had all been experienced pilots, Tereshkova and the other four women short-listed for Vostok missions—Valentina Ponomaryova, Tatyana Kuznetsova, Irina Solovyova, and Zhanna Torkina—were chosen from the ranks of more than 400 parachutists. Tereshkova soon rose to the top of the women in her training group, and gained the nod to fly on Vostok 5.

She trained hard for months in preparation for her mission, excelling at weightless flight training, isolation and centrifuge tests, engineering courses, more than 120 parachute jumps, and pilot training in MiG-15 jet fighters. Korolev considered her a particularly suitable candidate for the mission because of her humble background as a worker in a textile factory. Moreover, her father had been killed in the Winter War in Finland during World War II, and was therefore regarded as a hero.

Originally, Korolev planned to have two female cosmonauts in orbit simultaneously, with Tereshkova in Vostok 5 and Ponomaryova in Vostok 6, but this plan was altered in March 1963. The two spacecraft would still be launched days apart, in order to be in orbit simultaneously, but the male cosmonaut Valery Bykovsky (1934–) was inserted into the Vostok 5 mission slot, while Tereshkova was transferred to Vostok 6. Tereshkova watched Bykovsky's launch on June 14, and followed him into orbit two days later.

Like Gherman Titov (p. 109), Tereshkova also experienced nausea during much of the flight, even vomiting, but successfully completed 48 orbits of the Earth over three days. This was more time in space than the combined total of all of the American astronauts up to that point.

Tereshkova returned to a hero's welcome and, like Yuri Gagarin before her, she was sent around the world as a goodwill ambassador. She later went on to establish a career for herself as a politician, serving first as a deputy in the Supreme Soviet, and later as a member of the Russian State Duma in the post-Soviet era.

But while her feat seemed to signal a greater equality between men and women, within the Soviet space program the reality was somewhat different. Tereshkova's mission had been essentially a publicity stunt to achieve another space first for the

ABOVE Valentina Tereshkova undergoing tests on her reaction to colored lights during training for spaceflight.

BELOW LEFT Tereshkova practises feeding in simulated flight conditions for her spaceflight.

BELOW RIGHT After three days in space, Tereshkova returns to Earth in the Vostok 6 on 19 June, 1963.

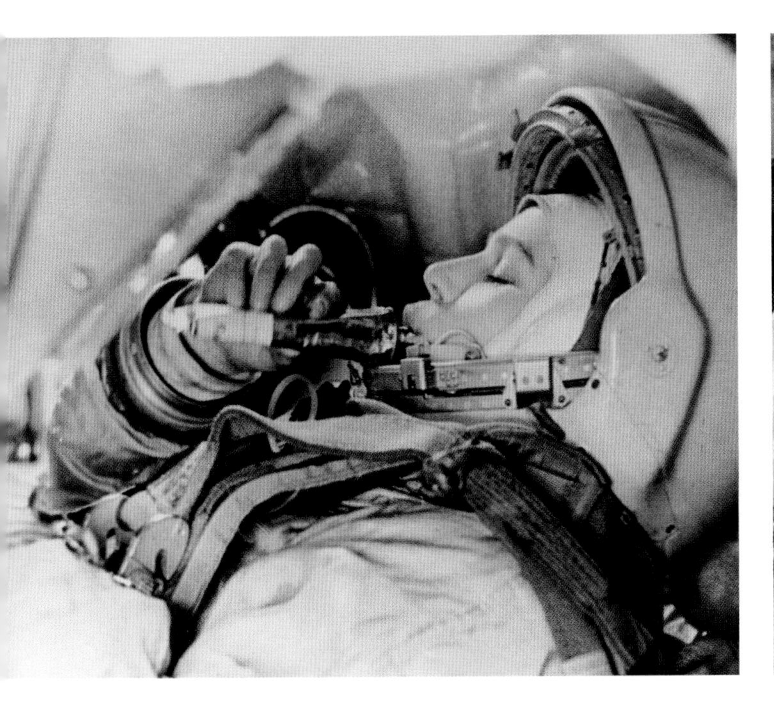

The Second Woman in Space

Cosmonaut Svetlana Savitskaya working aboard the *Salyut 7* space station in 1982 with Anatoly Berezovoy. Savitskaya was the second woman in space, the first woman to fly on a space station, the first woman to perform a space walk, and the first woman to participate in multiple space missions.

Soviet Union. None of the other women who had trained alongside her were given the opportunity to fly a space mission, and in 1969 the whole cadre of women cosmonauts was disbanded.

It was not until August 19, 1982, more than 19 years after Tereshkova's first and only flight, that a second woman, fellow Soviet Svetlana Savitskaya (1948–), went into space. As of January 2018, a total of 60 women from around the world have followed Tereshkova's flight path over the Kármán line and on to become cosmonauts and astronauts. That figure will doubtlessly increase in the coming years, as more and more women bring their considerable skills and expertise to the endeavor of space exploration—that 50 percent of the trainees on NASA's 2016 program were women is a sign of things to come.

Mercury 13 astronaut Jerrie Cobb during training being tested for resistance to stress in 1960.

The "Mercury 13"

Shortly after the public unveiling of the Mercury Seven astronauts in the United States (p. 105), William Lovelace II, one of the scientists who had been involved in their selection, began wondering if women could perform in similar selection tests as well as their male counterparts. After meeting American aviator Geraldyn "Jerrie" Cobb in 1959, Lovelace invited her to take the same tests as the Mercury astronauts—and was astounded by her aptitude. Uncertain whether Cobb might be an anomaly, Lovelace secured private funding from veteran pilot Jackie Cochran, the head of the Women Airforce Service Pilots (WASP) program in World War II, to bring 18 other experienced female pilots to his clinic for secret testing.

The women arrived alone or in pairs for four days of tests, including experiments on centrifuges to simulate the strains of launch and reentry. All of the women were skilled airplane pilots with commercial ratings, and fared well on all the tests. When word of the experiments leaked to the press, the top 12, along with Cobb, were dubbed the "Mercury 13."

When NASA's leadership learned of the experiments, they soon made it clear that they had no plans to employ women astronauts.

Cobb lobbied extensively for NASA to accept female astronauts, and eventually secured a dramatic congressional hearing in July 1962 to explore the possibility. NASA officials insisted that astronauts must be military test pilots, and that training women would slow down the astronaut program.

It was another 21 years before an American woman was eventually sent into space, with Sally Ride (1951–2012) serving as a mission specialist on Space Shuttle flight STS-7 (June 18–29, 1983). It was not until 1995 that a woman of any nationality first commanded a space mission, when Eileen Collins (1956–) piloted Space Shuttle mission STS-63.

ABOVE LEFT Gene Nora Jessen, Wally Funk, Jerrie Cobb, Jerri Truhill, Sarah Rutley, Myrtle Cagle, and Bernice Steadman, seven of the "Mercury 13," visit the Kennedy Space Center to witness astronaut Eileen Collins become the first woman to pilot a Space Shuttle.

ABOVE RIGHT Cobb undergoing tests on the Multiple Axis Space Test Inertia Facility used to train astronauts to control the spin of a tumbling spacecraft.

5

The Race to the Moon

The early years of the space race between the Soviet Union and the United States saw the rivalry between the two nations play out on a largely ad hoc basis of one-upmanship, with little planning or thought given to longer-term goals. That changed after May 25, 1961, when President John F. Kennedy announced his goal of sending a human to the Moon and returning him safely to the Earth (p. 132).

Both the Soviets and the Americans had already sent robotic missions to the Moon (p. 161), but a human mission was something else entirely. It was an objective that presented a great number of technical challenges, and one that need not necessarily have turned into a race (p. 134). Nevertheless, a race is exactly how many people saw it, even though the Soviet leaders were at first reluctant to engage in such a competition, and never publicly admitted their nation's involvement.

This period saw the two major space powers develop a range of new spacecraft, from Voskhod (p. 136) and Soyuz (p. 139) to Gemini (p. 136) and Apollo (p. 146). Their scientists and engineers would work to build some truly colossal launch vehicles, such as N-1 (p. 138) and the Saturn V (p. 138), as well as a host of other new technologies, from spacesuits to lunar landing vehicles (p. 140).

There were disasters along the way as well as triumphs, and some missions that were a mixture of both. But while it was astronauts from the United States who succeeded in taking the first small steps on the lunar surface (p. 148), it was ultimately humanity, and science, that benefited most from the whole endeavor. In the new age of space exploration that was to follow the race to the Moon, scientific enquiry would be the key.

OPPOSITE Apollo 11 mission commander Neil Armstrong took this iconic photograph of his crewmate Buzz Aldrin standing on the lunar surface just behind the leg of their lunar lander on July 20, 1969.

The First Moon Race: Robotic Precursors

RIGHT A contemporary illustration of the Soviet spacecraft Luna 3 taking pictures of the far side of the Moon in October 1959.

BELOW The first blurry image of the far side of the Moon taken by the Soviet satellite Luna 3 on November 7, 1959.

Soon after the Soviet Union and the United States had successfully placed satellites in orbit, the two superpowers turned their attention to reaching the Moon. Landing anything on the Moon would represent a significant publicity coup and, just as in the early space race, it was robotic probes that blazed the trail for human space explorers.

After some false starts in the fall of 1958, when three attempted launches each ended in catastrophic failure, the Soviets successfully launched their Luna 1 probe on January 4, 1959. It was the first human-made object to escape Earth's orbit but, due to an error in the ground-based control systems for the probe's launch vehicle, it missed the Moon by 3,666 miles (5,900 km), and went on to become the first human-made object to go into orbit around the Sun. Sergei Korolev, the chief architect of the Soviet space program, recognized that there was significant scientific knowledge to be gained from exploring the Moon, but the effort also suited the purposes of Soviet leader Nikita Khrushchev, who sought to amass spectacular space firsts as a means to embarrass his nation's Cold War rivals. The key to the early Soviet success rested mainly on the powerful launch vehicles it employed, adapted from the proven R-7 rocket. These launch vehicles had larger payload capacities than American rockets because Soviet nuclear and scientific payloads were both cruder and heavier.

In 1958, eager to respond to Soviet leadership in space technology, the United States began an accelerated effort to send a series of spacecraft to the Moon under the auspices of the newly created National Astronautics and Space Administration (NASA). In late 1958, the United States made several attempts to send a probe to the vicinity of the Moon. None reached its intended destination. Indeed, none succeeded in escaping Earth's orbit, but two of them did manage to transmit some data about the outer regions of the Van Allen radiation belts. In March 1959, Pioneer 5 came within 36,650 miles (58,983 km) of the lunar surface. The Soviets had greater success with their Luna 2 probe, which became the first spacecraft to impact on the Moon, or indeed any celestial body, on September 13, 1959. On October 7, 1959, Luna 3 became the first probe to photograph and transmit pictures of the far side of the Moon, thereby giving the Soviets another important first. This initial phase of lunar exploration ended with the Soviet Union looking like the clear winner.

In December 1959, stung by the failure of its first batch of lunar probes, NASA's Jet Propulsion Laboratory (JPL) started the Ranger project to send a probe to take photographs of the Moon's surface. Early Ranger launches were beset by failure, but Ranger 7, launched on July 28, 1964, and Ranger 8, launched on February 17, 1965, both returned photos, while Ranger 9, launched on March 21, 1965, broadcast live footage from its television camera until it smashed into the Moon.

These early robotic lunar exploration programs succeeded in gathering important scientific data about Moon, but it was more than four more years before the first humans set foot on Earth's sole natural satellite.

Ready for Impact

At the time the first probes were launched toward the Moon, neither the Soviet Union nor the United States had the technical ability to land their spacecraft. To get around this problem, Soviet engineers simply designed their probe to impact onto the lunar surface instead. It was a mission profile that scientists at NASA would later adopt in their design of their own probes for Project Ranger.

BELOW LEFT A scale model of the Ranger 1 space probe. It was the first NASA probe designed to reach the Moon, but was destroyed in the catastrophic failure of its Atlas-Agena launch vehicle.

BELOW RIGHT The Moon photographed by Ranger 9 moments before the spacecraft impacted on to the lunar surface on March 24, 1965.

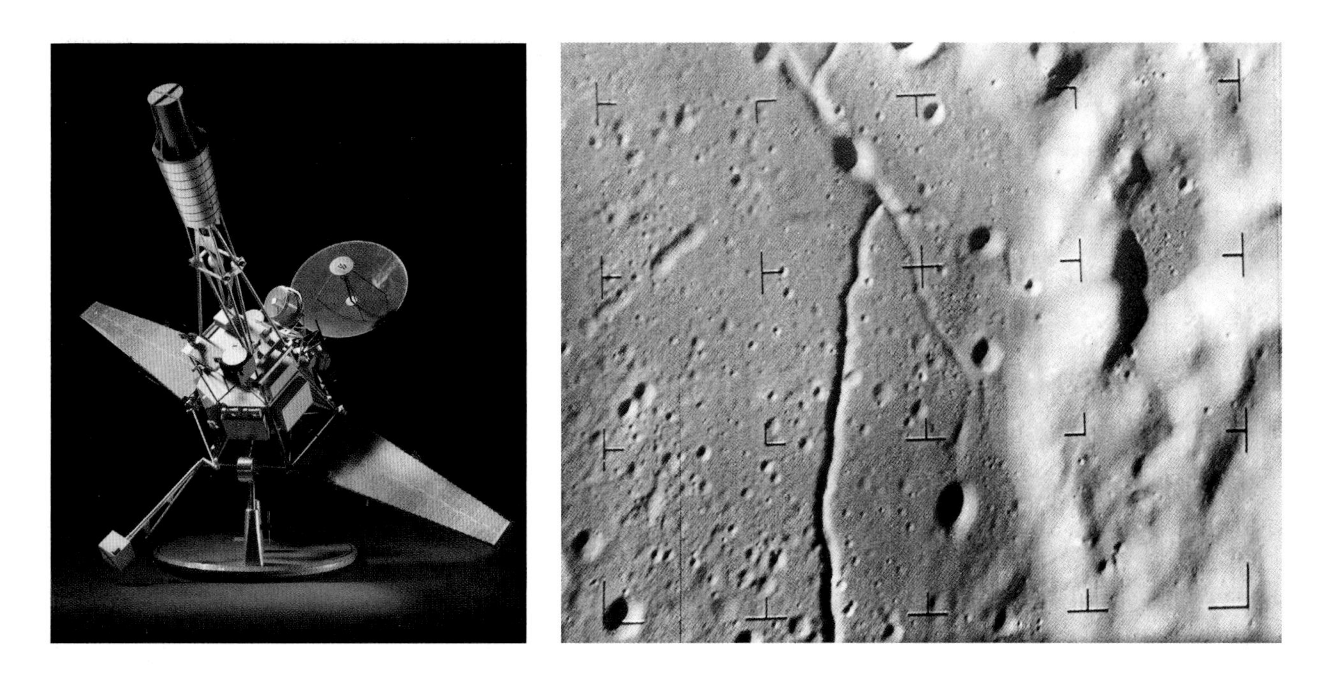

The Decision to Go to the Moon

President John F. Kennedy delivering his May 25, 1961, speech to Congress announcing the plan to send humans to the Moon.

The space race between the United States and the Soviet Union stepped up a gear on May 25, 1961, when American President John F. Kennedy announced that his country would land an astronaut on the Moon by the end of the decade. It was an ambition that might have appeared lofty to many observers, but it was in fact rooted in very Earthly concerns. The rampant success of the Soviet space program up to this point, and in particular the orbital flight of Yuri Gagarin on April 12, 1961 (p. 122), coupled with the bungled U.S.-backed invasion of Cuba at the Bay of Pigs, served as significant political embarrassments, and made the Communist system look stronger on the world stage.

Soon after the Gagarin flight, Kennedy discussed the possibility of a lunar landing program with his Vice President, Lyndon Johnson. He then followed their discussion with a memorandum on April 20, 1961, which asked: "Do we have a chance of beating the Soviets by…a trip around the Moon, or by a rocket to land on the Moon, or by a rocket to go to the Moon and back with a man? Is there any other space program which promises dramatic results in which we could win?"

Kennedy unveiled his commitment to what became Project Apollo in a speech to Congress on "Urgent National Needs," in which he outlined the extraordinary challenges facing the United States, and the extraordinary response that those challenges would imminently require. Announcing the Moon landing commitment, he said: "We go into space because whatever mankind must undertake, free men must fully share."

Kennedy's speech, and the goal it outlined, captured the public imagination and received immediate political support. At the time, no one seemed concerned about the difficulty or the expense of this undertaking. Congressional debate was perfunctory, and NASA found itself struggling to spend the funds committed to it during the early 1960s. Project Apollo gave the Americans the chance to set their own space exploration agenda. Now with a clear goal, they were no longer stuck playing catch-up with the Soviets. Nevertheless, in 1961, a lunar landing was far beyond the capabilities of either the United States or the Soviet Union.

In the Soviet Union, Sergei Korolev, chief architect of the Soviet space program, was quick to use the American commitment to obtain further resources for his own space exploration program, but Soviet leader Nikita Khrushchev stopped short of approving a Soviet Moon mission. Korolev tried to structure his nation's space efforts so that they could be extended into a lunar landing program later, and lobbied heavily on the mission's behalf, finally securing backing in 1963. By this time it was clear that a race to the Moon was underway, whether the Soviet leadership cared to acknowledge that fact or not.

> *"Is there any other space program which promises dramatic results in which we could win?"*
>
> PRESIDENT JOHN F. KENNEDY, MEMORANDUM, APRIL 20, 1961

BELOW LEFT NASA's Wernher von Braun (left) explains the Saturn rocket system to President John F. Kennedy on a tour of the Cape Canaveral Missile Test Annex on November 17, 1963.

BELOW RIGHT President Kennedy and his vice president, Lyndon Johnson, being briefed on Project Apollo at the Cape Canaveral Missile Test Annex on September 11, 1962.

Could the Moon Landing Have Been an International Program?

The public commitment of U.S. President John F. Kennedy to send Americans to the Moon (p. 132) was an overt challenge to the Soviet Union to demonstrate technological supremacy in space exploration. The winner would undoubtedly gain massive prestige around the world, perhaps swaying the allegiance of various unaligned nations within the wider Cold War. For the Americans, still smarting from the seemingly unlimited success of the Soviet space program, the competition was intended as a real challenge, prompting their government to further open up its treasury to help NASA achieve the goal of setting foot on another body in the Solar System. After some initial hesitancy, the Soviets took the bait, apparently accepting the terms of the American challenge to a race to the Moon.

But what if the United States and the Soviet Union had undertaken the Moon landing program cooperatively, rather than as a competition? Perhaps surprisingly, this is more than an academic question. There were genuine efforts to make the Moon mission a joint program. In his inaugural address in January 1961, Kennedy spoke directly to Soviet leader Nikita Khrushchev, and asked him to cooperate in exploring "the stars." In his State of the Union address ten days later, Kennedy asked the Soviet Union to join the United States "in developing a weather prediction program, in a new communications satellite program, and in preparation for probing the distant planets of Mars and Venus, probes which may someday unlock the deepest secrets of the Universe."

President John F. Kennedy meets Soviet leader Nikita Khrushchev in Vienna, Austria, in June 1961.

After Yuri Gagarin's first orbital flight (p. 122) and the failed American-backed invasion of Cuba in the early months of 1961, Kennedy asked his brother, Attorney General Robert Kennedy (1925–68), to assess the Soviet leadership's inclinations toward taking a cooperative approach to human space exploration. On the very same day that he gave his Apollo speech, President Kennedy instructed some of his key advisors to "offer the Soviets a range of choices as to the degree and scope of cooperation." Within two weeks of giving his bold May 25 speech to the Joint Session of Congress, Kennedy met Khrushchev at a summit in Vienna, Austria, and there

proposed making Apollo a joint mission between their two nations. The Soviet leader reportedly first said no, then replied "Why not?" before then apparently changing his mind again, saying that disarmament was a prerequisite for cooperation in space. Then, on September 20, 1963, Kennedy made his famous speech before the United Nations, in which he again proposed a mutual human mission to the Moon. He closed by urging, "Let us do the big things together." But his call for collaboration went unheeded.

In public, the Soviet Union was noncommittal. The official government newspaper *Pravda*, for example, dismissed its nation's own 1963 proposal for a Moon mission as premature. Some historians have suggested that Khrushchev viewed the American offer as a ploy to open up Soviet society and compromise Soviet technology. Behind the scenes, Khrushchev asked his son, Sergei, an engineer at the Experimental Design Bureau OKB-1 set up by Sergei Korolev, to come to the Kremlin and discuss this proposal. The two men agreed that such a collaborative project would enable the Soviet Union to reduce significantly the cost of space exploration, and such an agreement would ensure that Soviet prestige would remain intact. It would, in essence, take the Moon out of the Cold War rivalry. But no cooperative venture materialized. Kennedy was assassinated on November 22, 1963, and Khrushchev was deposed in the following year. But had both leaders remained in place or simply committed their nations to a single joint Moon mission, one of the most significant achievements in space exploration to date might have been very different.

ABOVE LEFT President John F. Kennedy rises to address the United Nations on September 20, 1963, in which he directly proposed a joint U.S.-Soviet mission to the Moon.

ABOVE RIGHT Astronaut Vance Brand (left) and cosmonaut Valeri Kubasov train together at the Johnson Space Center in Houston, Texas, in 1975 for the first joint mission between the space superpowers. But could such collaboration have succeeded a decade earlier?

Spacecraft Built for Two (or Three): Voskhod and Gemini

After the success of a series of Vostok flights in the early 1960s, Sergei Korolev set his OKB-1 design bureau the task of creating the Voskhod spacecraft, a capsule that could accommodate two or three cosmonauts. The result was a space vehicle that was the same size as Vostok, but with many of its duplicated systems and safety features removed in order to make room for extra crew members. Because of the weight of the additional cosmonauts, Voskhod required a new launch vehicle, the Molniya 8K78M, a further modified version of the R-7 rocket.

The Voskhod program was designed to explore further how the human body reacted to space travel, but it was relatively short-lived, with only two launches. Voskhod 1, launched on October 12, 1964, was the first multi-cosmonaut mission, with its crew of Boris Yegorov (1937–94), Konstantin Feoktistov (1926–2009), and Vladimir Komarov (1927–67). Crewed by Pavel Belyayev (1925–70) and Alexey Leonov, Voskhod 2, launched on March 18, 1965, saw the first extravehicular activity (EVA), or space walk. Nikita Khrushchev was removed from power in October 1964. The new Soviet leadership that replaced him was less focused on using the nation's space program to achieve prestigious scientific firsts. They allowed Korolev to cancel the Voskhod program and shift his focus toward developing the larger, more versatile Soyuz capsule (p. 139) as part of the Soviet Union's own lunar program.

A keen artist, cosmonaut Alexey Leonov took drawing materials with him into space. This painting, a collaboration between Leonov and fellow artist Andrei Sokolov, depicts Leonov's space walk on March 18, 1965. The tube visible on the side of the Voskhod spacecraft was the inflatable airlock through which Leonov exited and reentered the capsule.

Meantime, the Americans developed the two-astronaut Gemini capsule, which flew in 1965 and 1966. Project Gemini had three main goals, all of which were viewed as steps along the way to the larger Moon mission. Firstly, it gave NASA astronauts a platform on which to learn how to maneuver, rendezvous, and dock with another spacecraft. Secondly, it was designed to help astronauts practice working outside a spacecraft. Thirdly, it was to be a resource for the collection of physiological data about long spaceflights. Astronauts Virgil "Gus" Grissom and John Young (1930–) flew on Gemini 3, the first crewed mission in Project Gemini, on March 23, 1965. The next mission, in which astronaut Edward White II (1930–67) became the first American to perform a space walk, flew in June 1965. Eight more missions followed through to November 1966. Despite problems, most of which involved in-orbit docking practice using an Agena rocket, the program achieved all its goals and, moreover, packed 52 additional science experiments in to the ten Gemini missions, preparing the way for the Apollo missions to come (p. 146).

FAR LEFT Astronauts trained for rendezvous and docking missions on the Gemini simulator. The simulator was suspended from the roof of an aircraft hangar at Langley Research Center in Hampton, Virginia.

LEFT This time-exposure photograph shows the configuration of the launch of Gemini 10 on July 18, 1966.

Space Walking

Both the Soviets and Americans realized early on that an important element of true space exploration would be the ability of astronauts and cosmonauts to step outside of their spacecraft. Cosmonaut Alexey Leonov (1934–) made the first space walk on March 18, 1965, floating outside the Voskhod capsule to which he was tethered for a few minutes. When Leonov attempted to return to the capsule he found that his suit had ballooned up, making it difficult to grip onto things. He was forced to release some of the air from his suit before he was able to get back aboard the capsule and close the airlock's outer hatch.

The next space walk occurred on June 3, 1965, when astronaut Edward White stepped out of his Gemini 4 capsule. White's space walk, and those by Eugene Cernan (1934–2017) and Buzz Aldrin (1930–) during Project Gemini, demonstrated the flaws of the early NASA spacesuits. The biggest problems were the limited range of movement the suits allowed and that the visor had a tendency to fog up. These problems were addressed by the new spacesuit designs used in the Apollo Moon landings (p. 148).

Ed White becomes the first American to walk in space; June 3, 1965.

Lunar Exploration Technology

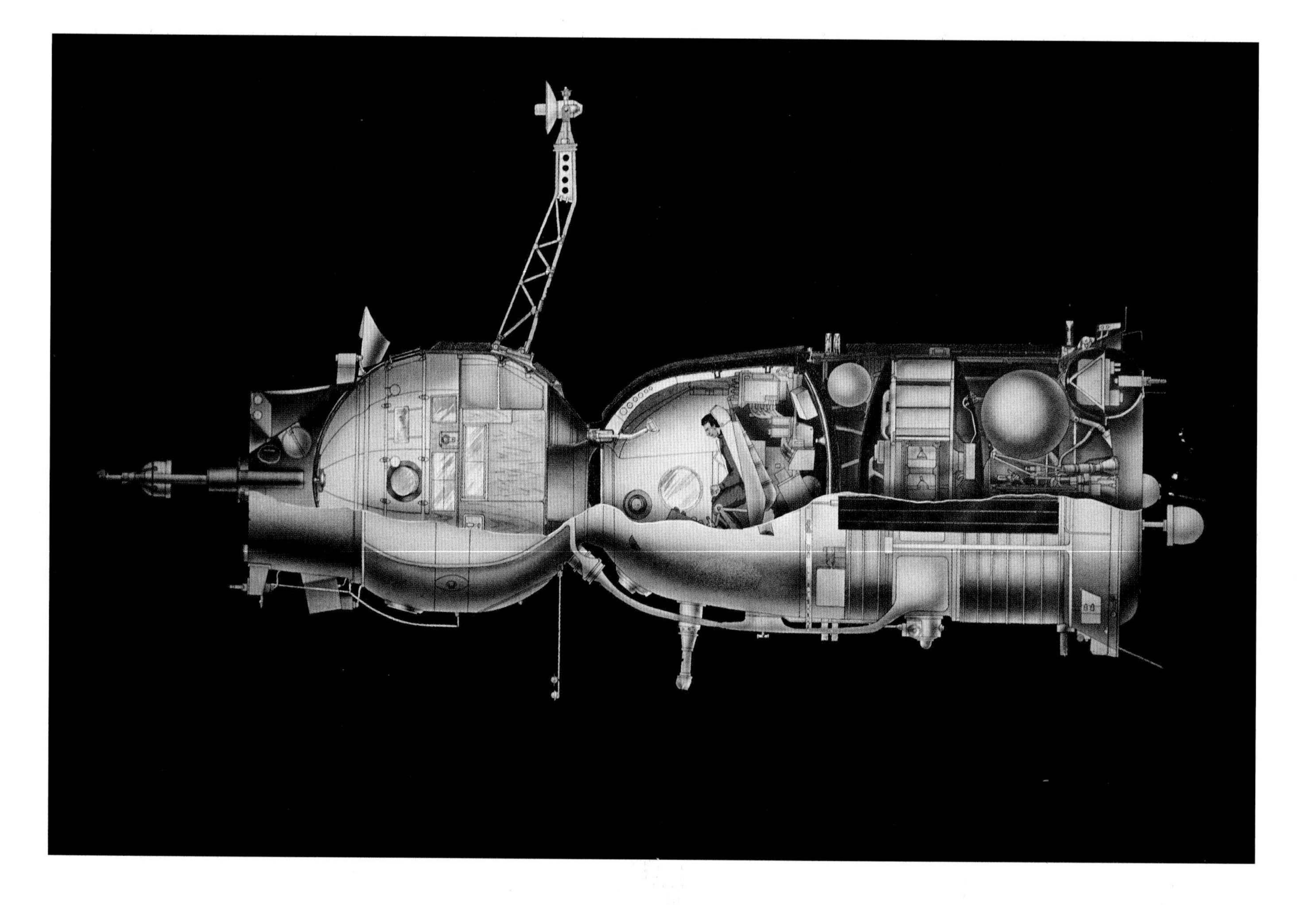

The Soyuz spacecraft represented a significant progression in Soviet space design, and, unlike earlier Soviet spacecraft, it did not require cosmonauts to eject before landing.

Engineers at OKB-1, the rocket design bureau set up by Sergei Korolev, had been clamoring for their own lunar program practically from the moment the Americans declared their intention to land on the Moon by the end of the 1960s. Korolev himself argued for a circumlunar flight to take place in time for the 50th anniversary of the Bolshevik Revolution in October 1967, followed by a Moon landing to take place in 1968. The Soviet Communist Party Central Committee Command decree 655-268, issued on August 3, 1964, officially established a number of secret programs to produce the technology for circumlunar flights and lunar landings. Unfortunately, all these programs were competing for the same limited financial resources. The circumlunar program, named Zond, was assigned to a Korolev acolyte-cum-rival Vladimir Chelomey (1914–84), who led design bureau OKB-52. Chelomey's main task was to design and build a larger and more powerful launch vehicle capable of reaching the Moon. This was the N-1.

For his part, Korolev was assigned the task of designing and building the Soviets' overall lunar landing program, comprising the Soyuz spacecraft and its lunar lander. Powered by the N-1, Korolev's newly developed spacecraft would take two cosmonauts and the lander to the Moon. Once at their destination, one crew member would take the lander down to the lunar surface, exit the spacecraft to explore, return to the lander, and fly it back into lunar orbit to reunite with the Soyuz capsule for the return trip to Earth. The four-stage N-1 rocket would be the launch vehicle that would send the Soyuz capsule to the Moon. However, the rocket's development was plagued with technical problems, and at the end of 1964 Korolev took over responsibility for the Zond program in addition to his own project. One problem that the Soviet Moon mission never resolved was how to build large rocket engines. NASA could obtain 7,500,000 pounds (33,000,000 N) of thrust from the five engines on the first stage of its Saturn V rocket, but the N-1 required 30 separate engines to accomplish the same feat. Getting these engines to work effectively together eluded Chelomey's best engineers.

The complexity and pressure of the Moon mission were heavy burdens for anyone to shoulder alone, and when Korolev died on January 14, 1966, apparently as a result of complications from a botched hemorrhoid operation (p. 73), he left a massive hole in the Soviet space program. Competing factions of engineers, including one led by Chelomey, struggled against rocket engine designer Valentin Glushko (1908–89) and Korolev's longtime ally Vasily Mishin (1917–2001) for control. But no one else had the same grasp of the technical details of the Soviet space program as a whole as Korolev had demonstrated, much less the gravitas to hold it all together. Difficulties continued to dog the N-1 rocket. After four test launch failures, its development was halted. The whole mission to send cosmonauts to the Moon was finally abandoned in 1974, by which time the Soviet space program had turned its attention to the development of space stations (p. 207).

The Enduring Soyuz

First flown in automated tests in 1966, Soyuz is the most enduring design in space exploration, and is still in use today to ferry space travelers up to the International Space Station. But while Soyuz now has a reputation for safety and reliability, its first piloted mission, Soyuz 1, ended in the death of its pilot, cosmonaut Vladimir Komarov (1927–67). Komarov died when his descent module crashed to the ground after its parachutes failed to open correctly.

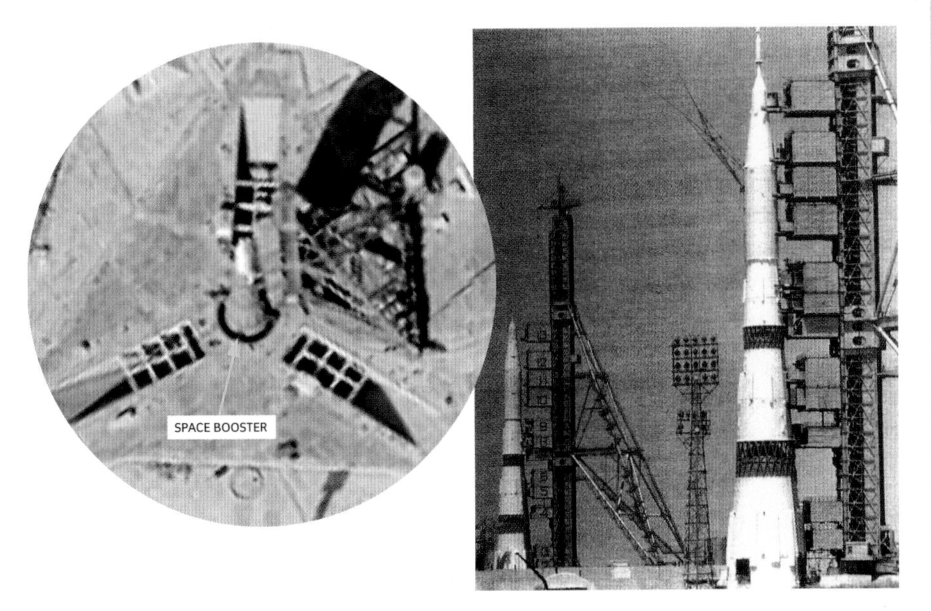

FAR LEFT This image of the Soviet N-1 on the launchpad at the Tyuratam missile (now Baikonur) test center in the Soviet Union was taken by the Gambit 3 reconnaissance satellite on September 19, 1968.

LEFT Two N-1 Moon rockets on the launchpad at Tyuratam in early July 1969.

While the N-1 might have been the downfall of the Soviet Moon mission, the Saturn V launch vehicle played a huge role in the success of the Apollo program. In many ways, the Saturn V represented the pinnacle of NASA's rocket know-how. It stood 363 feet (110 m) tall, and comprised three stages. Its five powerful first-stage engines required the development of new metal alloys and construction techniques to produce combustion chambers that could withstand the extreme heat and shock they would experience in firing. The rocket's second stage used liquid hydrogen for fuel with a liquid oxygen oxidizer, the first NASA rocket to employ such propellants. Liquid hydrogen is more efficient than the aviation fuel used in Saturn V's first stage, but it is difficult to work with because it has to be kept at very low temperatures to maintain its liquid form. The third stage was an enlarged and improved version of the earlier Saturn IB rocket.

This cutaway illustration shows the Apollo A7L spacesuit with backpack.

In addition to their launch vehicles, both the Soviet and American Moon programs invested heavily in the development of spacesuits for their respective lunar missions. Soviet engineers built the Krechet-94 extra-vehicular activity (EVA) suit. Weighing 198 pounds (90 kg) and worn over a liquid-cooled undergarment, the suit could operate continuously for 15 hours, and had a life cycle of 48 hours. It had a semi-rigid structure, with the hard upper torso section constructed from aluminum, but its most important innovation was the hatch opening in the suit's backpack, which made it easier for cosmonauts to get into and out of it. The Krechet-94's chest-mounted instrument panel offered cosmonauts excellent control over their own life support systems. While it was never used on the Moon, the design of Krechet-94 exerted a massive influence on many later Russian and American EVA suits.

NASA's lunar suit, the A7L, was far more complicated. Like the Krechet-94, it required liquid-cooled undergarments, but it also had an additional two-piece pressure suit and a further one-piece outer coverall. It was versatile: designed to function both on the Moon and during EVAs in space. It also had a helmet, gloves, and boots that were all attached individually. However, its greatest benefit was probably the fact that it had a longer operational life than the Krechet-94: it could withstand as much as 115 hours of use before being retired.

While the A7L was certainly an effective piece of equipment, one which kept all the Apollo astronauts who used it safe on their various lunar excursions, it had one main drawback: it was incredibly laborious to put on and take off. This problem would later be addressed in the development of simpler NASA spacesuits, which would be just as safe and reliable but much easier to use.

Lunar Landers

Both space superpowers developed their own lunar landers for their respective Moon missions. NASA's Lunar Module (LM) took two astronauts from the main Apollo spacecraft, dubbed the Command/Service Module (CSM), down to the Moon's surface. For NASA's engineers and contractors, the LM represented one of the most serious design challenges of the whole Moon mission, not least because it was required to convey a crew safely to a soft landing on the surface of another world, and later return them to a lunar orbit where they could rendezvous with the CSM for the trip back to Earth.

Throughout the early phase of the Apollo program, the LM was consistently behind schedule and over budget. Most of the project's difficulties surrounded the demands of devising two separate spacecraft components—one for descent to the Moon's surface, and one for ascent back to the orbiting CSM. Both engines had to work perfectly, or the astronauts might be stranded on the lunar surface without any means of getting home. Guidance, maneuverability, and spacecraft control also caused no end of headaches for the LM engineers.

The module's landing structure also presented problems; it had to be light, sturdy, and shock-resistant. After various engineering problems, an ungainly-looking vehicle emerged that was finally declared flight-ready in early 1968.

The Soviet lunar lander, known as LK, was a far simpler vehicle than the LM, but its mission profile was broadly the same as the one developed by NASA. The key difference between the LM and the LK was that the Soviet vehicle lacked any cargo-carrying capacity, and when NASA's astronauts finally reached the Moon the LK was still not ready for service.

ABOVE LEFT As seen in this 1:48 scale model, the Soviet N-1 launch vehicle (right) was approximately the same size as the Saturn V rocket beside it.

ABOVE RIGHT The Lunar Module being tested in Earth orbit during the Apollo 9 mission in 1968.

Hidden Figures in Space Exploration

> *"I was so involved in what we were doing, technically, that I was oblivious to the fact that I was outnumbered by men."*
>
> MARGARET HAMILTON, COMPUTER SCIENTIST

Throughout the 1960s, almost everyone who traveled into space was both white and male. But despite the racial prejudice and inequality that were widespread in both the United States and the Soviet Union at the time, neither of the nations' space programs was as monocultural as it may at first have appeared.

Many key facilities belonging to NASA were located in the southern United States, where racial segregation was still practiced. Notwithstanding that, the organization recruited a number of black mathematicians, engineers, and technicians throughout the decade, including Julius Montgomery. When he first arrived at NASA's Marshall Space Flight Center in Huntsville, Alabama, in 1965 to work as an engineer, Montgomery had to deal with NASA colleagues who were members of the white supremacist Ku Klux Klan. They refused to acknowledge him or shake his hand. But whether the Klansmen liked it or not, Montgomery represented a vanguard of progress toward diversity at NASA.

In 1961, United States Air Force pilot Captain Edward Dwight, Jr. (1933–) became the first black astronaut trainee, which catapulted him to instant fame. However, he faced severe discrimination from many of his fellow trainees, as well as from government officials, and this eventually prompted him to resign from the Air Force. Thereafter, the Air Force selected Major Robert Lawrence (1935–67) to become the first black pilot to join its military astronaut program, but he died soon afterward in a training accident. It was another 11 years before NASA recruited its first black astronaut, Guion "Guy" Bluford (1942–).

Ethnic discrimination worked differently in the Soviet Union but it was no less divisive. Yet in spite of that, some Soviet citizens from various ethnic minorities still rose to important positions in the Soviet space program. Three excellent examples include Andrian Nikolayev (1929–2004), of the Chuvash ethnic group of Siberia; ethnic Lithuanian Aleksei Yeliseyev (*né* Kuraitis, 1934–), who took his mother's more Russian-sounding family name in an effort to avoid discrimination; and Boris Volynov (1934–), who on July 6, 1976, became the first Jewish person to travel into space.

United States Air Force pilot Robert Lawrence.

LEFT Katherine Johnson, one of the human computers at NASA's Langley facility.

BELOW Margaret Hamilton demonstrating the extent of her Apollo code.

Women at NASA during the Race to the Moon

Although they were seldom seen in NASA's press conferences during the early years of the space race, a large number of women played important roles in the organization.

Katherine Johnson (1918–), the black mathematician whose NASA career was the focus of the 2016 movie *Hidden Figures*, calculated rocket trajectories for Project Mercury. She went on to make further significant contributions to the space agency's work until her retirement in 1986.

Margaret Hamilton (1936–) served as Director of the Software Engineering Division of MIT's Instrumentation Laboratory, and played a key role in developing the onboard flight software for the Apollo spacecraft. Her contribution to the software for the Apollo guidance computer was crucial to the success of the Moon landings.

Meanwhile, Carolyn Huntoon (1940–) worked as a physician in the Apollo program, ultimately rising to become director at the Johnson Space Center from 1994 to 1996.

2001 and Space Pop Culture

RIGHT A movie poster for the iconic 1968 feature film.

BELOW Astronauts from *2001: A Space Odyssey* investigating on the lunar surface with a Moon base in the background.

In 1968, legendary director Stanley Kubrick (1928–99) released an epic science fiction film, *2001: A Space Odyssey*, the story of humanity's transformative intergalactic evolutionary experience. Based on a story by science fiction author Arthur C. Clarke (1917–2008), it was at the time the most accurate presentation of spaceflight depicted in a movie to date. Among its many highlights is the presence of a great rotating space station in Low Earth Orbit (LEO), which uses centrifugal rotation to produce artificial gravity. It is serviced by a reusable winged spacecraft ferrying passengers up from the planet's surface and back. Space travel is shown as a routine commercial enterprise, with the winged shuttle operated by the now-defunct carrier Pan American World Airways. A Hilton hotel is located on the space station, and videophone services are provided by the American telecoms company AT&T. In addition to all this, humans have established several outposts on the Moon and, because of the discovery of an alien-built artifact that is sending signals toward Jupiter, a further interplanetary exploration mission to that planet is underway.

Upon its original release, *2001* thrilled viewers with its suggestion that an exciting future of commercially available space travel was on the horizon. Unfortunately, this vision has yet to materialize in reality. The single greatest obstacle that still stands in the way of regular passenger flights into space is the enormous cost of getting into orbit reliably.

As Clarke himself said in a 1993 interview: "You can't make much of a case for man in space until you've got efficient and reliable propulsion systems. Once we've got that, everything else will follow automatically." It seems that, if space is ever to be opened up for widespread commercial operations, humanity must first make radical improvements in its launch capabilities, significantly improving the efficiency of the propulsion methods employed, while also dramatically reducing the weight of the spacecraft and increasing their payload capacity.

One of the film's most celebrated characters is not a person at all, but the HAL 9000 computer that controls all systems on board Discovery One, the spacecraft sent to investigate the extraterrestrial signals. While basic computer-based personal assistants are now a commercially available reality, they lag far behind Hal 9000's powers of autonomy, not to mention his tendency to murder his crewmates when he thought they had become a threat to his mission. However, in recent years, scientists have had great success in creating increasingly complex thinking machines—computers that have the ability to learn.

But perhaps Kubrick's and Clarke's greatest achievement in *2001* was their portrayal of space exploration as a true odyssey: a long and heroic voyage of discovery and learning. Soon after the film had left the cinemas, humanity would finally reach the Moon in reality (p. 148).

Lifelong space travel enthusiast, author Arthur C. Clarke (right) glancing through a book of illustrations of spacecraft with cosmonaut Alexey Leonov.

An Odyssey of Sequels

Arthur C. Clarke wrote three sequels to *2001*, the first of which, *2010: Odyssey Two*, was also adapted into a movie by director Peter Hyams in 1984. This film depicted Jupiter's transformation into a second Sun, and explored the possibility of life evolving on Jupiter's moon Europa, a prediction that may yet prove to be feasible. The other two sequels, *2061: Odyssey Three*, and *3001: The Final Odyssey*, further explore life on Europa and the nature of the alien-built artifacts that first prompted humanity to explore Jupiter and its satellites.

The Early Apollo Missions and the Road to "Earthrise"

Despite the successes of the Gemini missions, NASA's Moon mission received a heavy blow at the very start of the Apollo program proper on January 27, 1967. Astronauts Virgil "Gus" Grissom, Edward White, and Roger Chaffee had been running a simulated flight for their scheduled Apollo 1 launch when a fire broke out in the spacecraft. The pressurized high-oxygen environment of the command module, in which all three members of the crew were located, quickly fueled flames and toxic smoke. Because the module had been pressurized, it took the NASA ground crew even longer to get the capsule's hatch open, and by the time they succeeded it was too late. The astronauts died of asphyxiation in a matter of seconds. These were the first fatalities directly linked to America's space program.

NASA immediately put its Moon mission on hiatus, and swiftly conducted a thorough investigation into the incident. It found that the pressurized oxygen-rich capsule environment, coupled with an abundance of combustible materials and possible ignition sources, all contributed to the disaster, which was exacerbated by inadequate emergency preparedness procedures. As a consequence of the tragedy, improvements were made, and after several automated test flights and the crewed test flight of Apollo 7 in October 1968, George Low of the Manned Spacecraft Center, Houston, Texas, and Samuel Phillips, the Apollo Program Manager, pressed the rest of the NASA leadership to give the green light to a circumlunar flight. Their plan promised both to gather valuable technical and scientific knowledge, and to be an important public demonstration of America's space exploration capabilities. Convinced by these arguments, and the safety of their improved procedures, NASA's leadership agreed to the mission.

On December 21, 1968, Apollo 8 launched atop a Saturn V booster with its three-astronaut crew: Frank Borman (1928–), James Lovell (1928–), and William Anders (1933–). Four days later, on Christmas Eve, the spacecraft's crew offered a special holiday greeting to the world in a broadcast from their Apollo Command/Service Module (CSM) in lunar orbit. They read the first few verses from the Bible's Book of Genesis and sent back TV images of the Earth.

They also took the powerful still image now known simply as "Earthrise," which immediately captured the popular imagination worldwide. Journalist Archibald MacLeish

ABOVE The crew of Apollo 8—(left to right) James Lovell, William Anders, and Frank Borman—pose on a simulator at the Kennedy Space Center.

RIGHT The iconic Earthrise image gave humanity a new perspective on life on the planet. It shows the Moon, gray and lifeless in the foreground, with the blue and white Earth, teeming with life, hanging in the blackness of space.

wrote in the *New York Times* that Apollo 8 had allowed all of humanity "to see the Earth as it truly is, small and blue and beautiful in that eternal silence where it floats." That, he continued, is "to see ourselves as riders on the Earth together, brothers on that bright loveliness in the eternal cold—brothers who know now that they are truly brothers." Earthrise made clear that our planet was a tiny raft, adrift in the vastness of the cosmos, with no lifeboats to rescue its population should catastrophe strike. This realization helped to galvanize environmental movements across the world.

But more importantly for NASA, Apollo 8's successful completion of 10 orbits of the Moon, and the safe return of its crew to Earth, demonstrated that its wider lunar mission was now back on track.

Taking up the Gauntlet

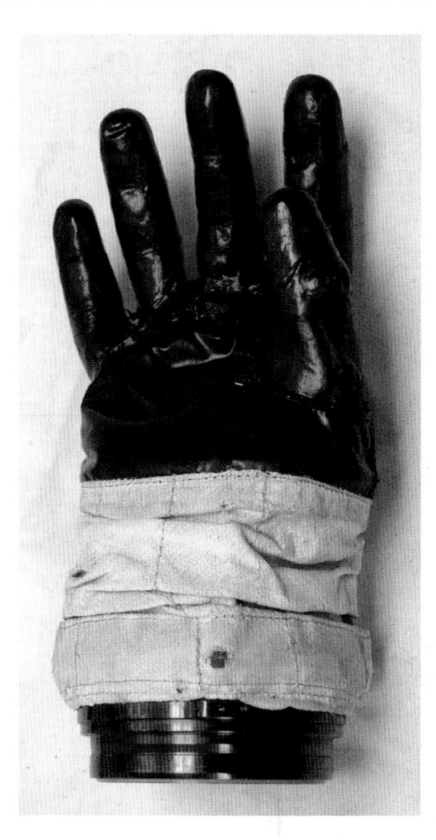

This right-hand intra-vehicular glove was worn by astronaut James Lovell during the launch of Apollo 8. It included a pressure bladder, which was dip-molded from a cast of Lovell's hand and intended to minimize the effects of acceleration associated with the launch of Saturn V. The anodized red aluminum connector at the wrist was designed for attachment to the arm of Lovell's spacesuit.

The launch of Apollo 8 to the Moon on December 21, 1968. This was the first crewed flight using the Saturn V launch vehicle.

Apollo 11: "One Giant Leap for Mankind"

After the return of Apollo 8 (p. 146), NASA launched a further two missions to test the Lunar Module (LM) and other hardware essential to the lunar landings. Apollo 9 tested docking maneuvers with the LM in Earth orbit, and Apollo 10 did the same in lunar orbit. With these tests successfully completed, the NASA mission leadership felt ready to attempt a landing on the Moon with Apollo 11.

Astronauts Neil Armstrong (1930–2012), Buzz Aldrin, and Michael Collins (1930–) were thoroughly prepared and well rehearsed for their mission. NASA scientists had found a suitable landing site; engineers had drilled the crew on every aspect of the weeklong return flight to the Moon. The trio practiced space walks in a deep swimming pool dubbed the Neutral Buoyancy Simulator (NBS), donned and doffed spacesuits repeatedly, and trained for entry and exit into and out of both their capsule and the LM. Armstrong also practiced landing the LM in the Moon's lower gravity, roughly one-sixth that of the Earth, on the Lunar Landing Research Vehicle (LLRV) simulator.

A 1967 flight of the Lunar Landing Research Vehicle, a platform designed to simulate the handling characteristics of the Lunar Module (LM) on the Moon.

On July 16, 1969, Apollo 11 launched from the Kennedy Space Center without incident, and began the three-day trip to the Moon. On July 20, the LM, dubbed The Eagle, separated from the Command/Service Module (CSM) to begin its descent toward the lunar surface. The landing was difficult for the LM crew of Armstrong and Aldrin. As they neared the surface, Armstrong realized that the automatic landing system was poised to set them down in the middle of a boulder field, so he took manual control and searched for another landing spot. As he slowed the descent over the lunar surface, the LM used more and more of its fuel, setting off low-fuel alarms in the spacecraft. Aldrin reset the alarms, and called out the altitude and the status of the fuel. With just 11 seconds of fuel left, and amid rising tension at Mission Control, Armstrong finally set the Eagle down on the lunar surface, announcing, "Contact light. Houston, Tranquility Base here. The Eagle has landed." Charlie Duke (1935–), the astronaut at Mission Control responsible for communicating with the crew, responded in a flustered voice: "Roger, Tranquility, we copy you on the ground. You got a bunch of guys about to turn blue. We're breathing again. Thanks a lot!"

After the landing, Armstrong and Aldrin were scheduled for a five-hour rest period, but they decided to skip it, reasoning that they would be too excited to sleep anyway. After some final checks, the pair suited up. Armstrong left Eagle to set foot on the Moon's surface, telling millions on Earth that it was "one small step for man—one giant leap for mankind." Aldrin soon followed, and while he was not the first on the Moon, he paused for a moment to relieve himself just as he reached the surface, claiming a different kind of first. Thereafter, the two lumbered around the landing site in their bulky spacesuits, planting an American flag, collecting soil and rock samples, and setting up scientific experiments.

Moon Menu

Among the food that was sent up with the Apollo 11 astronauts were such delicacies as dehydrated compressed beef and vegetables, dehydrated compressed beef hash, and dehydrated chocolate pudding. Possibly due to the excitement of the mission, not all of these rations were consumed during the Moon trip. The food that made it back to Earth uneaten was eventually presented to the National Air and Space Museum by NASA officials.

ABOVE Pineapple fruitcake cubes flown on the Apollo 11 mission in July 1969. The cubes could be eaten without the addition of water, and were "bite-sized" to minimize the possibility of crumbs.

LEFT Buzz Aldrin standing next to the American flag on the lunar surface, July 20, 1969. For a generation of Americans, this photograph symbolized the sense of national pride in the success of the Moon landings.

Reactions around the World

1969

Just as the first man and woman in space had excited crowds across the globe, so did the world tour of the first men to land on the Moon.

The Apollo 11 astronauts wear sombreros and ponchos to greet the large and enthusiastic crowds during the Mexico City stop of their world tour in September 1969.

The success of the Apollo 11 mission produced an ecstatic reaction around the planet, as everyone shared in the sheer joy of the achievement. The ticker-tape parades within the United States were followed by a world tour for the astronauts, with speaking engagements and public relations events, all of which served to increase American prestige abroad.

The Apollo program demonstrated American technological prowess, but the mission swiftly became much more than an American feat: it was a triumph for humanity. Newspapers worldwide featured the Moon landing prominently on their front pages. NASA estimated that, due to blanket radio and television coverage, more than half of the planet's population had followed the progress of the Apollo 11 mission. Even in the Soviet Union, which tried to jam the radio broadcasts of the U.S. government-financed Voice of America, many citizens attentively followed the American astronauts' lunar adventure.

In the United States, police reports noted that streets in many cities were quiet during the Moon walk, as residents watched television coverage at home or in bars and other public places.

Not long afterward, President Richard Nixon told an assembled TV and radio audience worldwide that the flight of Apollo 11 marked the most significant week in the history of Earth since its creation. The successful Moon landing captured the world's attention. As the legendary American journalist Walter Cronkite commented in 2003, those who witnessed the Apollo Moon landing were part of "the lucky generation" that witnessed as humanity "first broke our Earthly bonds and ventured into space. From our descendants' perches on other planets or distant space cities, they will look back at our achievement with wonder at our courage and audacity and with appreciation at our accomplishments, which assured the future in which they live."

The ceremonial flag planting, which conjured images of European explorers seizing various territories across the world, notably did not include any attempt to claim the Moon for the United States. Instead, the astronauts proclaimed that they "came in peace for all mankind." It was an important move, signifying to the world a break with imperial Earthbound exploration of the past.

After completing their assigned tasks, the astronauts returned to the LM and, the next day, launched it back into lunar orbit to rendezvous with the Apollo CSM before heading back toward to Earth.

For a brief moment in the summer of 1969, the Apollo 11 mission managed to unify an American nation deeply involved in the Vietnam War and bitterly divided by political, social, racial, and economic tensions. Civil rights leader Reverend Ralph Abernathy led a protest to Cape Canaveral to call attention to the plight of the poor of the United States. The protest served to remind all those watching the Apollo 11 launch that poverty still ravaged the lives of too many Americans. But Abernathy was not protesting the mission itself, as he soon made clear. He wished nothing but success for Apollo 11, insisting that he would pray for the safety of the astronauts and that, as an American, he was as proud of his nation's space achievements as anyone else in the country.

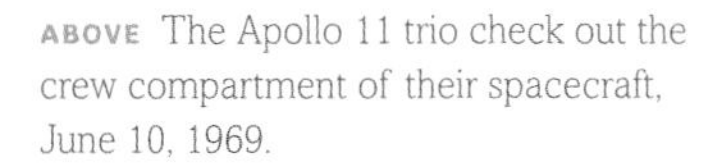

ABOVE The Apollo 11 trio check out the crew compartment of their spacecraft, June 10, 1969.

LEFT In a Hong Kong park, crowds stop to watch the broadcast of the Apollo 11 Moon landing, July 1969.

LEFT Astronaut Buzz Aldrin during the Apollo 11 extravehicular activity on the Moon. He has just deployed the Early Apollo Scientific Experiments Package. In the foreground is the Passive Seismic Experiment Package; beyond it is the Laser Ranging Retro-Reflector; in the center background is the American flag; in the left background is the lunar surface TV camera; to the right is the Lunar Module.

BELOW Photograph of Aldrin's bootprint on the lunar surface during the Apollo 11 mission to the Moon, July 20, 1969.

Apollo 12 and Precision Landing

RIGHT Alan Bean removing parts of Surveyor 3. The Apollo 12 Lunar Module is visible in the distance.

BELOW The Apollo 12 crew—Pete Conrad, Dick Gordon, and Alan Bean.

The second lunar landing mission, Apollo 12 in November 1969, really cemented the United States' position of dominance in lunar exploration. While Apollo 11 had experienced some difficulties in its landing stage (p. 148), the crew of Intrepid, the Apollo 12 lander, achieved a largely automated precision touchdown in the southeastern part of the Ocean of Storms, just 600 feet (182 m) from the robotic Surveyor 3 spacecraft, which had reached the lunar surface two years earlier, on April 20, 1967.

Although their flight was of less historic importance than the Apollo 11 mission, the crew of Intrepid, Charles "Pete" Conrad (1930–99) and Alan Bean (1932–2018), were no less excited, with Conrad exclaiming "Whoopee! Man, that may have been a small [step] for Neil [Armstrong], but that's a long one for me."

During their time on the lunar surface, Conrad and Bean both took two Moon walks, each lasting just under four hours. They collected rocks, and set up experiments that measured the Moon's seismic activity, solar wind flux, and magnetic field. Meanwhile, Richard Gordon (1929–2017) on board the Command/Service Module (CSM) took multispectral photographs of the Moon's surface. After Conrad and Bean returned from the surface, the Apollo 12 crew stayed an extra day in lunar orbit taking additional photographs. When they were finally ready to begin the trip back to Earth, the crew detached the Lunar Module (LM) ascent stage, and let it drop down onto the Moon. The seismometers the astronauts had left on the lunar surface registered the vibrations of its impact. This demonstrated that seismic activity could be produced on the Moon.

The Apollo 12 mission also produced some unexpected science by way of its recovery of the camera and a variety of other parts from the Surveyor 3 probe. The camera had sat exposed on the lunar surface for 31 months before the Apollo 12 astronauts retrieved it and took it back to Earth. NASA's scientists were anxious to examine it for signs of any micrometeoroid damage it might have suffered while on the Moon; this would give some indication of the level of such activity on the lunar surface. But what the scientists were not prepared for was the discovery of the terrestrial bacteria *Streptococcus mitis* on the camera. Humans cannot survive exposure for more than a few seconds in the vacuum of space, but the bacteria had apparently lasted for two and a half years in such conditions.

The scientists went on to check whether the other 33 samples from various other parts of Surveyor 3 also harbored the bacteria. The question then arose as to whether the bacteria had survived in space at all, or whether they had resulted from accidental contamination following the samples' return from the Moon.

Whether or not the Surveyor 3 samples were contaminated after their return from the Moon remains in doubt, because subsequent experiments have revealed that some forms of microbial life can go into hibernation while in space, reviving once they reach a hospitable environment. This discovery raised some profound questions about the possibility of microbial life forms traveling on asteroids and other bodies, reaching Earth and perhaps seeding it with a basic form of life billions of years ago.

A Probing Visit

Launched on April 17, 1967, Surveyor 3 was the third NASA probe to perform a soft landing on the lunar surface. The probe's original mission was cut short by the fall of the lunar night, which left it unable to generate electricity from its solar panels. NASA's scientists were subsequently unable to reactivate the probe at the next lunar dawn. It went on to become the first robotic probe to be visited by human space explorers during the Apollo 12 mission.

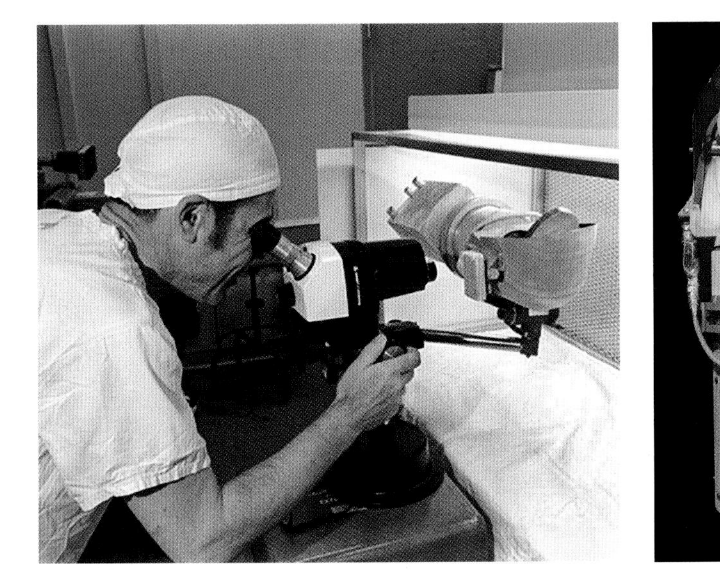

FAR LEFT The Surveyor 3 camera, returned from the Moon, is analyzed by a scientist at the Jet Propulsion Laboratory in 1969.

LEFT The Surveyor 3 camera on display at the Smithsonian's National Air and Space Museum in 2014.

Triumph from Disaster: The Apollo 13 Story

Three of the four Apollo 13 Flight Directors—(foreground, left to right) Gerald Griffin, Eugene Kranz, and Glynn Lunney—applaud the successful splashdown of the spacecraft while behind them Robert Gilruth, Director of the Manned Spacecraft Center, and his deputy Christopher Kraft light up cigars in celebration and relief.

If the successes of the Apollo 11 and 12 lunar landings made Moon missions look easy, then the flight of Apollo 13, launched on April 11, 1970, served to underline just how dangerous such journeys could still be.

Apollo 13 was to have been NASA's third trip to the lunar surface, and astronauts James Lovell, Fred Haise (1933–), and John Swigert (1931–82) practiced for it enthusiastically. The mission profile would have had the astronauts exploring the Moon's Fra Mauro highlands, a mountainous area that NASA scientists suspected was formed by material thrown up from a crater caused by a meteor impact. But the Apollo 13 crew never reached Fra Mauro. Just 56 hours into the mission, an oxygen tank in their spacecraft's Service Module (SM) ruptured and damaged several of the vehicle's power, electrical, and life support systems. NASA engineers quickly determined that the SM had been damaged beyond repair, and now had limited ability to sustain the astronauts in the attached Command Module (CM).

Fortunately, the Lunar Module (LM) had been unaffected by the accident. As a self-contained spacecraft, the LM had its own power supply and life support systems that could keep the astronauts alive while mission controllers at the Johnson Space Center and NASA facilities across the United States frantically sought the best way to bring the crew home safely.

As the NASA teams worked, people around the world watched the drama of the mission unfolding in real time on the TV news. One of the biggest problems the crew faced was the need to remove the carbon dioxide they were all exhaling from the enclosed environment of the LM. The limited supply of lithium hydroxide in the LM usually employed for this task was not enough to sustain the three astronauts for the whole time it would take to get them back to Earth on the trans-lunar flight path that offered the best chance of the crew's safe return. While the CM contained additional lithium hydroxide canisters, these were cube-shaped and were not compatible with the equipment in the LM. Using a piece of hose from one of the spacesuits, and guided by their mission controllers, the Apollo 13 crew managed to improvise a device, dubbed "the mailbox," which allowed them to feed lithium hydroxide from the CM canisters into the LM's carbon dioxide filtration system.

Having found this solution to the problem, the Apollo 13 crew completed its journey around the Moon and returned to the CM for a successful reentry and splashdown on April 17, 1970.

Ironically, even more than the successes of the previous two Moon landings, NASA's ability to keep its astronauts alive and bring them back home safely after such a catastrophic mission-ending failure helped to boost the organization's reputation for competence and capability internationally. NASA truly had snatched triumph from the jaws of disaster.

"Failure Is Not an Option."

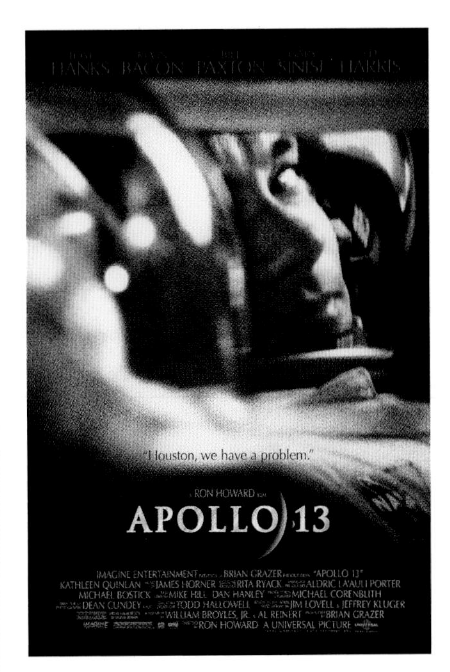

Based on astronaut James Lovell's bestselling autobiography and directed by Ron Howard, the 1995 movie *Apollo 13* proved to be an unconventional Hollywood blockbuster that detailed the near disaster of the mission. NASA Flight Director Eugene Kranz, played in the movie by Ed Harris, never actually uttered the words "Failure is not an option" during the crisis, but he freely admits that he wished he had said them, because the phrase so neatly summed up NASA's response to what could easily have been a tragedy.

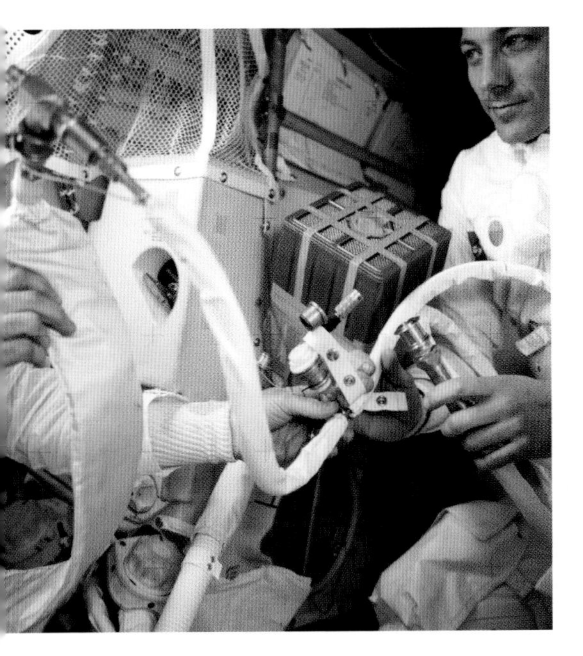

FAR LEFT "The mailbox" device used to connect the cube-shaped lithium hydroxide canisters into the Lunar Module's systems and keep its limited air supply breathable on Apollo 13's trip home.

LEFT Apollo 13 astronauts James Lovell and Fred Haise practice for geological investigations on the Moon at Kilauea, Hawaii, in 1969.

Scientific Harvest: Apollo 14 – Apollo 17

Following the triumph of the first Moon landing, the United States government began to reassess and scale back the extent of the Apollo program. The proposed Apollo 20 mission was the first to be scrapped and, shortly after the dramatic failure of the Apollo 13 mission (p. 156), so too were Apollo 19 and Apollo 18. NASA also saw dramatic cuts to its budget, as the government sought to limit expenditure at a time when the nation was still heavily embroiled in the Vietnam War.

OPPOSITE Apollo 14 Commander Alan Shepard stands by the Modular Equipment Transporter, called the "rickshaw" by astronauts, which provided a means to move tools, cameras, equipment, and samples on the lunar surface.

BELOW Deploying the Apollo Lunar Surface Experiments Package on Apollo 14, January 31–February 9, 1971.

Despite this, NASA worked hard to reap a bountiful harvest of scientific knowledge about the Moon's origins and evolution from its last four landings on Apollo missions 14 through 17. Three of these missions also involved a new piece of equipment in the form of the Lunar Roving Vehicle (LRV), which enabled the astronauts to travel farther on the Moon's surface than any previous mission, and brought within reach several of the Moon's most interesting surface features—mountains, and rilles (long, narrow, channel-like depressions)—that had previously been part of a distant, unreachable backdrop.

NASA's scientists were keen to exploit the improved capabilities of the later Apollo missions. By the end of the last Moon mission, astronauts had deployed more than 50 experiments on the lunar surface. Perhaps the most important of these was the Apollo Lunar Surface Experiments Package (ALSEP), a set of instruments to measure various aspects of the lunar environment, such as soil mechanics, seismic activity, heat flow, magnetic fields, and solar wind. To date, data from these science packages has yielded more than 10,000 scientific papers and helped to provide evidence for our modern understanding of the origins and evolution of the Moon.

Only one of the 12 astronauts to walk on the Moon, Apollo 17's Harrison Schmitt (1935–), was a doctorate-level geology specialist, but NASA ensured

Old Theories on the Origin of the Moon

Prior to the Apollo program, there were three principal theories to explain the origins of the Moon:

Fission
A theory that asserted that the Moon naturally split off from the Earth at some point.

Co-accretion
A theory that proclaimed that the Moon and the Earth formed at the same time from the same solar nebula debris.

Capture
A theory that held that the Moon formed elsewhere, and was subsequently drawn into orbit around the Earth.

that every crew member knew enough to undertake useful geological fieldwork on the lunar surface.

Indicative of the approach taken by some of the non-geologist astronauts was that made by David Scott (1932–) on Apollo 15. He trained enthusiastically for his mission, and, once on the Moon, threw himself into his work with immense energy. As he later recalled: "Most of my thoughts on the Moon were of the geology involved. Our mission was especially heavy in science, trying to understand the geology of the local site and the Apennines—why things occurred as they did."

Together, Scott and crewmate James Irwin (1930–91) found a 4-billion-year-old mineral sample formed in the early stages of the solar system's evolution. The sample, later dubbed the "Genesis Rock," provided a window into the conditions and origins of the Moon, the Earth, and the rest of the planets in the solar system.

The astronauts on the six missions that landed on the Moon returned almost 881 pounds (400 kg) of lunar samples. Since then, more than 60 research laboratories throughout the world have conducted continuing studies on these samples. After a decade of analysis of lunar samples gathered on the Apollo missions and some of the automated Soviet Luna probes, a new scientific consensus emerged that the Moon had been formed by debris from a massive collision that occurred around 4.6 billion years ago between an early Earth and an object roughly the size of Mars. Dubbed "the big whack," the full implications of this theory have yet to be determined, and lunar scientists are anxious to return to the Moon to obtain additional samples that might reveal further details about this probable event.

RIGHT Astronaut David Scott, commander of the Apollo 15 mission, gets a close look at the "Genesis Rock" at the Lunar Receiving Laboratory of the Manned Spacecraft Center in Houston, Texas.

OVERLEAF Scientist-astronaut Harrison Schmitt during the Apollo 17 mission, at the Taurus-Littrow landing site in December 1972. The crew returned with 242 pounds (110 kg) of rock and soil samples, more than was returned from any of the other lunar landing sites.

Soviet Sample Return Missions and Robotic Rovers

1970–76

Soviet scientists collected their lunar samples and pioneered remote exploration of the Moon using unmanned probes and rovers.

While the United States remains the only nation to have sent human explorers to the Moon, the Soviet Union did succeed in completing three robotic missions to retrieve samples from the lunar surface. On September 24, 1970, Luna 16 returned 3.6 ounces (101 g) of soil that it had gathered with its extendable drilling apparatus. This mission was followed by Luna 20, which returned a 1.9-ounce (55-g) sample of material from the Apollonius Highlands near the Sea of Fertility, and Luna 24, which collected 6 ounces (170 g) of soil from the Sea of Crises in 1976. These samples played a significant role in expanding acceptance of "the big whack" theory.

The Soviets also sent two eight-wheeled robotic rovers to the lunar surface, the first of which, Lunokhod 1, arrived on November 17, 1970. It traveled around the Sea of Rains for 11 months before ceasing operations on October 4, 1971. Lunokhod 1 blazed a trail for robotic space exploration: it was the first remote-controlled rover to operate on another world, and it took readings of the lunar environment using an x-ray spectrometer, an x-ray telescope, and cosmic-ray detectors.

Lunokhod 2 landed on the Moon on January 15, 1973. It took photographs of the lunar surface, analyzed solar x-rays, measured the Moon's magnetic field, and tested the properties of various kinds of lunar surface material. It operated for roughly four months, during which it covered 23 miles (37 km) of terrain, sending back 86 panoramic images. It eventually ceased to function after it rolled into a crater and had its power supply disrupted by a layer of dust that obscured the surface of its solar panels.

Aside from the data they collected, the true legacy of the Lunakhod missions can be seen in the success of the various rover expeditions to Mars that have followed (p. 279).

ABOVE LEFT Soviet scientists examine Moon rock samples returned by the three Luna probes.

ABOVE A computer-generated image of the Lunokhod 1 Moon rover.

6

New Nations, New Missions

In the years that followed the Moon landings, the first two nations to journey into space, the Soviet Union and the United States, undertook a range of new and ever more ambitious missions to travel farther outside the Earth's orbit to study the Sun (p. 172) and Venus (p. 178).

Meanwhile, other nations eagerly followed their path of space exploration into orbit and beyond. A significant number of the countries took their first tentative steps toward space exploration in cooperative projects with the established space powers.

Often, the choice of which of the two superpowers these other nations turned to first reflected their allegiances in the wider Cold War. America's allies in Europe and elsewhere mainly entered into cooperative agreements with NASA. These collaborations typically resulted in satellite launches. Meanwhile, nations that looked more toward the Soviet Union enlisted its aid for access to rocket technology and seats for its citizens on various Soviet space missions. Some unaligned countries, such as India, even managed to achieve significant cooperation with both nations (p. 196).

As a result of their initial projects, many rising spacefaring nations, particularly China, developed greater technological expertise of their own. Several resolved to develop their own space launch capabilities, either alone or within the framework of regional space organizations, to reduce their reliance on others for access to space.

Every country that sent experiments, satellites, or people into space reaped rewards and benefits in areas as diverse as weather and climate monitoring, remote surveillance, and data gathering. In the age of space exploration, one thing was certain: the heavens were no longer solely the preserve of Soviet and American space explorers.

OPPOSITE This montage of planetary images taken by various space probes shows (top to bottom) Mercury, Venus, Earth and its Moon, Mars, Jupiter, Saturn, Uranus, and Neptune.

The Expanding Space Club

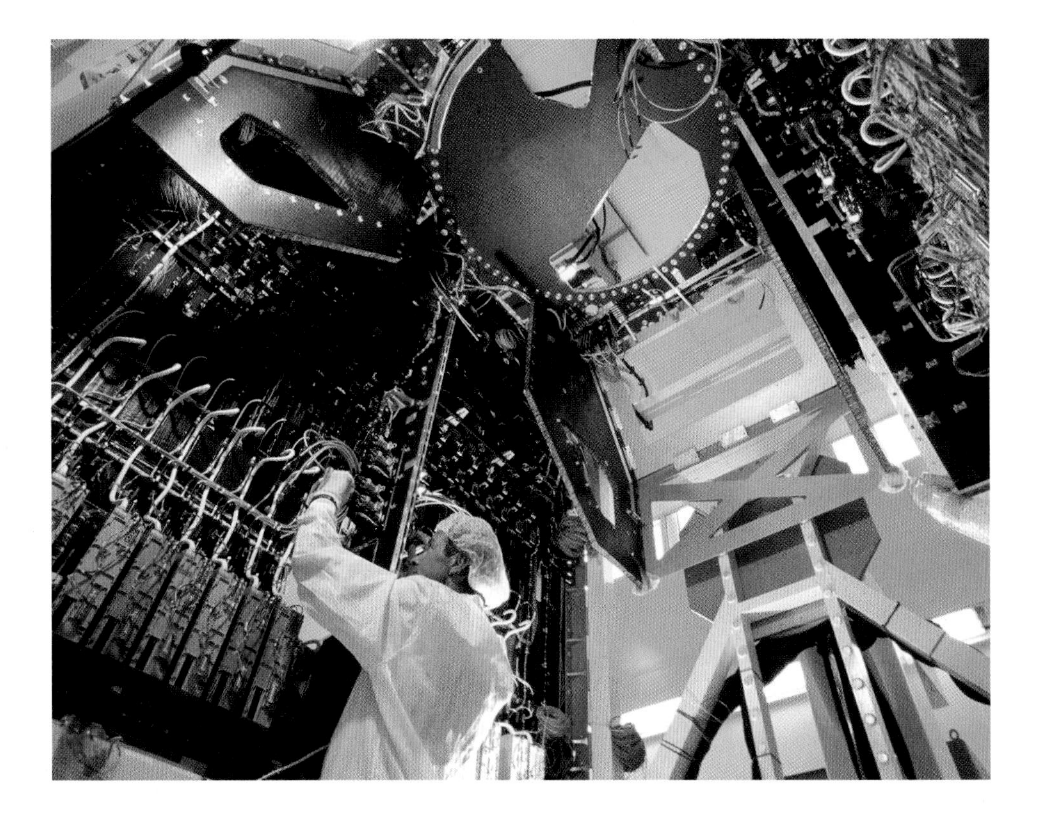

RIGHT The ability to design, build, and deliver communication satellites, such as the Eutelsat seen here under construction, represents a significant level of achievement in the "space club."

OPPOSITE LEFT A 1:40 scale model of the Shenzhou spacecraft that was used to send the first Chinese astronauts into space.

OPPOSITE RIGHT Japanese astronaut Koichi Wakata of the Japan Aerospace Exploration Agency prepares to enter a Soyuz spacecraft trainer on May 6, 2013, at the Gagarin Cosmonaut Training Center in Star City, Russia.

"A country that sees itself as a power deserving of a seat at the table of world governance is expected to race for space... Joining the space club is a legitimate and rational decision."

DEGANIT PAIKOWSKY,
ISRAELI SPACE POLICY ANALYST

The successes of the Soviet Union and the United States highlighted the prestige and opportunities that came with the development of spaceflight capabilities, and the desire for similar kudos inspired many nations to embark upon their own space exploration programs.

To date, more than 70 nations have become members of the "space club," a loose grouping of countries that have developed or invested in space technology. The club can be viewed as a pyramid-like structure with five basic levels. At each level there is an increased capability for space activities, and a smaller number of nations in the membership.

Level 1 Scientists and/or engineers from a particular nation engage in space projects overseen and funded by outside countries or entities. There are commercial benefits, either through space satellites that enhance existing infrastructure, such as telecommunications, or directly, through the hire of space launch services.

Level 2 Nations and organizations operating at this level have established a space agency of some type to manage concerted space exploration efforts and oversee basic space research and development. Such nations also typically have the capability to build and operate their own satellites.

Level 3 Nations and organizations operating at this level can build, operate, and fly their own satellites into Earth orbit, often from their own launch facilities.

Level 4 Nations and organizations operating at this level can typically send space probes to other bodies, such as the Moon and Mars. They also often engage in human spaceflight programs, although they may rely on crewed spacecraft operated by other nations to do so.

Level 5 Nations operating at this level have typically developed their own launch vehicles for human spaceflight, and can undertake the full range of space activities.

The nations at Level 1 include a broad array of states around the world. For example, Slovenia, which has no space agency of its own, has assisted higher-level nations and consequently in 2010 became an associate member of the European Space Agency (ESA). Before that, Slovenian space scientist Dušan Petra (1932–) worked on NASA's Infrared Astronomical Satellite (IRAS) and other projects sponsored by other national programs.

During most of the Cold War, the top of the pyramid was occupied exclusively by the Soviet Union and the United States, while both powers flew citizens from other countries into orbit. The Russian Federation inherited the former Soviet Union's space program after the collapse of Communism in 1991. Roughly a decade later, China developed its own human spaceflight capacity with its Shenzhou capsule (p. 194). Since the retirement of its fleet of space shuttles in 2011, the United States has not had the capacity to send astronauts into space by its own means. Until a replacement vehicle is developed, the United States relies on the Russian space agency, Roscosmos, to carry astronauts into space. This leaves Russia and China as the only two nations that can currently place people into orbit with their own spacecraft and launch vehicles.

Space Exploration Goals

Generally the goals of countries' respective programs mirrored those of the space superpowers:

- To secure national security benefits using space assets to aid military operations.
- To secure commercial benefits, either through space satellites that enhance existing infrastructure and capabilities, or directly through the hire of space launch services.
- To secure technology development benefits that give rise to a cascading multiplier effect when technology developed for space applications is adapted for other uses.
- To pursue pure science through curiosity-driven research, which often yields unforeseen benefits decades after the initial investment.
- To increase prestige, enhancing both national pride at home and respect abroad with a conspicuous display of technological prowess.

Global Launch Sites

At the start of 2018, there were 24 spaceports around the world capable of launching rockets that can take payloads into orbit. Some, such as NASA's Kennedy Space Center in Florida, are well known and open to the public. Others are top-secret installations shrouded in mystery, such as the Palmachim launch site in Israel. They are usually, but not always, located within the territory of the nation that operates them, and away from major population centers or international borders. They are also frequently located as close to the Equator as possible, and launch their projectiles in an easterly direction. This allows rockets launched from these sites to take advantage of the Earth's rotational velocity, which is at its greatest at the Equator. Rockets launched eastward from a site close to the Equator are already traveling faster than those launched from sites located closer to the poles, and consequently require less fuel to reach orbit. This advantage is lost on rockets that are designed to deliver their payloads into polar orbits, which are typically favored for weather monitoring or reconnaissance satellites. Israel's launch complex, located on the eastern shore of the Mediterranean Sea, is the only major spaceport in which launches are conducted to the west, but this is largely due to geographic and political considerations.

The busiest spaceports in the world tend to be operated by the nations, or groups of nations, most heavily involved in space exploration. The space programs of the United States and the Soviet Union/Russia have long been comparable in size and the scope of their operations, with launch sites at Cape Canaveral in Florida and Vandenberg Air Force Base in California in the United States, and at Plesetsk in Russia and Baikonur in a Russian-administered region within Kazakhstan. The next busiest spaceports are one launch site at Sriharikota in India and three in China, at Jiuquan, Tiayuan, and Xichang. Close behind these are the ESA launch site at Kourou, French Guiana, and the Tanegashima and Kagoshima launch sites in Japan.

While most of the facilities listed above specialize in orbital launches, various other sites have concentrated on non-orbital flights. Some of these establishments, such as NASA's Wallops Flight Facility on the eastern Atlantic coast in Virginia, specialize in flight and upper atmosphere research, while others, such as Spaceport America, located near White Sands, New Mexico, have been built to service the emergence of a much-anticipated space tourism industry. However, with commercial operator Virgin Galactic yet to commence a regular suborbital space service, it remains to be seen how many more tourism-driven spaceports will eventually be built.

OPPOSITE A world map of sites capable of launching rockets into orbit.

BELOW Media crews await the launch of the Epsilon-1 rocket at the Uchinoura Space Center in Kimotsuki, Kagoshima, Japan.

World Launch Sites

ABANDONED SITE ●

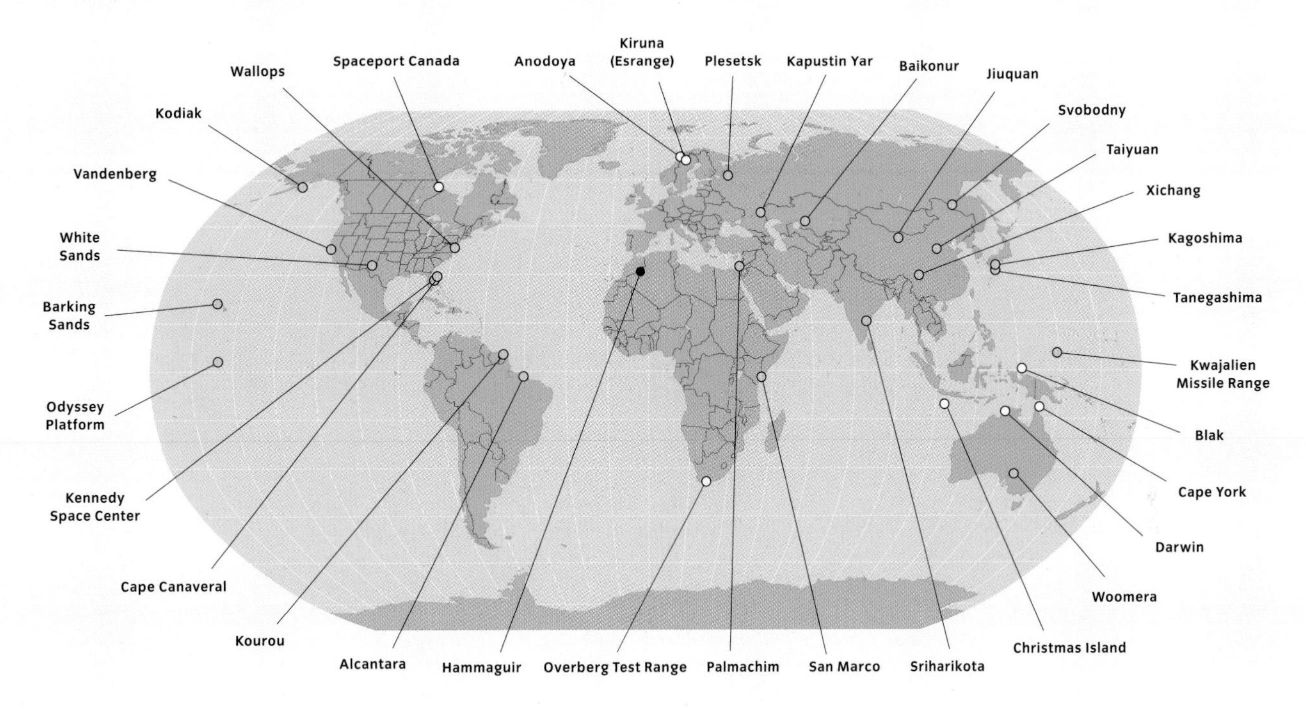

The Kourou Spaceport

FRENCH (1968–)

This transnational launch facility helped to pioneer the market for commercial space launch services.

First established as a launch site by France in 1968, the Kourou spaceport is located close to the Equator in French Guiana on the eastern coast of South America. After the creation of the European Space Agency (ESA) in 1975, the intergovernmental agency negotiated to use Kourou for its space launches. Since then, the European commercial satellite launch company Arianespace and, since 2011, the Russian Space Agency Roscosmos, have also used the site. Despite hosting several different rocket types, Kourou still most frequently uses Arianespace's Ariane family of rockets as launch vehicles. The first rocket in this series, Ariane 1, launched its payload from Kourou on December 24, 1979. Its success opened the way for the development of a new commercial market for the launch system. Since then, there have been more than 200 launches by Ariane rockets. The series' largest and most versatile iteration to date, Ariane 6, is currently planned for a test launch in 2020.

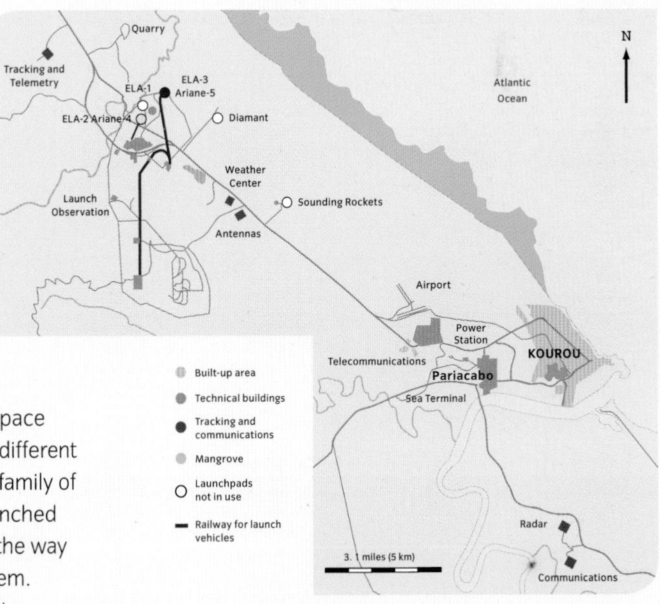

A map of the Kourou launch site.

Early Satellites around the Globe

RIGHT A Chinese-made Long March rocket of the type used to place the first satellite from the People's Republic of China into orbit in 1970.

BELOW The French Diamant A rocket. It was the first rocket to place into orbit a satellite not made by the Soviet Union or the United States.

Since 1957, there have been more than 5,000 rocket launches that have placed in excess of 7,000 satellites into orbit. At the start of 2018, more than two-thirds of these satellites had reached the end of their service lives and re-entered Earth's atmosphere, leaving approximately 2,500 satellites in orbit, of which around 1,070 are operational. The oldest satellite still in orbit is Vanguard 1, launched by the United States in 1958, although it ceased to function in the early 1960s.

After the Soviet Union and the United States, the third country to put a satellite into orbit was the United Kingdom, whose Ariel 1, designed to measure solar radiation and fluctuations in the Earth's ionosphere, was launched on April 26, 1962 using an American Thor-Delta two-stage rocket. The project also prompted the establishment of an international network of tracking sites based in the United States, Argentina, South Africa, and Australia to ensure that future space launches could be constantly monitored. This network would go on to play a vital support role in America's human space missions.

Through to the end of 1970, an additional eight other nations or groups of nations launched satellites, either using their own launch sites and vehicles or paying to use those of the United States. To date, more than 75 countries have operated artificial satellites in Earth orbit and beyond, placed there either by their own rockets or those of another spacefaring nation. That figure will likely increase as more nations seek to benefit from space exploration.

The Next Satellites in Space by Nation 1962–74

Country	Satellite	Operator	Launcher	Launch Site	Launch Date
United Kingdom	Ariel 1	Royal Aircraft Establishment	Thor-Delta	Cape Canaveral, USA	Apr 26, 1962
Canada	Alouette 1	Defence Research and Development Canada	Thor-Agena-B	Vandenberg Air Force Base, USA	Dec 15, 1964
Italy	San Marco 1	Italian Space Research Commission	Scout X-4	Wallops Island, USA	Dec 15, 1964
France	Astérix	National Center for Space Studies	Diamant A	Hammaguir, Algeria	Nov 26, 1965
Australia	WRESAT	Weapons Research Establishment	Sparta	Woomera, Australia	Nov 29, 1967
Europe	ESRO 2B	European Space Research Organisation	Scout B	Vandenberg Air Force Base, USA	May 17, 1968
Federal Republic of Germany	Azur	—	Scout B	Vandenberg Air Force Base, USA	Nov 8, 1969
Japan	Osumi	Institute of Space and Aeronautical Science	Lambda-4S	Kagoshima, Japan	Feb 11, 1970
People's Republic of China	Dong Fang Hong 1	—	Chang Zheng 1	Jiuquan, China	Apr 24, 1970
The Netherlands	Astronomical Netherlands Satellite	—	Scout	Vandenberg Air Force Base, USA	Aug 29, 1974

Ariel 1 and International Science Collaborations

A replica of the Ariel 1 satellite, assembled out of spare parts from the original spacecraft.

In late 1959, space exploration advocates in the United Kingdom, responding to an offer from the United States to provide assistance to nations seeking to develop their own space exploration programs, drafted a proposal to NASA to build a new satellite to gather data about the Earth's ionosphere that would be shared with the global scientific community.

In the resultant collaboration, which set the pattern for many subsequent projects, NASA engineers and contractors designed and built the satellite, while British scientists provided the experiments the vehicle carried.

Ariel 1 marked a significant milestone in the development of scientific satellites; it pioneered the orbital use of various metal alloys and ceramics, autonomous thermal regulation systems, electro-optical sensors, and data storage and processing systems.

Although the satellite failed in July 1962, just a few months after its launch, the experiments it carried were still able to return important data about the effects of solar energy on the ionosphere. The legacy of international scientific cooperation it fostered has been even more enduring.

Exploring the Sun

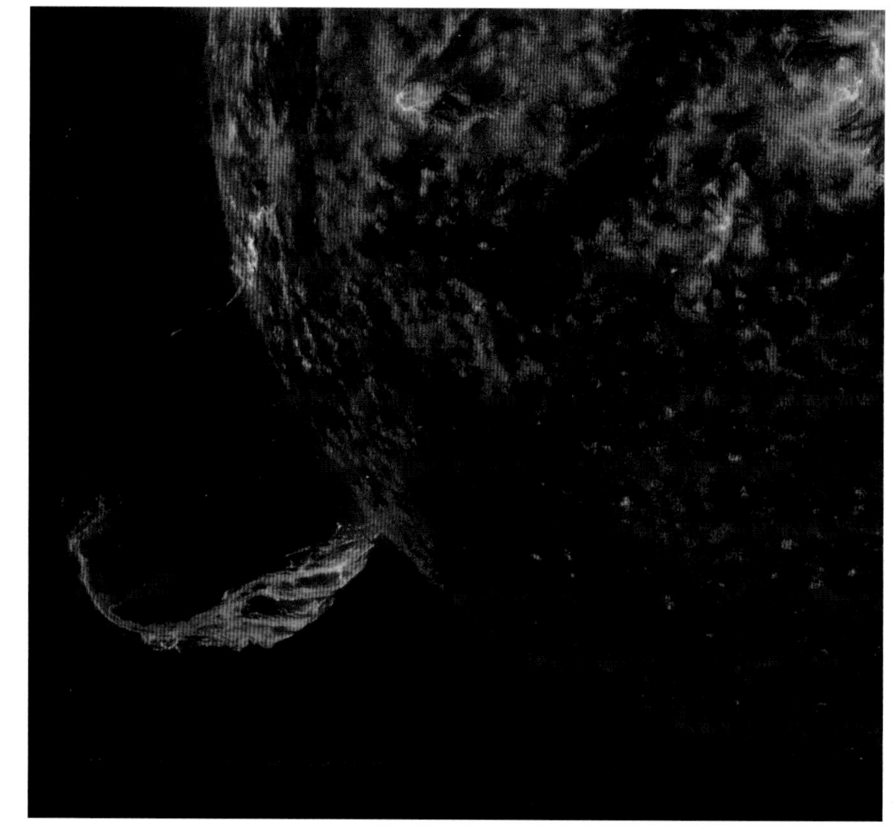

RIGHT The Sun in the throes of a coronal mass ejection along with part of a solar filament on February 24, 2015. While some of the strands fell back into the Sun, a substantial part of the ejection raced into space in a bright cloud of particles.

BELOW This image was taken on March 26, 2007, by the SECCHI Extreme UltraViolet Imager mounted on the spacecraft STEREO-B. STEREO-B is positioned on the side of the Earth that is beyond the Sun in order to observe the movement of our planet around its star.

People have been fascinated by the Sun throughout recorded history, but the advent of the space age enabled scientists to investigate our nearest star to an extent not previously imagined. Starting with sounding rockets that measured the effects of various kinds of solar radiation on the Earth's upper atmosphere, and proceeding to science-focused space probes that studied solar activity more directly, scientists sought to improve humanity's understanding of how the Sun works and the various different ways in which it affects life on Earth.

This new phase of solar investigation concentrated on solar flares and Coronal Mass Ejections (CMEs), the explosions that throw out millions of tons of very hot electrified material from the Sun. The violent effects of CMEs were first recorded on Earth by amateur astronomer Richard Carrington (1826–75) and others in 1859, but their propensity to disrupt Earth-orbiting satellites and Earth-based electronics became increasingly clear in the early 1960s, as telecommunication satellites and the ownership of consumer electronics became more common.

Some of the earliest space missions to concentrate on studying the Sun were the eight satellites that comprised the Orbiting Solar Observatory (OSO). These spacecraft performed their observations from a Low Earth Orbit (LEO), but later probes made use of Lagrange points, areas between two bodies in an orbital system where their respective gravities equalize, effectively canceling each other out. Some of the major satellites involved in exploring the Sun include:

Orbiting Solar Observatory (OSO): A series of eight NASA satellites built to study the Sun were launched between 1962 and 1975. International science teams analyzed the data produced.

Helios I and Helios II: Launched in 1974 and 1976, these probes were a joint project undertaken by NASA and the Federal Republic of Germany. The probes were designed to study solar processes.

Solar Maximum Mission: Launched in 1980, this probe was designed to study solar activity during the 1989 solar maximum, the point in the 11-year solar cycle during which the Sun is at its most active. The probe collected significant images and data for nine years before it lost altitude control in November 1989.

Geotail: Launched in 1992 by Japan's Institute of Space and Astronautical Science (ISAS) in collaboration with NASA, Geotail studied Earth's magnetosphere to monitor its interaction with the solar wind.

Yohkah: This Japanese satellite, launched in 1992, used x-ray instruments to collect data about the Sun.

"Earth's Moon is about 1/400th the diameter of the Sun, but it is also 1/400th as far from us, making the Sun and the Moon the same size in the sky—a coincidence not shared by any other planet–moon combination in the solar system, allowing for uniquely photogenic total solar eclipses."

NEIL DEGRASSE TYSON,
AMERICAN ASTROPHYSICIST

Technicians at Cape Canaveral Air Force Station, Florida, complete the preflight checkout and testing of the Ulysses spacecraft.

Josef Lagrange and Lagrange Points

Key locations in a planet's orbit where observations can be made.

In any orbital system in which a small body orbits a larger body, there are five points where the gravitational pull from each of the bodies cancels the other out. These areas are called Lagrange points after the Italian-French mathematician Josef Lagrange who, building on the work of Swiss mathematician Leonhard Euler (1707–83), identified the location of those five points. Lagrange points are extremely useful in space exploration because, in the absence of gravity acting upon it, a spacecraft can maintain a stable position in the vicinity of a Lagrange point while expending minimal amounts of fuel. Of all the Lagrange points, the L1, which occurs directly between the two bodies in an orbital system, is particularly important, because it provides a perfect, uninterrupted view from one body to the other. In the Earth/Sun system, the L1 point offers an unbeatable vantage point for probes studying the Sun; it is here that the SOHO probe is currently located.

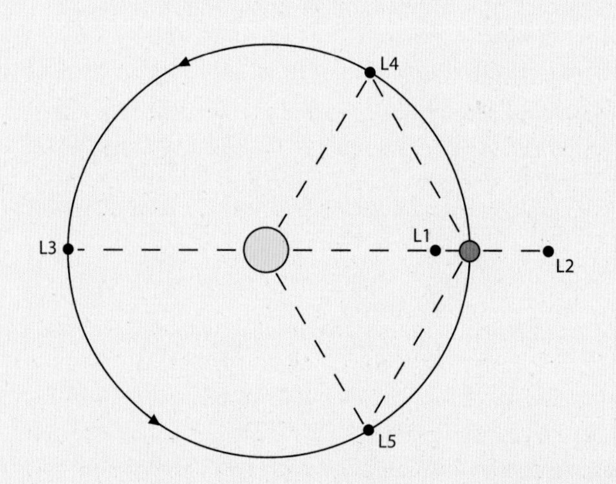

Of the five Lagrangian points in a simple orbital system, three, termed L1, L2, and L3, rest along a straight line running directly through the two bodies. The other two Lagrange points, L4 and L5, form the apex of two equilateral triangles, with edges that are equal to the distance between the two bodies.

Ulysses: Launched in 1990, this probe was part of a collaborative mission between NASA and the ESA. The probe completed its first polar orbit of the Sun in 1995, and continued to study high-altitude solar wind during the solar maximum in 2001.

Wind: This NASA probe, launched on November 1, 1994, was designed to analyze the physical and chemical properties of the solar wind.

Transition Region and Coronal Explorer (TRACE): A spacecraft launched in April 1998 to study magnetic fields in the Sun's photosphere, transition region, and corona during the 2001 solar maximum.

Imager for Magnetopause-to-Aurora Global Exploration (IMAGE): A spacecraft designed to collect data about the Sun's magnetosphere during magnetic storms in the solar maximum in 2001.

Solar TErrestrial RElations Observatory (STEREO) mission: A 2006 mission that used two nearly identical probes—one ahead of Earth in its orbit, and the other trailing behind—to study the structure and evolution of solar storms as they emerge from the Sun and move out through space.

Hinode: Japan's third solar probe was launched on September 23, 2006, and contained three telescopes to study the Sun's eruptive phenomena and space weather with the aim of improving scientists' ability to predict the Sun's influence on the Earth.

Solar Dynamics Observatory (SDO): An ongoing NASA mission with a large international science team that has been observing the Sun since 2010. Its goal is to understand the influence of the Sun on the Earth and near-Earth space by studying the solar atmosphere in small scales of space and time and in many wavelengths simultaneously.

The SDO is a good example of modern solar exploration missions. Since its launch, it has provided a continuous stream of images and data about the Sun, helping solar scientists better to understand the star's patterns of behavior with the aim of being able to predict its perturbations and the influence of those perturbations on life on Earth. It is hoped that analysis of SDO's key data will eventually help scientists to determine how the Sun's magnetic field is generated and structured, as well as how its stored magnetic energy is converted and released into the heliosphere to form the solar wind variations in the solar irradiance. Using three main instruments—the Atmospheric Imaging Assembly, the Extreme Ultraviolet Variability Experiment, and the Helioseismic and Magnetic Imager—SDO has already gathered significant data on the developing field of solar-generated space weather and how it affects Earth, and this fund of knowledge seems set to continue growing as more nations seek to learn the Sun's secrets.

Solar Filaments

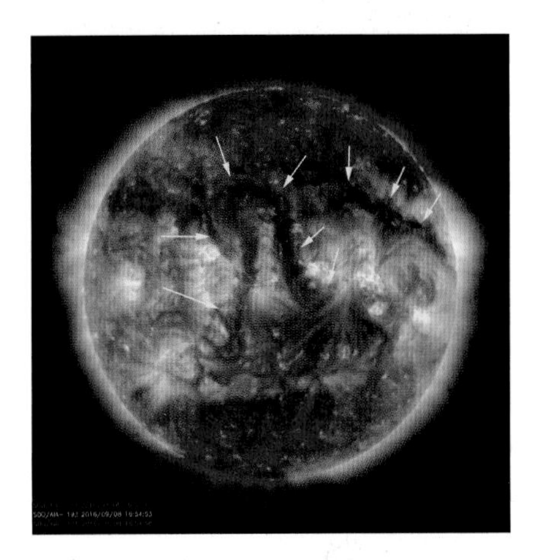

On September 8, 2016, the two most noteworthy features on the Sun were a pair of elongated solar filaments, as shown in the above composite image captured by the Solar Dynamics Observatory (SDO). Solar filaments are elongated strands of plasma suspended above the Sun by its own magnetic field. They are notoriously unstable, and usually break apart within a few days.

This composite image was assembled out of three separate images capturing different wavelengths of extreme ultraviolet light from the Sun. It shows a central filament that is twisted into the shape of an arch at the center of the Sun (highlighted by the yellow arrows). If this filament were straightened out, it would extend almost 1 million miles (1.6 million km), roughly the diameter of the Sun. The other, smaller filament (highlighted by the white arrows), might stretch out to about half that distance.

OPPOSITE LEFT This backup gamma ray detector was installed in the main section of the Orbiting Solar Observatory 1 satellite, launched on March 7, 1962.

OPPOSITE RIGHT An engineering prototype of the Orbiting Solar Observatory 1.

OVERLEAF The sun emitting a mid-level solar flare on January 12, 2015. Solar flares are bursts of radiation.

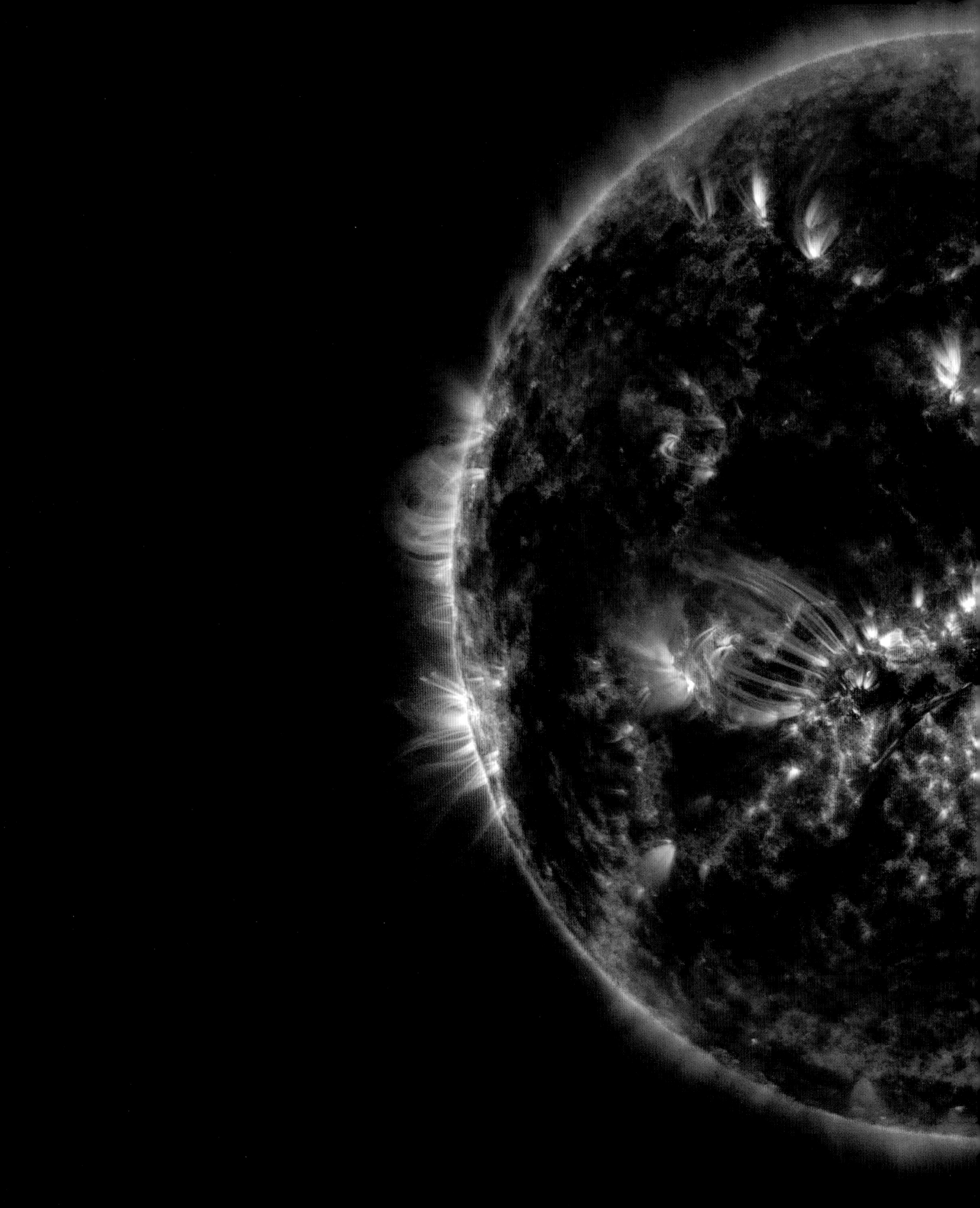

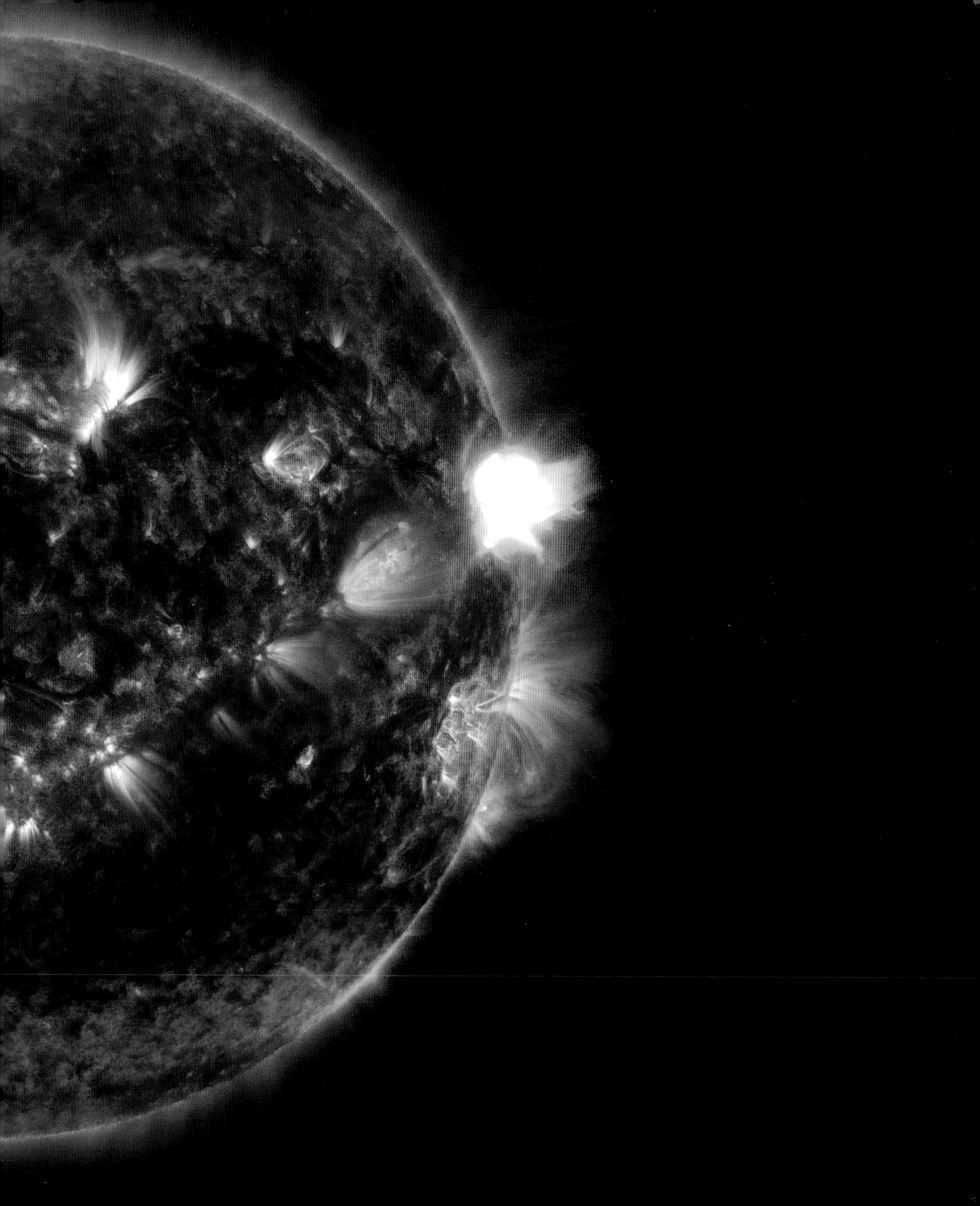

Venus and the International Dimensions of Space Science

While the race to put satellites and humans into orbit and onto the Moon dominated the headlines in the early years of the space age, scientists in the Soviet Union and the United States also had an eye on exploring the planets, and Venus was one of the first candidates for such an expedition. A prominent feature in the early evening and early morning skies, Venus is the closest planet to Earth, and a near twin in terms of size, mass, and gravitation. It has also long held fascination for astronomers, even more so once they realized that the planet was shrouded in a mysterious cloak of clouds that kept its surface hidden from view.

This led some to speculate about the nature of Venus, and the possibility that life existed there in some form. One popular theory during the first half of the twentieth century held that the Sun had gradually been cooling for millennia, and that, as it did so, each planet in the solar system had a turn as a haven for life of various types. Although it was now Earth's turn to harbor life, the theory suggested that Mars had once been habitable, and that life on Venus might be at a much earlier stage of evolution. This notion persisted well into the space age, and had supporters even among NASA scientists. One unnamed researcher at NASA's JPL stated that, "If Venus were covered by water, it was suggested that it might be inhabited by Venusian equivalents of Earth's Cambrian period of 500 million years ago, and the same steamy atmosphere could be a possibility."

In 1961, JPL researchers used ground-based radar systems in an attempt to penetrate the clouds of Venus and ascertain what lay beneath them. While this failed to deliver evidence of vast oceans, it did help to establish that Venus rotated in a retrograde motion, spinning in the opposite direction from its orbital direction.

Eager to discover more about the enigmatic Venus, both the Soviet Union and the United States sent robotic spacecraft to explore the planet in the early 1960s. The Soviets made the first attempt on

ABOVE Depictions of the surface of Venus in science fiction works such as *First Spaceship on Venus* (1960) have subsequently been proven to be wildly inaccurate by the findings of various probes.

RIGHT A computer-generated image of the landing of Venera 4 on the surface of Venus in 1967.

February 12, 1961, with the launch of their Venera 1 space probe. Unfortunately, the probe's telecommunications systems failed before it reached the vicinity of Venus, and consequently it returned no data. A similar setback in July 1962 resulted in the loss of NASA's Mariner 1 probe. NASA had greater success with its Mariner 2 probe, which flew by Venus on December 14, 1962, at a distance of 21,641 miles (34,800 km), gathering in the process some simple measurements of the planet's extremely hot atmosphere.

Although the Soviet Union made several further attempts to reach Venus, these missions were dogged by failure and bad luck. In 1965, the Venera 3 probe fell into the Venusian atmosphere without returning any scientific data. Then, in 1967, Venera 4 successfully deployed a landing capsule into the planet's atmosphere, which returned information about the harsh terrain on its surface. The same year, NASA launched its Mariner 5 flyby probe to investigate the Venusian atmosphere. Both spacecraft demonstrated that Venus was a very inhospitable place for life to

ABOVE LEFT NASA scientists at the Jet Propulsion Laboratory (JPL) poring over a wealth of data from the Mariner 2 probe in 1962.

ABOVE RIGHT This hemispheric view of Venus is a composite of images from the Magellan probe. The higher terrain is shown at the red end of the color spectrum. The lower terrain is close to the blue end of the spectrum.

Pioneer Venus Missions

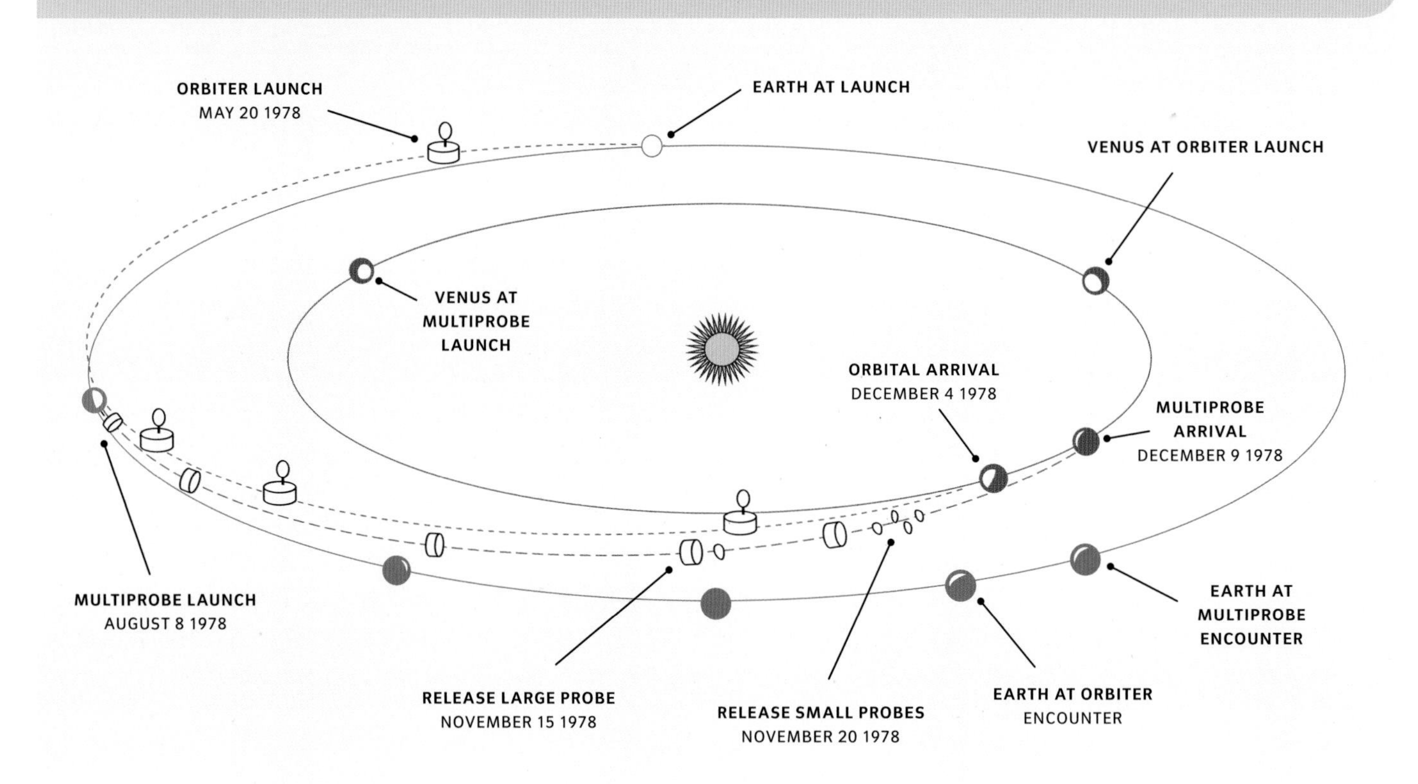

The mission profiles of NASA's two Pioneer Venus missions in 1978.

exist, with a superheated atmosphere caused by its dense clouds that trap the heat from the Sun, and a surface atmospheric pressure that was about 90 times greater than that on Earth.

To get past the thick, cloudy Venusian atmosphere, NASA scientists proposed the two-stage Pioneer Venus mission. The first stage was Pioneer Venus 1, launched in May 1978. This orbiter used its radar to assemble a low-resolution topographic map of the planet, the first of its kind. The second stage, Pioneer Venus 2, was a spacecraft that carried four probes designed to enter the Venusian atmosphere. The probes varied in size, with one significantly larger than the other three. Together, they recorded atmospheric temperatures of between 838°F and 858°F (448–459°C).

The Soviet Venera 15 and Venera 16 probes, both launched in 1983, provided high-resolution coverage over the northern reaches of Venus. But it wasn't until the launch of NASA's highly successful Magellan probe in 1989 that scientists were able to produce a comprehensive high-resolution map of 99 percent of the surface of Venus. The Magellan data provided a number of surprises; among them the discovery of evidence of lava flows and plate tectonics, suggesting that Venus was geologically active at some point in its history.

The Venus Express Mission

2005–15

This probe measured the Venusian atmosphere and detected traces of the planet's geological history.

The first probe to visit Venus that did not come from either the Soviet or American space programs was the Venus Express probe created by the ESA. Venus Express was launched from the Baikonur Cosmodrome aboard a Soyuz-FG-Fregat launch vehicle (a further adaptation of the R-7 rocket) on November 9, 2005. The spacecraft reached Venus in April 2006, where it promptly entered into a polar orbit.

The probe undertook a long-term observation of the planet's clouds and atmospheric dynamics. Scientists analyzing the data soon identified that the Venusian atmosphere was even more complex than had previously been suspected. Italian project scientist Giuseppe Piccioni (1953–) described the Venusian clouds as presenting "repetitive patterns and recurrent features, but they are very variable in position both on short and long time scales. Essentially variable in both thickness and velocity."

Venus Express made several important discoveries during its near decade-long period of operations. It completed the first infrared map of the surface temperature on the southern hemisphere of Venus, confirming that the planet once had a plate tectonics system and an ocean of water. The probe also provided data indicating that water is still being lost from the upper atmosphere, which suggests that Venus had more water in its early history.

While data from Venus Express has largely ended the theory of the planet being a cradle from which life might evolve, it has not yet ended the hope that life might still exist there. With uniform temperatures on the surface at nearly 791.6°F (422°C), and massive atmospheric pressure, the development of life similar to that on Earth is unlikely. However, the evidence that Venus once had more water makes it possible that hardy microorganisms, tolerant of the harsh Venusian conditions, might have evolved and continue to be present.

Concept art of ESA's Venus Express in orbit around the planet.

Establishing the European Space Agency

ABOVE LEFT French scientist Pierre Auger was the first Director General of the European Space Research Organisation.

ABOVE RIGHT The Europa launcher, developed by the European Launcher Development Organization, during a test launch at the Woomera launch facility in Australia.

The successful launch of Sputnik 1 in 1957 did not just galvanize the efforts of the United States to invest more heavily in space exploration; scientists, engineers, and politicians across Western Europe were equally quick to see the benefits of their nations developing independent space capabilities, but none could match the spending of the postwar superpowers, at least, not on their own. An interregional approach to space exploration was required. In 1958, French scientist Pierre Auger (1899–1993) joined with Italian Edoardo Amaldi (1908–89) to recommend that European governments establish a "purely scientific" space agency on the same model as the recently established European Council for Nuclear Research (CERN). After much negotiation, ten European countries—Belgium, West Germany, Denmark, France, the United Kingdom, Italy, the Netherlands, Sweden, Switzerland, and Spain—finally came together to found two new organizations in 1964; the

European Space Research Organisation (ESRO), for satellite development, and the European Launch Development Organization (ELDO), which would build rocket-based vehicles to send the satellites into space.

Immediately, ELDO began developing the multistage Europa rocket. A modified version of the British Blue Streak missile was to serve as the rocket's first stage, with the French Coralie rocket adapted for use as a second stage, and the West German-developed Astris rocket adapted for the third stage. Unfortunately, the politicians underestimated the engineering difficulties involved in successfully integrating the three different rocket systems. The plan was eventually scrapped.

In contrast to ELDO's failure, ESRO proved far more successful, conducting various experiments launched on a variety of British-, French-, and American-built suborbital sounding rockets from its purpose-built rocket-testing range, dubbed the Esrange, in Kiruna, Sweden. ESRO also developed two small satellites, which were launched by the Americans from Vandenberg Air Force Base using Scout B rockets. While ESRO 2A (also known as Aurorae) failed to reach orbit, ESRO 2B (also known as Iris), launched on March 6, 1968, succeeded, and went on to detect a variety of x-rays reaching Earth from non-solar sources. ESRO also flew the Highly Eccentric Orbiting Satellite-A (HEOS-A), launched on December 5, 1968, from Cape Canaveral on an American Delta rocket. This satellite carried instruments from nine scientific groups in five European countries for measuring interplanetary magnetic fields. But despite its achievements, ESRO was soon superseded.

BELOW LEFT The ESRO 2 was the first satellite developed by the European Space Research Organisation (ESRO). ESRO's first satellites concentrated on solar and cosmic radiation and their interaction with the Earth and its magnetosphere. This image shows the ESRO 2 being vibration tested in 1970.

BELOW RIGHT A Highly Eccentric Orbit Satellite being launched aboard a Delta rocket from Cape Kennedy's Launch Complex 17B, December 5, 1968.

In the early 1970s, ELDO and ESRO were integrated into a single regional space organization; the European Space Agency (ESA), established on May 30, 1975. Over the years, the original membership carried over from ELDO/ESRO has expanded to 22 states with full member status, including one non-European member, Canada. Alongside these nations are a number of cooperating members engaging in joint projects with other ESA members. As products of a transnational organization, ESA projects operate within a funding structure in which the percentage of costs contributed by any member state on a project equals the percentage level of participation that member has in the project. This structure lowers the cost of participation for member states, while ensuring that the bigger-spending nations see more of the benefits of their investment.

Although ESA has not yet developed its own launch vehicle capable of taking humans into orbit, this issue has been resolved through cooperation with nations that do have this capacity. One of the earliest examples of this kind of high-profile cooperative project was Spacelab, a venture undertaken with NASA.

Spacelab was a laboratory module that could be flown in the Space Shuttle's payload bay. Work on the module began in 1974, just before the formation of ESA, and the lab flew 25 times between 1983 and 1998. Part of the agreement between ESA and NASA was that European astronauts would fly on these missions. The first ESA astronaut, West Germany's Ulf Merbold (1941–), flew on Space Shuttle Columbia during the STS-9 Spacelab mission in 1983. Subsequently, ten further ESA astronauts flew on various Space Shuttle missions, some more than once, with many more subsequently going on to fly on Russian Soyuz spacecraft.

ESA also served as a full partner in the consortium that built the International Space Station (ISS) (p. 232). The orbital assembly of the ISS began in 1998 and the space station has been permanently occupied since 2000. ESA's principal contribution to its structure is the Columbus laboratory, which was added in 2009. ESA also plays an important role in keeping the station supplied through the use of its Automated Transfer Vehicle (ATV).

As with Spacelab, participation in the ISS gave a limited number of ESA astronauts the opportunity to serve as a part of its crew and to oversee experiments. To train these astronauts, ESA established its own European

TOP ESA astronaut Ulf Merbold aboard STS-9, Spacelab-1 mission, working at the Gradient Heating Facility on the Materials Science Double Rack.

ABOVE Concept art of ESA's Huygens probe descending on to Saturn's largest moon, Titan, in 2005.

Astronaut Centre (EAC) in Germany in 1998. The most recent intake of ESA astronauts came in May 2009, when the corps recruited six new members.

Away from human spaceflight, ESA's science program is one of the largest and most successful in the world, second only to NASA's. ESA's first major satellite, Cos-B, designed to measure gamma-ray emissions, was launched from Vandenberg Air Force Base on August 9, 1975, on a Delta rocket. Over a decade later, on July 2, 1986, ESA launched its Giotto probe on an Ariane 1 rocket from the Kourou launch facility in French Guiana (p. 169). Giotto was a deep-space mission to study Halley's Comet, with a secondary mission to Comet Grigg-Skjellerup. More recently, ESA has engaged in a variety of cooperative missions with NASA, particularly with the Solar & Heliospheric Observatory (SOHO) and Ulysses probes to study the Sun. ESA projects involving the Hubble Space Telescope (p. 331) have been enormously successful, and have significantly expanded the knowledge base of scientists and engineers. ESA took on a more equal partnership with NASA on the Cassini-Huygens mission to Saturn, which operated between 1996 and 2017. NASA's Cassini probe carried ESA's Huygens lander to Saturn's moon, Titan, where it was dropped into the atmosphere (p. 318). During its parachute descent, and for around 90 seconds after landing, Huygens recorded data on the composition of the moon's lower atmosphere, and captured images from Titan's surface that have left scientists around the world eager to mount another mission to explore and survey it in greater detail.

> *"For the first time in my life I saw the horizon as a curved line. It was accentuated by a thin seam of dark blue light—our atmosphere. Obviously this was not the ocean of air I had been told it was so many times in my life. I was terrified by its fragile appearance."*
>
> ULF MERBOLD,
> FIRST ESA ASTRONAUT

The Ariane Rocket Family

1979–

Opening up the commercial satellite launch market.

Drawing on the valuable lessons learned from the failure of the Europa rocket, ESA worked to design and build its Ariane family of launch vehicles from the ground up. From the first launch of Ariane 1 in 1979, the rockets have proved to be versatile and reliable launch vehicles.

Moreover, Ariane rockets freed ESA from reliance on NASA to launch its spacecraft, and allowed for the creation of Arianespace, a commercial multinational company owned and operated by a consortium composed of ten member nations. Arianespace swiftly established itself as a world leader in commercial space launches. Each subsequent iteration of the Ariane rocket up to the Ariane 5, the most recent, has advanced the launch capabilities of both Arianespace and ESA as a whole.

This concept art depicts the Ariane family of rockets in flight and on the way to place their various satellite payloads into orbit.

The Commonwealth in Space

Like the United States and the Soviet Union, the United Kingdom acquired V-2 rockets and rocket technicians from Germany at the end of World War II (p. 48), but it emerged from the conflict with a severely damaged economy and infrastructure. Britain did not have the same resources to expend on rocket development as its wartime allies, or to pursue much of a space program. As a result, British scientists were forced to rely on American rockets and engineers to get their experiments into space on Ariel 1 (p. 171) in 1962.

Working within a limited budget, the British military still managed to produce a range of atmospheric sounding rockets as well as short- and medium-range ballistic missiles using highly efficient, but highly unstable, liquid propellants. Britain's most prominent rocket in this period was Blue Streak. Originally designed as a delivery system for Britain's nascent nuclear weapons program, its effectiveness in this role was severely limited by the fact that its propellants could not be stored in its fuel tank for long periods, making it unsuitable for rapid deployment.

This Skylark high-altitude research rocket was just one of many rocket types that were tested at Australia's Woomera launch facility.

> *"We should be pushing our boundaries. After all, we Britons are explorers and adventurers."*
>
> HELEN SHARMAN, FIRST BRITISH ASTRONAUT

The British Black Arrow rocket lifting off from its launchpad in Woomera, Australia, in October 1971, on a mission to place the Prospero satellite in orbit.

However, the technology was more suitable for use in a satellite launch vehicle. While plans to incorporate Blue Streak as the first stage of the Europa multistage rocket were abandoned (p. 183), British engineers redeveloped the technology into a series of test rockets, including Black Knight and Black Prince, before building the Black Arrow launch vehicle. Black Arrow successfully placed Britain's Prospero satellite into orbit on October 28, 1971, but even before the flight had taken place, the UK government had already canceled all further funding for the rocket. Subsequently, Britain has been reliant on other nations to provide launch services for its satellites, while its first astronaut, Helen Sharman (1963–), flew to the Soviet/Russian space station Mir on Soyuz mission TM-12 on May 18, 1991.

Black Arrow was not launched from the United Kingdom, but from a launchpad at Woomera Range Complex in Australia, a member of the Commonwealth of Nations, which had its roots in the former British Empire. Established in 1947, Woomera was used as a rocket launch and testing area for a variety of British and Australian rockets in the years following the end of World War II. Australia's first

ABOVE This Canadarm patch was owned by NASA's first female astronaut, Sally Ride.

RIGHT These pages from the Operations Checklist for Space Shuttle Mission STS-7 include a drawing of Space Shuttle Challenger, with Canadarm in place.

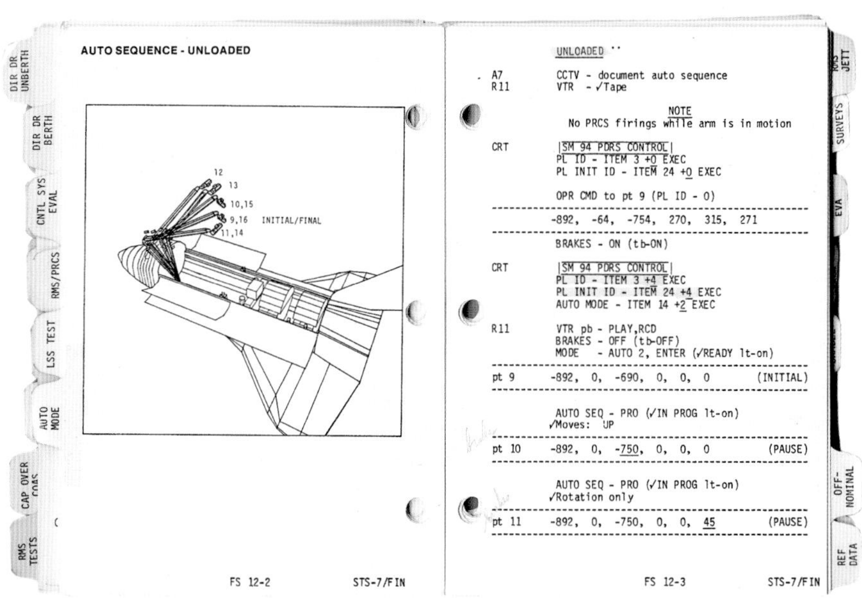

DIR DR UNBERTH | DIR DR BERTH | CNTL SYS EVAL | RMS/PRCS | LSS TEST | AUTO MODE | CAP OVER COAS | RMS TESTS

AUTO SEQUENCE - UNLOADED

FS 12-2 STS-7/FIN

UNLOADED

A7 CCTV - document auto sequence
R11 VTR - ✓Tape

NOTE
No PRCS firings while arm is in motion

CRT |SM 94 PDRS CONTROL|
PL ID - ITEM 3 +0 EXEC
PL INIT ID - ITEM 24 +0 EXEC

OPR CMD to pt 9 (PL ID - 0)

-892, -64, -754, 270, 315, 271

BRAKES - ON (tb-ON)

CRT |SM 94 PDRS CONTROL|
PL ID - ITEM 3 +4 EXEC
PL INIT ID - ITEM 24 +4 EXEC
AUTO MODE - ITEM 14 +2 EXEC

R11 VTR pb - PLAY,RCD
BRAKES - OFF (tb-OFF)
MODE - AUTO 2, ENTER (✓READY lt-on)

pt 9 -892, 0, -690, 0, 0, 0 (INITIAL)

AUTO SEQ - PRO (✓IN PROG lt-on)
✓Moves: UP

pt 10 -892, 0, -750, 0, 0, 0 (PAUSE)

AUTO SEQ - PRO (✓IN PROG lt-on)
✓Rotation only

pt 11 -892, 0, -750, 0, 0, 45 (PAUSE)

FS 12-3 STS-7/FIN

RMS JETT | SURVEYS | EVA | OFF-NOMINAL | REF DATA

The Canadarm Remote Manipulator

1981–2011

In space, astronauts sometimes need a robotic arm to give them a hand with unwieldy payloads.

One of the Canadian space program's most significant achievements has been the development of robotics for space exploration missions. Its Canadarm, which first flew on the Space Shuttle Columbia mission STS-2 on November 13, 1981, improved astronauts' ability to manipulate large satellites and other spacecraft. After a long and successful space career, the Canadarm was retired along with the Space Shuttle program after mission STS-135 in 2011, which marked the robotic arm's 90th flight. The CSA also contributed a larger and more capable Canadarm to the ISS, and followed it with another similar arm known as Dextre. These instruments support ISS's maintenance, and also perform "cosmic catches" to capture and dock the various unpiloted spacecraft that carry supplies and equipment to the ISS.

The Space Shuttle's Canadarm in action, enabling this ultimate space selfie during the STS-114 mission in July 2005.

satellite, named WRESAT after the Weapons Research Establishment responsible for its design, was launched from Woomera on a modified American Redstone launch vehicle on November 29, 1967. The site also hosted the failed launch of the Europa rocket on June 12, 1970.

Australia was not the only Commonwealth country to pursue its own space exploration activities. India began developing its own space program in 1962 (p. 196), while Canada worked in close collaboration with NASA and successfully launched its first satellite, Alouette 1, on September 29 that same year, from Vandenberg Air Force Base using an American Thor-Agena rocket. The Canadian government subsequently invested in a variety of international cooperative space projects. It is a full partner in the ISS (p. 232) and its astronauts have flown on multiple Space Shuttle missions, starting with Marc Garneau (1949–), who flew on Space Shuttle mission STS-41-G in 1984.

But despite their space successes, most of these Commonwealth nations did not develop their own space agencies until relatively recently. The Canadian Space Agency (CSA) was founded on March 1, 1989, a decade after Canada first joined ESA as a cooperating member. Australia consolidated its space exploration activities together in the Australian Space Research Institute (ASRI) in the early 1990s, while Britain itself officially formed the UK Space Agency on April 1, 2010, although it had been a charter member of ESA since its formation in 1975 (p. 183).

Among the UK Space Agency's most exciting projects currently under development is Skylon, a space plane that uses an innovative air-breathing rocket propulsion system; when the present volume went to press, Skylon was scheduled to enter flight tests by 2025 (p. 353).

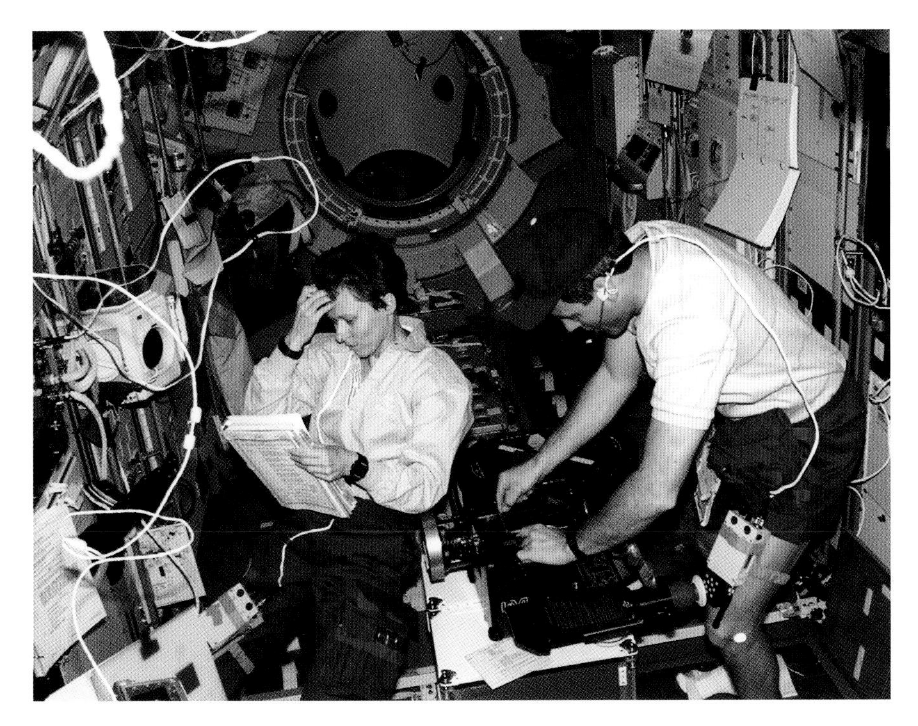

Canadian astronaut Roberta Bondar (left) works alongside colleague Steve Oswald in the International Microgravity Laboratory during Shuttle mission STS-4.

Get a Grip

While the original Canadarm retired with the Space Shuttle, its successor, Canadarm2, is a key component of the ISS. Its functionality has been greatly enhanced by the Mobile Remote Base System (MBS) depicted here. Delivered in June 2002 on Space Shuttle Endeavour during mission STS-111, the MBS enables the Canadarm2 to travel the length of the station.

The Orbital ATK Cygnus cargo spacecraft is released by Canadarm 2 from the ISS on December 6, 2017.

China Undertakes Space Exploration

The launch of a Long March 2C rocket from the Jiuquan Space Centre, China, on August 19, 1983.

The People's Republic of China's space program had its origins in the work of one man, American-educated Chinese engineer Qian Xuesen, also known as Hsue-Shen Tsien (1911–2009). Qian was a leading figure in the foundation of the JPL in Pasadena, California (p. 54). He served in the United States Army during World War II, and was a key member of the military's Scientific Advisory Group under the leadership of Theodore von Kármán. But in the early years of the Cold War, the United States was convulsed with a virulently anti-Communist hysteria. The Chinese Communist Revolution, also known as the War of Liberation, took place between 1946 and 1950 and resulted in Mao Zedong's Communist forces taking control of the Chinese mainland and there establishing the modern People's Republic of China (PRC).

American anti-Communist sentiment led to Qian being stripped of his security clearance; he was then placed under house arrest before he was eventually allowed to return to mainland China in 1955.

Upon his return, and due to his expertise and experience, Qian was given a prominent position presiding over the Fifth Academy of the Chinese National Defense Ministry. As the organization's director, he worked on his country's nuclear weapons and ballistic missile programs.

After the launch of Sputnik 1 by the Soviet Union in 1957 (p. 90), Qian persuaded Mao to commit to developing a similar capability. Accordingly, China adopted what was termed Project 581, a space exploration program with the objective of placing a satellite in orbit by 1959 to celebrate the tenth anniversary of the Communist revolution. While Qian's team failed to achieve this goal, they did gain access to advanced ballistic missile design from the Soviet engineers, which aided the development of the Dongfeng family of ballistic missiles. From 1966 onward, these missiles were the main delivery system for China's nuclear weapons.

A few years later, on February 20, 1968, the Chinese Academy of Space Technology (CAST) was formed with Qian under its auspices. Soon after, his engineers succeeded in redeveloping the two-stage Dongfeng 4 rocket into the three-stage Long March 1 launch vehicle that placed China's first simple satellite,

Propulsion and Expulsion

Qian Xuesen (1911–2009) first came to the United States In August 1935 on a scholarship to study at Massachusetts Institute of Technology (MIT). A brilliant and gifted student, he eventually moved to California to study for his doctorate at Caltech, where he became an enthusiastic participant in the rocket experiments of Frank Malina before World War II. During the conflict, Qian worked on various rocket projects for the United States Military, and was a founding member of the Jet Propulsion Laboratory. But postwar anti-Communist paranoia in the United States effectively compelled Qian to return to the country of his birth.

This satellite image depicts the Jiuquan Space Launch Complex, China, shortly before the launch of China's first human spaceflight mission on October 3, 2003.

Sophisticated Shenzhou

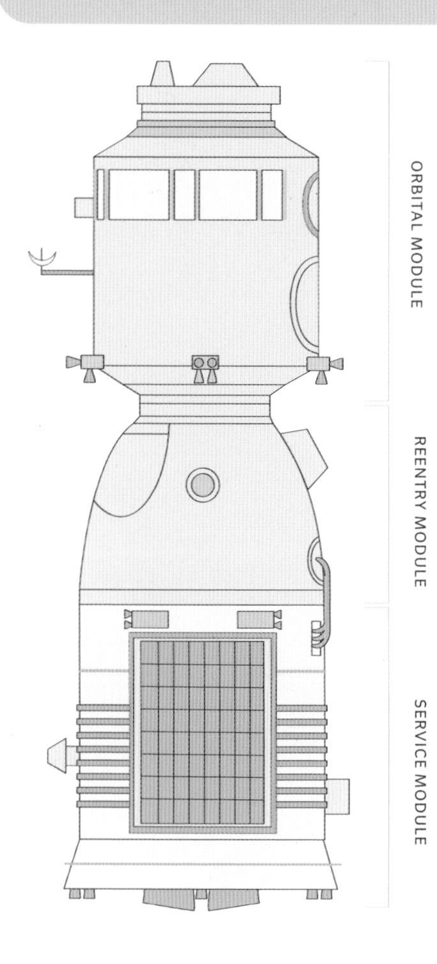

The Shenzhou capsule is the principal spacecraft of the Chinese space program. Like the Soyuz spacecraft that it resembles, the Shenzhou capsule has an orbital module, a descent module, and a service section, but it is larger and more versatile, with an additional airlock module to facilitate space walks and other extravehicular activities.

Dong Fang Hong 1, into orbit from the Jiuquan launch site on April 24, 1970. On March 3, 1971, CAST engineers launched a more complex satellite, Shijian XI-01, which used a magnetometer as well as cosmic ray and x-ray detectors to measure conditions in the ionosphere.

In 1993, CAST was succeeded by two bodies: the China Aerospace Science and Industry Corporation (CASIC), which oversaw international space projects and commercial satellite launch services; and the China National Space Administration (CNSA), which oversaw the national space program.

CASIC's operations faced a significant early setback early on February 14, 1996, when a short circuit in its guidance systems caused a Long March 3B heavy launcher carrying a telecommunications satellite for the American Intelsat consortium to crash into a village close to the Xichang Satellite Launch Center. Six villagers were killed, and a further 57 injured. The crash prompted the Chinese to improve the reliability of their rockets, and after August 1996 they carried out 75 consecutive successful launches without a failure.

In addition to its human space program, the CNSA has achieved significant success with its lunar exploration program. The lunar orbiters Chang'e 1 and Chang'e 2 captured high-definition images of the lunar landscape that allowed for the creation of detailed 3D maps of the Moon's surface. They also prospected for chemical elements that might be useful at some point in future missions.

The Yutu Moon rover as photographed by the camera on the Chang'e-3 lunar orbiter on December 16, 2013.

China's Human Spaceflight Program

2003–

China's twenty-first-century human space program is proceeding apace with the goal of returning humans to the Moon.

In 1967, Mao Zedong gave his approval for China's pursuit of a human space exploration program, but for decades little concrete progress was made toward this goal. Chinese scientists did produce designs for a two-person spacecraft in 1968, along with a heavy-lift launch vehicle, Feng Bao 1 (FB-1), but these plans were consistently underfunded and considered a low priority. The program was eventually canceled after many key scientists on the project became caught up in the political turmoil of the Cultural Revolution (1966–76).

For decades, human exploration was just a background goal for those involved in the Chinese space program, but the idea eventually reemerged in 1992 as Project 921-1. This set out a clear list of goals, including a human spaceflight mission, followed by human missions to the Moon.

The Project 921-1 spacecraft was eventually renamed Shenzhou, and Yang Liwei (1965–) became the first Chinese astronaut to go into orbit on the Shenzhou 5 spacecraft on October 15, 2003. A follow-up mission, Shenzhou 6, saw two more astronauts making 75 orbits in October 2005. During the three-person Shenzhou 7 mission, launched on September 25, 2008, Zhai Zhigang and Liu Boming became the first Chinese citizens to walk in space.

China's space program entered a new phase in 2011, when work began toward establishing a Chinese space station. The initial step was the launch of Tiangong-1, the first of two small prototype space stations. Shenzhou 9 took the first humans to visit Tiangong-1—Jing Haipeng (1966–), a veteran of the Shenzhou 7 mission, Liu Wang (1969–), and Liu Yang (1978–), the first Chinese woman in space. Tiangong-1 has since been superseded by the larger and more capable Tiangong-2 orbital station. Future space station missions have already been announced for 2019 through 2022, while the longer-term goal of a human mission to the Moon is as yet unscheduled.

Chinese astronaut Yang Liwei sits in the reentry capsule of his Shenzhou-5 spacecraft during a training exercise on September 27, 2003.

Vikram Sarabhai and India's First Steps into Space

RIGHT Before launch in 1981, the Indian Space Research Organisation's (ISRO) APPLE communications satellite was transported between locations on a bullock cart.

BELOW ISRO's Vikram Sarabhai and the National NASA Administrator Thomas Paine sign a cooperative satellite agreement between India and the United States in 1969.

The existence, and early success, of the Indian space program is largely credited to the efforts of wealthy Indian industrialist and scientist Vikram Sarabhai (1919–71). Enthralled by the early success in space exploration achieved by the Soviet Union and the United States, Sarabhai sought to establish relationships and cooperative space efforts with both powers. He pressed the founding Indian Prime Minister Jawaharlal Nehru to establish first the Indian National Committee for Space Research (INCOSPAR) in 1962, and later its successor, the Indian Space Research Organisation (ISRO) in 1969.

Sarabhai also championed the idea that India should develop its own satellite. The first Indian satellite, Aryabhata, was eventually launched by the Soviet Union on April 19, 1975, some four years after Sarabhai's death. Aryabhata conducted experiments in x-ray astronomy and solar physics, and measured ionized gas in the upper atmosphere. Although it failed after five days, it was only the beginning of India's active space program. Applying lessons learned from its collaboration with the Soviets, ISRO developed its own rocket family, the Satellite Launch Vehicle (SLV). An SLV-3 rocket launched the Rohini Satellite-1, on July 18, 1980. This achievement signaled that, like a number of other emerging space powers, India was no longer reliant on others to provide its satellite launch services.

ISRO has subsequently developed its launch capabilities. Its rockets have launched more than 100 satellites for applications as diverse as communications, Earth observation, space science, and navigation, as well as providing launch services for other nations.

ISRO has not yet developed its own launch capacity for human spaceflight, but two Indian nationals have flown into space. Rakesh Sharma (1949–) was the first, flying on the Soviet Union's Soyuz T-11 mission in April 1984 as part of the cooperative Intercosmos program. Kalpana Chawla (1962–2003) was the second. An Indian-American astronaut who became a naturalized American citizen in 1991, she flew into space on Space Shuttle mission STS-87 in 1997. She was also a mission specialist on Space Shuttle Columbia's mission STS-107, which disintegrated on reentry on February 1, 2003.

ISRO has also launched a lunar orbiter, Chandrayaan-1, which conducted scientific investigations of the Moon from 2008 to 2009, and Mangalyaan, a Mars orbiter which reached the Red Planet on September 24, 2014. This is a particularly significant achievement, since ISRO is only the fourth space agency in the world to succeed in sending a probe to Mars (p. 274).

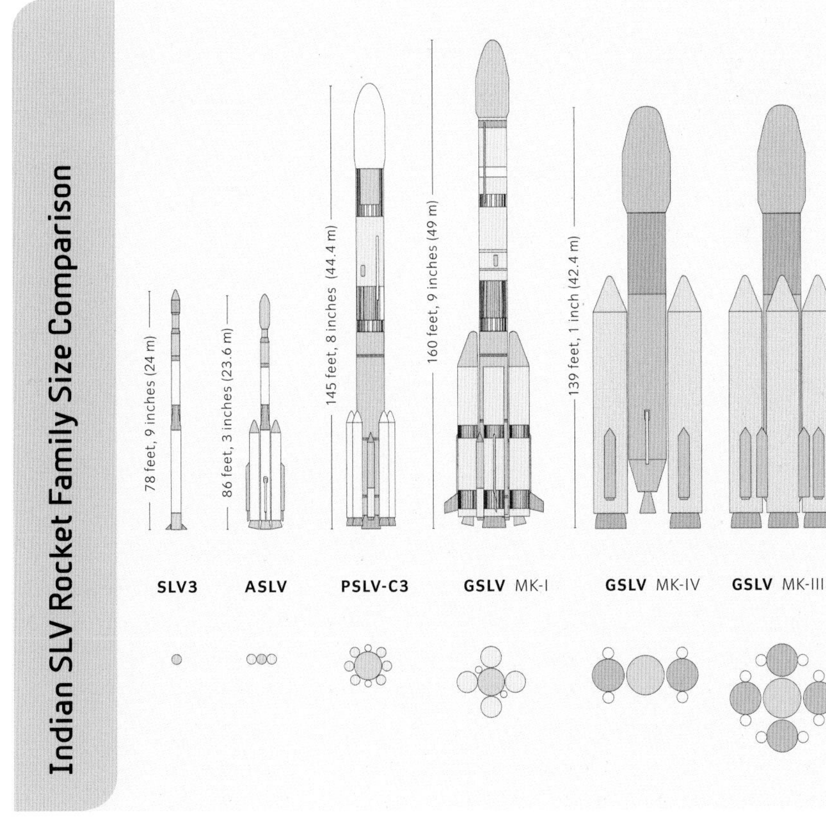

ABOVE Astronaut Kalpana Chawla floats beside a rack as she configures a wiring harness during the ill-fated STS-107 mission in January 2003.

LEFT This illustration depicts the current ISRO launcher family.

BELOW Scientists from the Soviet Union and the Indian Space Research Organisation perform remote tests of the systems of Aryabhata, India's first satellite, before launch in 1975.

Japan and the Frontiers of Space Exploration

Huyabusa Mission Specifications

Launch Mass: 1,120 pounds (510 kg)

Dry Mass: 840 pounds (380 kg)

Launch Date: May 9, 2003

Rocket: M-V

Launch Site: Uchinoura Space Center

Mission End: June 13, 2010

Sample Recovery: June 14, 2010

Japan was another nation that was spurred into pursuing space exploration by the successes of the Soviet Union and the United States. In 1964, Tokyo University established its Institute of Space and Astronautical Sciences (ISAS). Like many other early organizations studying space science, ISAS focused its initial exploration efforts on experiments to measure various astrophysical phenomena in the upper atmosphere, using a mixture of weather balloons and suborbital sounding rockets. ISAS later developed the Lambda family of solid propellant rockets, which were capable of reaching orbit. ISAS employed a Lambda L-4S launch vehicle to take Japan's first satellite, Ohsumi, into orbit in 1970.

While this was an important step forward, the Japanese government sought to expand its space exploration program, and established the National Space Development Agency (NASDA) on October 1, 1969, to help achieve this goal. Led by renowned Japanese railway engineer Hideo Shima (1901–98) from its inception until 1977, NASDA began an assertive effort to make Japan a significant space power. ISAS largely oversaw the scientific aims of the program, but NASDA was responsible for building the satellites, launch vehicles, and the Tanegashima Space Center and launch facilities on the island of Tanegashima, 24 miles (40 km) south of the island of Kyushu. A third entity, the National Aerospace Laboratory of Japan

(NAL), focused more on aviation research, but also contributed valuable expertise to solving the engineering problems that NASDA faced in developing its rockets.

After first relying on American rocket designs, NASDA eventually developed its own range of versatile launch vehicles. Of these, the H-II family of rockets is probably the most prominent, with some variants capable of sending payloads of up to 4,000 pounds (1,800 kg) into orbit. These rockets and the Tanegashima launch facility have been used for scientific and commercial launches on behalf of numerous nations around the world.

The close relationship between Japan and the United States after World War II has helped to foster a culture of ongoing collaboration between the Japanese space program and NASA. Since the 1970s, there has hardly been a NASA science mission that has not involved Japanese technology and the participation of Japanese scientists, and a total of seven Japanese astronauts have also flown on the Space Shuttle.

On October 1, 2003, the Japanese government merged NASDA, ISAS, and NAL into a single organization, the Japanese Aerospace Exploration Agency (JAXA). The new entity was given an elevated position within the Japanese government, and greater authority to undertake space activities. Possibly JAXA's most prominent collaborative project undertaken to date has been its work on the ISS. JAXA's Kibo Science Module has been a key component of that space station since its installation in 2009.

OPPOSITE The pressurized Kibo module at its manufacturing facility in Nagoya, Japan, prior to its transportation to the ISS.

ABOVE Engineer Hideo Shima, the first director of NASDA, February 25, 1993.

The Voyage to Asteroid 25013 Itokawa

2005–10

The ultimate asteroid sample return mission.

One of the most demanding missions ever undertaken by JAXA was the flight of the probe Huyabusa to encounter asteroid 25013 Itokawa in its irregular orbit that regularly crosses between Earth and Mars. Huyabusa flew to the tiny asteroid, twice touched down on its surface, collected a sample, and returned it to Earth for analysis. The mission also facilitated detailed study of the asteroid's irregular shape, spin state, topography, color, composition, density, physical properties, and geological history, but it did not result in any major new discoveries, and served mainly to demonstrate the technology required for long-range sample return missions. Such missions may yield more significant results in the future.

Artist's conception of the Huyabusa encounter with asteroid 25013 Itokawa.

HUGHES

The New Asian Space Race

After the race to the Moon, the intensely competitive aspect of the Soviet and American space programs began to wane. The two nations would eventually partner up to perform a joint mission (p. 231) in 1975. But for other emerging space powers, particularly those in Asia, regional competition remained a powerful driving force toward developing space technology, sparking what was in effect a new space race. In addition to the space programs of the People's Republic of China (p. 192), India (p. 186), and Japan (p. 198), Pakistan, the Republic of Korea, and Indonesia all looked to develop their own space science capabilities.

These countries all had slightly different reasons for getting more involved in the development of space technology, but largely their motivations echoed those of previous entrants into the space club. Ballistic missile technology used in rocket-powered launch vehicles was often important for national security. The ability to launch, or at least to design and build, communications and other types of satellites enhanced commercial opportunities, and helped to advance the technological capability of each nation. National pride and international prestige were also factors.

Pakistan began its pursuit of space technology with the formation of its Space and Upper Atmosphere Research Commission (SUPARCO) on September 16, 1961. Using suborbital rockets obtained from various other nations, it conducted upper atmosphere research from 1962. SUPARCO's first orbital satellite, Badr-I, was launched on a Chinese rocket from the Xichang Satellite Launch Center on July 16, 1990. The Republic of Korea established the Korea Aerospace Research Institute (KARI) in 1989, launching its first satellite, Arirang-1, on an American Taurus rocket on December 21, 1999. KARI has also developed its own launch vehicle, the NARO-1 rocket, first flown in January 2013.

Indonesia formed its space agency, the National Institute of Aeronautics and Space (LAPAN), on November 27, 1963. Since Indonesia occupies an archipelago, one of LAPAN's early priorities was improving the telecommunications infrastructure between the nation's component islands. To this end, LAPAN commissioned a number of American companies to design and build the Palapa series of communications satellites. Operated by the Indonesian company Indeosat, the first Palapa satellite was launched in July 1976.

Significantly, while a number of Asian countries now offer commercial satellite launch services to other nations, this does not seem to have diminished the appetite for the development of indigenous launch systems across the region. For some nations, it seems, the quest for space is still a race.

OPPOSITE Astronaut Dale Gardner works to secure the failed Palapa B-2 communications satellite in the payload bay of Space Shuttle Discovery as part of mission STS-51A, on November 12, 1984. The satellite was subsequently brought back to Earth for repair before being returned into orbit.

BELOW Three leading commercial launch vehicles of the emerging space powers of Asia.

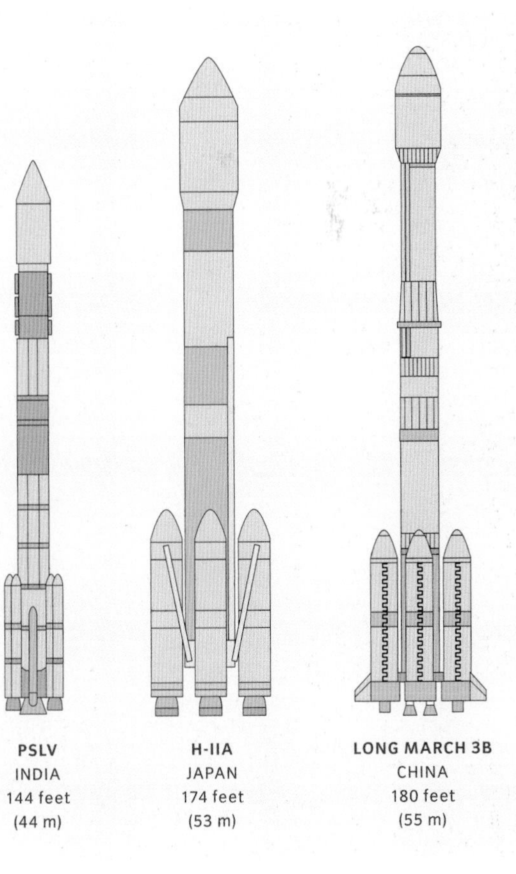

Space Exploration in the Global South

Nations need not necessarily be dominant global or even regional powers to undertake significant space exploration programs. Various countries across the global south, a term often used to collectively describe the emerging nations of Latin America and Africa, have also developed considerable space exploration capabilities of their own.

In common with many space powers, Argentina's space program began at the dawn of the space age. Established in 1960, the National Space Activities Commission (CNIE) was initially tasked with developing multistage rockets and missiles, but it later focused on satellite development. In 1997, Argentina became one of the first nations in Latin America to have its own communications satellite, Nahuel 1A, placed in orbit, a task achieved by an Ariane rocket. CNIE's other most prominent mission to date was the launch of the SAC-D Earth monitoring satellite. Built in collaboration with NASA, SAC-D's primary instrument, Aquarius, measures ocean salinity, and thus helps to show how ocean, atmosphere, and sea ice influence Earth's climate.

Established on August 3, 1961, Brazil's Organizing Group for the National Commission on Space Activities (GOCNAE), spent most of its first decades conducting extensive upper atmosphere research. In the 1980s, the organization oversaw the construction of the Alcântara Launch Center in Maranhão province, Brazil. Because it is a mere 2.3° south of the Equator, the Alcântara facility is one of the most cost-effective launch sites in the world.

Soon after Alcântara opened for business in October 1990, the Brazilian space program entered a new phase. GOCNAE was reorganized into the National Institute for Space Research (INPE), and the new organization initiated a program of satellite launches. The first Brazilian-made satellite, SCD-1, was placed into orbit in 1993 by an American Pegasus rocket, which is "air-launched" from a B-42 bomber in flight. SCD-1 was designed to collect environmental data on the country's vast rain forests and other geographical features. It was followed in 1994 by the first in the Brazilsat series of communications satellites launched by an Ariane rocket.

In 2006, Marcos Pontes (1963–) became the first Brazilian astronaut when he flew on a mission to the ISS aboard Soyuz TMA-8. He stayed on the ISS for a week, and conducted eight experiments.

The Mexican government formed its first space agency, the National Commission of Outer Space (CONEE), on August 31, 1962, to pursue rocket development, telecommunications, and space science experiments. The nation launched its first satellite, Morelos I, aboard the Space Shuttle Discovery on mission STS-51G on June 17, 1985. CONEE was later superseded by the Mexican Space Agency (AEM), a reorganization that underlined the nation's continued commitment to developing the scope of its space program.

OPPOSITE LEFT The Soyuz TMA-8 lifts off from Baikonur Cosmodrome in Kazakhstan on March 30, 2006, toward the ISS. Among the mission's crew was Marcos Pontes, Brazil's first astronaut.

OPPOSITE RIGHT The Mexican Morelos-D communications satellite is deployed from the cargo bay of the Space Shuttle Atlantis in 1985 during the STS 61-B mission.

BELOW Engineers inspect the Aquarius instrument for the SAC-D observatory ahead of the spacecraft's launch in 2011.

Latin American Launch Vehicles

While most Latin American countries relied on more established space powers to provide launch vehicles for their initial satellite launches, both Argentina and Brazil have developed their own rockets, with varying degrees of success.

Argentina's Tronador II and III rockets are currently still in development, but they are expected to make their first flights sometime in 2019.

Brazil began developing its own suborbital Sonda sounding rockets in the 1960s. More recently, the nation developed its own four-stage VLS satellite launch system, but after a series of costly failures between 1997 and 2003, this program was terminated in 2016.

Space Junk

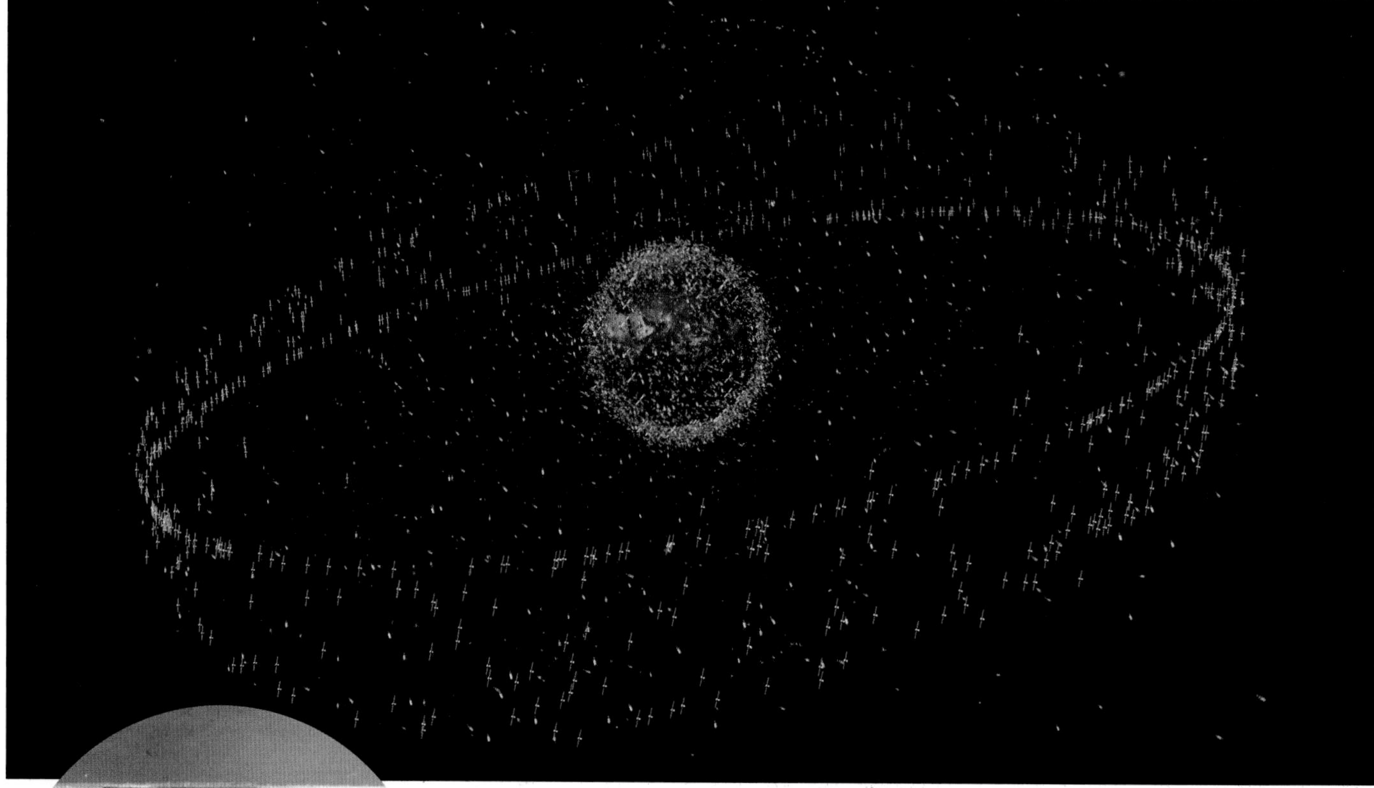

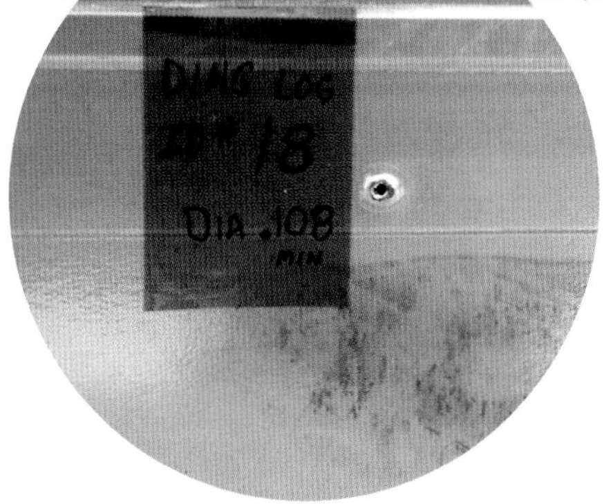

ABOVE LEFT A "ding" from a Micrometeoroid Orbital Debris (MMOD) impact on the right-hand payload bay door of Space Shuttle Atlantis.

ABOVE RIGHT This artist's depiction of the orbital objects circling the Earth is not to scale, but it gives a sense of the complexity of the issue of space junk.

In the first dozen or so years of the space age, little attention was paid to the issue of space junk, the discarded human-made objects left in orbit from various space missions. But since the Apollo Moon landings (p. 148), the proliferation of satellites and other space vehicles circling the Earth, not to mention their orbital debris, has been a growing problem. There are currently more than 2,000 working satellites in Earth orbit. There are also more than 12,000 other human-made objects sufficiently large to be tracked in orbit. These include spent rocket boosters, old satellites, bolts, filled garbage bags, and broken hardware. Most objects in low orbit eventually reenter the atmosphere, where they typically burn up. However, when a satellite or other large space object breaks apart, it can leave a cloud of debris in its wake that may remain in orbit longer than the larger object of which it was originally a part. While there are currently large areas above the Earth that are relatively free of space junk, concentrations of such detritus are a source of serious concern to space agencies around the world. Moreover, when one piece of space junk collides with another, they may break apart again, releasing even more debris. Scientists have estimated that 21,000 pieces of space trash larger than 4 inches (10 cm) currently circle the planet.

Fortunately, the density of objects in Earth orbit currently remains relatively low. In the most densely occupied region of Low Earth Orbit (LEO), there is an average of about one object per 12 million square miles (50 million km^2). While that might sound like a lot of space, the real difficulty with such debris is its relative speed. Space junk, like everything else in orbit, is traveling at 17,500 mph (28,000 km/h). Because of its high velocity in orbit, even a small object can cause serious damage to anything with which it comes into contact.

United States Strategic Command and other national observation sites attempt to track all space objects larger than 4 inches (10 cm) in diameter. Together, they are currently keeping tabs on more than 8,000 objects. This information can be used to maneuver spacecraft out of the path of the larger pieces of junk. Nevertheless, in spite of all precautions, accidents do happen. A small military satellite operated by France was hit by rocket debris in July 1996. The satellite briefly lost stability, but it remained in orbit and under control.

However, the single event that has generated the greatest quantity of orbiting space debris was not an accident. In January 2007, the Fengyun-3A Chinese weather satellite was deliberately destroyed by a Chinese ground-launched rocket as part of an apparent test of an anti-satellite weapon. The explosion put more than 2,000 new fragments into LEO, increasing by approximately one-third the total amount of space debris that was known to be in orbit at that time. Not to be outdone, the Americans responded by destroying one of their own spent satellites, demonstrating their own anti-satellite capability, although they targeted a lower-altitude satellite to ensure that the resulting debris field entered the atmosphere.

In the future, this problem will only get worse, and eventually some kind of regulation will have to be agreed by all the spacefaring nations to minimize the pollution caused by space junk.

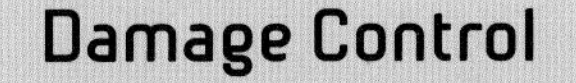

Damage Control

Space Shuttles and Soyuz capsules have been hit many times by orbiting debris on their various missions; the ISS has also suffered several such impacts. NASA recorded more than 300 impacts on the outer windows of various Shuttles that required the windows' replacement. Replacing Space Shuttle insulation tiles such as the one pictured above, damaged by space junk, was a routine duty after each flight.

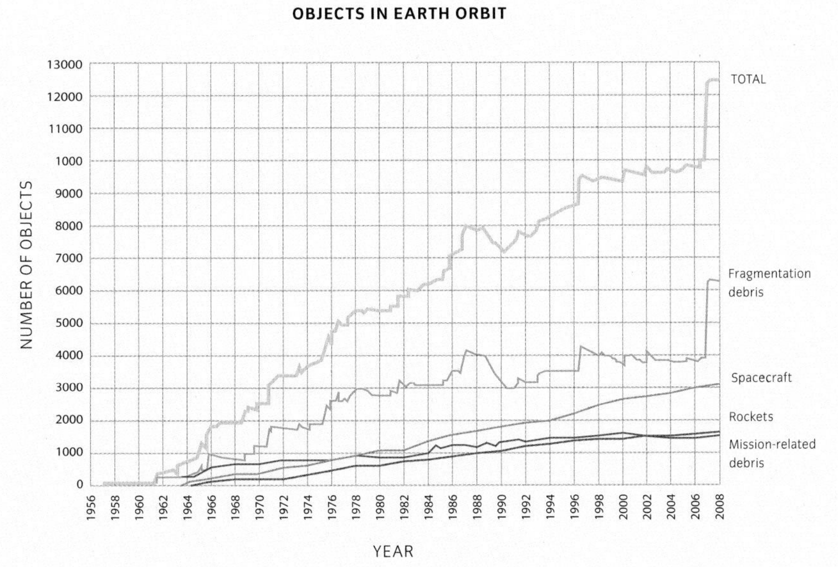

Chart showing the number of operational satellites and pieces of debris in orbit. "Mission-related debris" was intentionally left in orbit. "Fragmentation debris" is the result of explosions, collisions, or malfunctions.

Atlantis

7

Space Planes and Orbital Stations

Since the invention of powered flight on winged aircraft at the beginning of the twentieth century, the idea of a reusable winged vehicle flying from Earth into orbit has dominated popular thinking about space travel. However, when the goal of getting into space became a competitive race, the development of space plane designs was put on the back burner in favor of ballistic capsules, which were much less technologically complex but could deliver results faster.

Once the race to the Moon concluded, NASA revisited the notion of a space plane. Its scientists recognized that the idea could result in a more efficient means of reaching orbit, but the concept they eventually came up with was the rocket-powered Space Shuttle. This was still a far more versatile space vehicle than anything that had traveled into orbit previously, but it still fell short of being a truly reusable space plane.

Despite developing their own version of the shuttle, dubbed Buran (p. 228), Soviet space scientists and engineers instead largely focused their resources on a succession of space stations. From Almaz/Salyut (p. 208) to Mir (p. 230), Soviet stations set the standard for others, and led the way toward the international consortium of space agencies that would eventually build and operate the ISS (p. 232).

This new era saw the whole enterprise of space exploration transformed from a race into a joint venture in which all participants had a stake. The increase in human missions was matched by an increasing diversity of nations and peoples going into orbit, and by the unprecedented willingness of all these participants to share their expertise and discoveries. Before long, this democratization of space would lead to the advent of the first space tourists (p. 242).

OPPOSITE The Space Shuttle Atlantis over the Bahamas prior to a perfect docking with the International Space Station, July 10, 2011. Part of a Russian Progress spacecraft that is docked to the station is visible in the foreground.

Almaz and Salyut

ABOVE Salyut 7 in Earth orbit with a Soyuz-T spacecraft attached in the mid-1980s.

BELOW Soviet cosmonauts Vladimir Vasyutin, Georgy Grechko, Victor Savinykh, Aleksandr Volkov, and Vladimir Dzhanibatov aboard the Salyut 7 space station in Earth orbit.

In the 1960s, both the Soviet Union and the United States began working on designs for their first orbital space stations. Remarkably, in this process both nations independently abandoned the giant wheel-shaped space station design emphasized by more than 50 years of studies and subsequent popular fiction in favor of an entirely different layout and structure. Their designs made no provision for artificial gravity, which could be provided through the centrifugal force of a rotating space station. Neither were their efforts aimed at creating orbital base camps for further exploration of the Moon or Mars. Instead, they were developed as military assets for reconnaissance purposes, or as orbital laboratories in which humanity could harness microgravity to assist scientific research.

The earliest space station designs devised in the Soviet Union were for military use. The Almaz military station program envisioned an orbiting facility weighing approximately 40,000 pounds (18,000 kg) that had a crew of three cosmonauts. On April 19, 1971, the Soviet space program achieved a spectacular success with the launch of Salyut 1, a simplified version of the Almaz design designated for non-military use.

Unfortunately, Salyut 1 and other first-generation Soviet space stations were plagued with technical problems. The crew of Soyuz 10, the first mission sent to rendezvous with the station, was unable to reach its intended destination because of difficulties with the docking mechanism. The next visitors, the crew of Soyuz 11, docked successfully, and lived aboard the station for three weeks in 1971, but died of asphyxiation during a return to Earth when a malfunction on their spacecraft caused all its air to leak out. A further three first-generation stations either failed to reach orbit, or broke up in orbit before their crews could reach them.

The first fully realized Almaz military station to fly was given the designation Salyut 2, but soon after achieving orbit it lost altitude, depressurized, and became unusable. However, Salyut 3, Salyut 4, and Salyut 5, all based on the Almaz design, were more successful, and collectively supported a total of five crews during their time in service.

Salyut 6 incorporated several improvements to the original Almaz design that enabled its crews to remain on the station for longer periods. The most significant of these improvements was the addition of a second docking port that permitted the station's refuel and resupply by automated Progress freighters. Based on the Soyuz spacecraft design, Progress docked automatically at the unoccupied port. Once docked, the craft was then unloaded by the cosmonauts on the station. Another larger experimental transport logistics spacecraft, Cosmos 1267, docked with the station in 1982. Originally designed for the Almaz program, it was larger than either Soyuz or Progress, and proved that even bulky modules could be docked automatically, a major advance in space station technology.

The second docking port also allowed long-term resident crews to receive visitors. These crews often included cosmonaut-researchers from countries allied with, or at least sympathetic to, the Soviet Union. Cosmonaut Vladimir Remek (1948–) of Czechoslovakia (now the Czech Republic) became the first person from a nation other than the United States or the Soviet Union to go into space when he flew to Salyut 6 in 1978.

Czechoslovak cosmonaut Vladimir Remek (front) and Soviet pilot Alexei Gubarev board the Soyuz 28 spacecraft to fly to the Salyut 6 in 1978.

Salyut's International Visitors

1977–82

Opening opportunities for countries without human space programs of their own to go into orbit.

Over the course of its operational life, Salyut 6 received a total of 16 cosmonaut crews, including six on long-duration missions, and hosted cosmonauts from Hungary, Poland, Romania, Cuba, Mongolia, Vietnam, and East Germany. Visiting crews often traded the Soyuz spacecraft they arrived in for the one that was already docked at the station, because Soyuz craft had only limited service lives in orbit. Over time, estimates for this service life were gradually extended from 60 to 90 days for the Soyuz 7K-T capsules, and to more than 180 days for the Soyuz-TM capsules. The Soviet space program moved on to launch Salyut 7 in 1982. While it remained in orbit until 1991, its last crew departed in 1986. A near twin of Salyut 6, the last Salyut station was home to 10 cosmonaut crews, including six long-duration crews. The station saw further international visitors, with space travelers from France and India, before it was superseded in 1996 by the first long-term space station, Mir (p. 230).

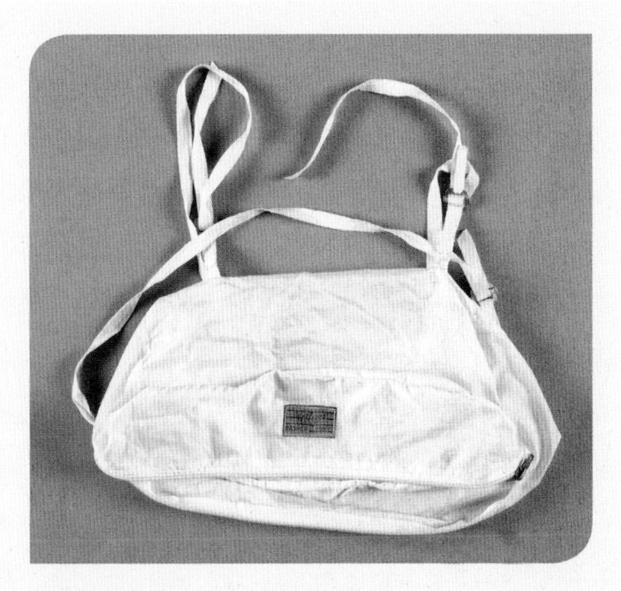

Safe storage of essential equipment is important in space. This stowage bag was used to store the flight suit worn by Soviet cosmonaut Yury Malyshev on the way to the Salyut 6 space station.

Repurposing Apollo Hardware

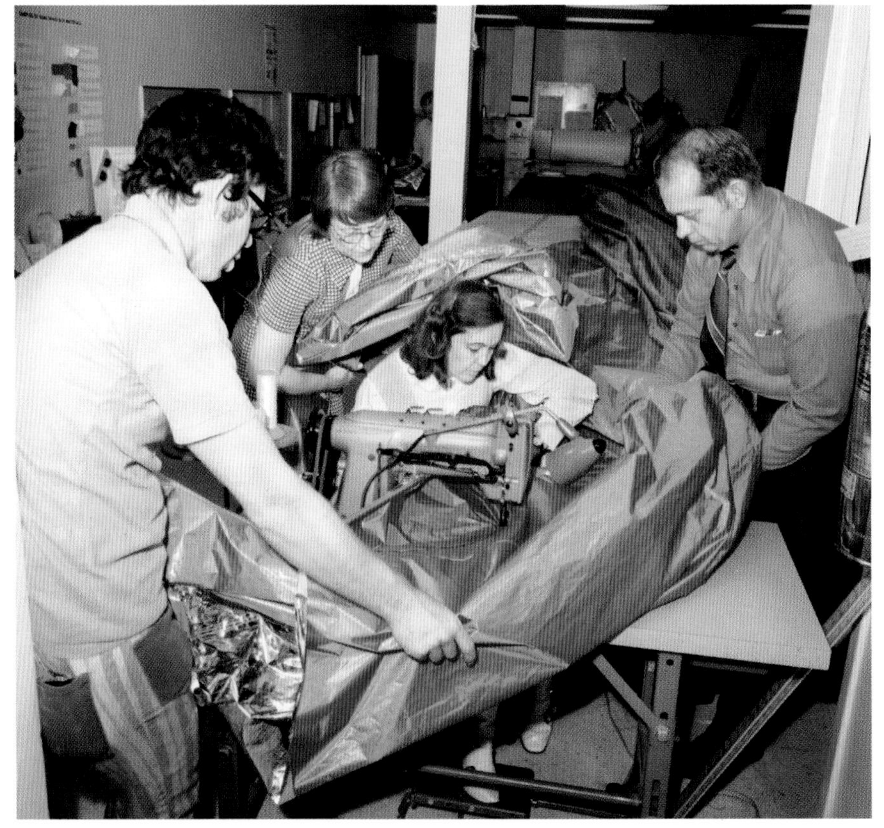

ABOVE LEFT Skylab commander Pete Conrad leads the parade of astronauts as they walk out for a ride to the launch pad on May 25, 1973. Behind Conrad are Paul Weitz and Joseph Kerwin.

ABOVE RIGHT To protect from the heat of the Sun in orbit, technicians at General Electric, including skilled industrial seamstress Alyene Baker seen here hard at work, fashioned this sunshade for use on Skylab.

As the 1970s wore on, political and financial support for Moon missions began to diminish. The United States was heavily involved in a costly war in Vietnam, and could ill afford to keep sending astronauts to the lunar surface when the benefits were purely scientific, and had little impact on the geopolitics of the ongoing Cold War. As more and more planned Apollo missions were scrapped, NASA considered possible alternative uses for the Apollo/Saturn hardware that had already been produced. Ideas abounded, the most foreseeably practicable of which was to construct an orbital workshop out of the third stage of a Saturn V rocket, and crew it with teams of astronauts flown there in an Apollo Command/ Service Module (CSM).

The first section of NASA's experimental orbital workshop, dubbed Skylab, was launched on May 14, 1973 in a mission that also marked the final flight of the giant Saturn V rocket. Soon after launch, the mission began to experience technical problems caused by vibrations during lift-off. Sixty-three seconds into the flight, the meteoroid shield—designed to protect the workshop from damage in orbit, as well as to shade Skylab from the Sun's rays—ripped off the rocket, taking with it one of the workshop's two furled solar panels. Despite this, the station successfully achieved a near-circular orbit at the desired altitude of 270 miles (434 km). NASA's mission control personnel maneuvered Skylab so that the solar panels attached to the Apollo Telescope Mount (ATM) faced the Sun, thus compensating for the missing solar panel, but the station still faced a significant reduction in its power generation capability. The loss of the meteoroid shield caused temperatures inside the station to rise to 126°F (52°C) when it was in direct sunlight.

NASA technicians worked on a solution to the problem, and swiftly assembled a plan to rescue the workshop and make it habitable. On May 25, 1973, the Skylab 2 mission carried astronauts Pete Conrad, Paul Weitz (1932–), and Joseph Kerwin (1932–) to rendezvous with the orbital workshop. After substantial extravehicular repair work, including the deployment of a large space parasol to keep the interior temperature of the station to a more manageable 75°F (23°C), Skylab became both habitable and operational. The crew went on to conduct solar astronomy, medical studies, and various other scientific experiments over the course of 404 orbits before returning to Earth on June 22, 1973, leaving the space station unoccupied until the arrival of the Skylab 3 crew on July 28. After a 60-day mission, the Skylab 3 crew returned to Earth on September 25, 1973. The last mission to the space station, Skylab 4, lasted for 84 days, from November 16, 1973, to February 8, 1974.

Fixing Skylab

The principal aim of the Skylab 2 mission was to repair Skylab and make it habitable for future crews. The orange sunshade seen in this image from the Skylab 4 mission in 1974 served as an unorthodox yet effective means of keeping the temperature inside at a manageable level.

Skylab astronauts in the Marshall Space Flight Center's Neutral Buoyancy Simulator rehearse procedures to be used during the Skylab space walks.

The Apollo-Soyuz Test Project

1975

The international space mission that paved the way for future space collaborations.

ABOVE LEFT These Apollo-Soyuz cigarettes were manufactured as part a joint venture between the American cigarette manufacturer Phillip Morris and the Soviet Yava cigarette factory. It was thought the Apollo/Soyuz mission could lead to wider industrial cooperation between the Soviet Union and the United States.

ABOVE RIGHT Concept art depicting the first rendezvous and international docking between Apollo and Soyuz spacecraft.

The mid-1970s saw a period of improved relations between the United States and the Soviet Union and a notable example of this was the Apollo-Soyuz Test Project that took place in July 1975. It was the first international human spaceflight mission and was specifically planned to test the compatibility of rendezvous and docking systems for American and Soviet spacecraft in the hope that developing such a capability might open the way for future joint missions. NASA engineers designed and built a universal docking module that would serve as an airlock and transfer corridor between the American Apollo Command/Service Module and the Soviet Soyuz spacecraft, which had become the Soviets' primary spacecraft since its introduction in 1967.

While NASA's missions had been more focused on pure scientific research since the end of the Moon landings, the Apollo-Soyuz flight was primarily a diplomatic mission. The two spacecraft both launched on July 16, 1975, rendezvousing in orbit some 45 hours later. After docking, the two crews, consisting of Americans Thomas Stafford (1930–), Vance Brand (1931–), and Donald Slayton (1924–93) and Soviets Alexey Leonov and Valeri Kubasov (1935–2014) were able to swap gifts and conduct experiments.

After separation, the two vehicles each remained in space for a few days before returning to Earth. Although there were no similar joint missions for the rest of the Cold War the mission led the way for greater international cooperation in human spaceflight in the years that followed.

Despite the recurring mechanical difficulties that plagued Skylab, NASA was delighted with the scientific return from the station. The three visiting crews occupied the Skylab workshop for a total of 171 days and 13 hours, during which time they conducted nearly 300 experiments.

Following the final occupied phase of the Skylab mission, ground controllers performed some remote engineering tests for several of the station's key systems. These tests, which the ground personnel were reluctant to perform while astronauts were aboard the station, positioned Skylab into a stable attitude and shut down its systems. It was expected that Skylab would remain in this orbit for eight to ten years, by which time NASA hoped to have developed a new launch vehicle that would enable the organization to return to the station and reactivate it. However, in the fall of 1977, NASA scientists determined that, as a result of greater than predicted solar activity, Skylab had entered a rapidly decaying orbit, and would reenter the Earth's atmosphere within two years.

Accordingly, NASA began to put together a plan for Skylab's final descent through the atmosphere toward an impact in a remote area of the Indian Ocean. Unfortunately, this descent did not go fully according to plan. Shortly before its final impact on July 11, 1979, Skylab broke apart. This greatly altered the trajectory of the resulting cloud of debris. A significant amount of this space junk was scattered across a chunk of sparsely populated land in Western Australia. No humans were injured in the incident, but a jackrabbit reportedly died after being struck by a chunk of Skylab. NASA's leadership, and the United States more generally, faced strong international criticism over the incident. Subsequently, NASA made it a priority to avoid any situation in which orbital debris from its spacecraft would have any chance of reaching the Earth's surface.

"There's plenty of unearthly-looking things moving around in my refrigerator, so there's always a chance of life springing up almost anywhere."

PETE CONRAD,
APOLLO AND SKYLAB ASTRONAUT

Astronaut Alan Bean undertakes some science in space as he operates the Ultraviolet Stellar Astronomy experiment during the Skylab 3 mission.

USA
NASA

Building the Space Shuttle

The Space Shuttle that NASA flew for 135 missions between 1981 and 2011 was strikingly different from the kind of reusable space plane the organization originally proposed in the 1960s. However, the objectives that shaped its design evolution remained largely the same as those outlined in that early proposal. NASA's core idea was to move away from the costly single-use rockets of the Apollo program, and to look for a way to lower the cost of spaceflight that would facilitate a more proactive space exploration effort.

NASA officials came to the view that both a reusable space capsule and some kind of space station were necessary to support future human space missions to explore the rest of the solar system. They recognized that the most difficult part of spaceflight was leaving the Earth's surface. Finding a way to do this easily, flexibly, safely, and, above all, economically, were the core requirements. The agency's answer was the Space Shuttle.

Much of the planned cost savings were to be achieved through the Space Shuttle's reusability and its large and versatile payload bay, which could be used to haul scientific, military, and commercial satellites of various types into orbit for whichever organization required them. The shuttle would essentially be a one-size-fits-all space vehicle, providing all orbital services required by any users. It would also be able to serve as a space breakdown and recovery service, retrieving broken satellites and either repairing them in orbit, as in the case of the Hubble Space Telescope (p. 331), or returning them to Earth. The potential benefits were clear, but getting the shuttle built and off the ground was not easy because President Richard Nixon was at first not minded to give NASA the requisite political backing.

What first prompted Nixon to change his mind was the impassioned memorandum he received in August 1971 from Deputy Director of the Office of Management and Budget Caspar Weinberger. In his note, the future Secretary of Defense wrote that to reduce U.S. activity in space "would be confirming in some respects, a belief that I fear is gaining credence at home and abroad: that our best years are behind us, that we are turning inward, reducing our defense commitments, and voluntarily starting to give up our superpower status, and our desire to maintain world superiority." In a handwritten scrawl on Weinberger's memo,

OPPOSITE Atop the Mobile Launcher Platform and crawler-transporter, the Space Shuttle Atlantis begins its journey to Launch Pad 39A from the Vehicle Assembly Building on August 20, 1996.

BELOW Dimensions and configuration of the Space Shuttle orbiter.

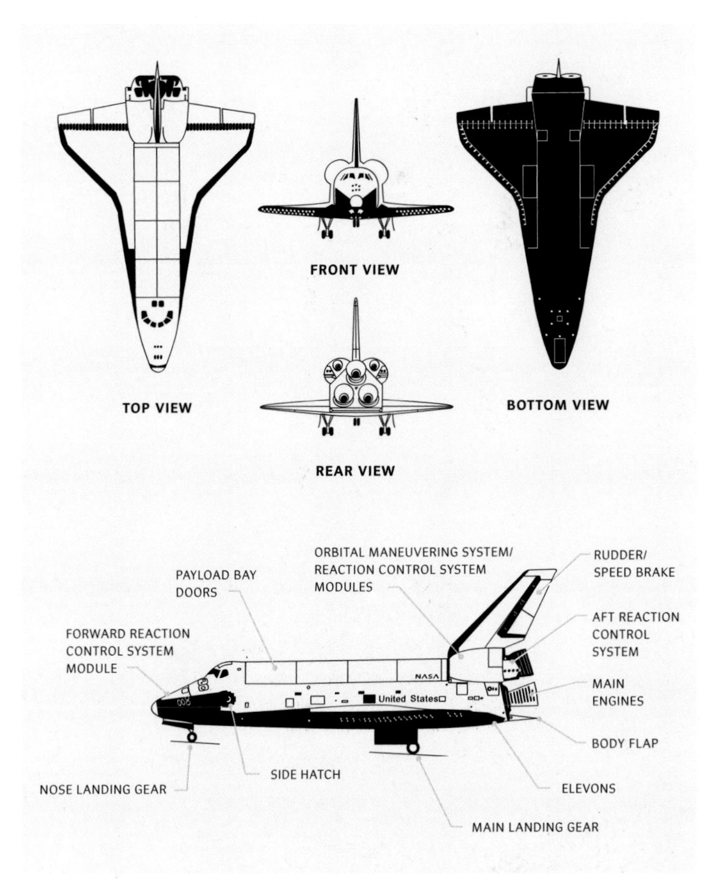

Go with the Flow

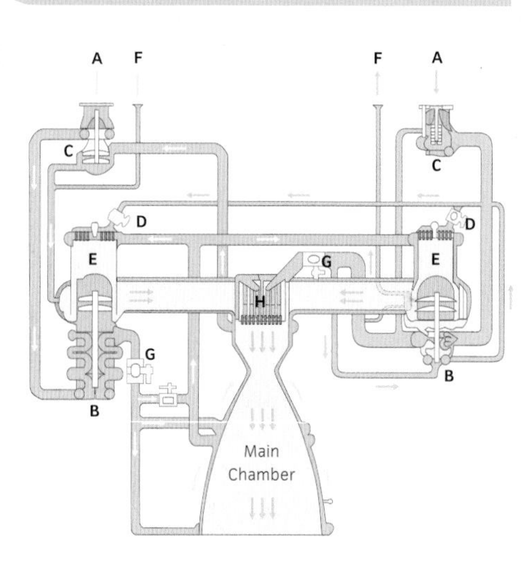

This schematic for one of the Space Shuttle's three main engines shows both its complex design and the flow of its liquid fuel and oxidizer propellants from the external fuel tank. Without an efficient flow of these propellants, the Shuttle would never have been able to achieve orbit.

- Fuel
- Oxidant
- Hot gases
- Combustion zone

A Inlet
B High-pressure turbopump
C Low-pressure turbopump
D Preburner OX valve
E Preburner
F Outlet to external tank
G Main valve
H Main injector

Nixon responded, "I agree with Cap." Consequently, on January 5, 1972, Nixon issued a statement announcing his decision to support the Space Shuttle. The spacecraft he endorsed consisted of three primary elements: a delta-winged orbiter vehicle with a large crew compartment; a cargo bay measuring 15 feet by 60 feet (4.5 m by 18 m); and three main engines supplied by an external fuel tank. The tank and orbiter combination was further supported by two solid fuel rocket boosters.

The original NASA design brief also set out the operational parameters of the Space Shuttle. It would be able to transport approximately 45,000 tons (40,823 t) of cargo into near-Earth orbit, up to 250 miles (400 km) above the Earth. It could also accommodate a flight crew of up to 10 astronauts (although a crew of seven would be more common) for a basic space mission of seven days. During its return to Earth, the orbiter was designed so that it had a cross-range maneuvering capability of 1,265 miles (2,000 km) to meet requirements for lift-off and landing at the same location after only one orbit.

The shuttle met and surpassed each of these requirements. Its extensive service history is a testament to its flexibility and adaptability to all tasks that it was given.

Contracted engineers assembling the Space Shuttle Discovery at the Rockwell International assembly plant in Downey, California, on October 5, 1982.

Propulsion and Thermal Protection

1981–2011

Getting the Space Shuttle into orbit and back to Earth safely were the two greatest design challenges NASA would face in building its first reusable spacecraft.

Engineers faced two major challenges in building the Space Shuttle: the propulsion system, and the thermal protection system. The shuttle launched like a rocket, with two booster rockets that were discarded once spent, and an external fuel tank, which was also jettisoned when no longer required. When it had performed its mission, the orbiter reentered Earth's atmosphere like a glider, before landing on a conventional aircraft runway. Perhaps the most important design issue, after the vehicle's overall configuration, concerned the decision that the boosters should burn solid fuel. Alone, such boosters would not be sufficient to hoist the shuttle into orbit, but they would provide enough thrust when operating in concert with the three liquid-fueled main engines on the shuttle orbiter, which could be vectored and throttled to allow for a gentler more controlled ride for crew and cargo. The boosters were reusable, and consequently had to be built, and refurbished, to standards never attempted before, but they proved to be among the most important innovations of the entire program.

The second challenge, and the key to the reusability of the Shuttle orbiter, was the development of a new method of reentry. The question was how best to pass through the ionosphere. The designers opted for a blunt-body reentry, similar to that employed by previous NASA space capsules. This required the use of a thermal protection system comprising special ceramic tiles on the underside and nose of the orbiter that could withstand prolonged exposure to reentry heat.

Both the Shuttle's main engines and its thermal protection system proved to be vital factors in the orbiter's success and longevity. Without them, a reusable space vehicle would not have become a reality.

ABOVE LEFT Manufactured by Boeing Rocketdyne, a Space Shuttle main engine displayed at the Smithsonian Institution, National Air and Space Museum.

ABOVE RIGHT A Space Shuttle main engine undergoes test firing at the National Space Technology Laboratories (now the Stennis Space Center) in Mississippi.

First Flights

On April 12, 1981, Columbia, the first Space Shuttle orbiter, took off from the Kennedy Space Center, Florida. The launch was greeted with much public excitement. It had been six years since the previous American astronaut had returned from orbit following a cooperative Apollo-Soyuz mission in 1975 (p. 212). This first shuttle flight, mission STS-1, was piloted by astronauts John Young and Robert Crippen (1937–). Young was a spaceflight veteran, having flown on two Gemini and two Apollo missions, while Crippen was a key member of the team that had conducted atmospheric flight tests on early versions of the spacecraft.

BELOW Space Shuttle Columbia blasts off from Launch Pad 39A at the Kennedy Space Center on mission STS-1 on April 12, 1981.

At launch, the orbiter's three liquid-fueled Space Shuttle main engines (SSMEs), which drew propellant from a large external fuel tank, and the two solid rocket boosters (SRBs), generated approximately 7 million pounds (31 million N) of thrust. After about two minutes, at an altitude of 31 miles (50 km), the two boosters were spent, and separated from the external tank. Ships waiting below in the Atlantic Ocean recovered them for refurbishment, assessment, and eventual reuse on subsequent missions. The spacecraft's main engines continued to fire for around eight minutes, before shutting down just as the Shuttle entered orbit. As they did so, the external tank separated from the orbiter, and followed a ballistic trajectory back to the ocean. The tank was not recovered or reused.

After 36 orbits over two days in space, Columbia landed like an aircraft at Edwards Air Force Base in California, successfully accomplishing the first such landing of an orbital vehicle. After three further flights had been completed without incident, President Ronald Reagan declared the system operational.

The Space Shuttle was to be used for all of the United States' space launch needs, both military and scientific, while NASA went ahead selling rides to the space programs and private corporations of other nations.

High hopes abounded that NASA had finally delivered the goal of creating low-cost, routine access to space. The 24 flights that followed through mid-January 1986 were, in the main, as successful as their launches were spectacular. During the fifth flight (STS-5) on November 11, 1982, Space Shuttle Columbia carried the first commercial payloads delivered to space by a piloted spacecraft, the ANIK-C3 and SBS-C communication satellites. Over the following three years, 24 commercial satellites were deployed from the shuttle.

Not all missions went as planned. For example, on STS-41-B in 1984, the boost engines of both satellites to be deployed failed to fire correctly, leaving the Palapa-B2 and Westar-6 communications satellites in orbits that were lower than intended. But the shuttle offered NASA a unique capability to retrieve and repair and/or replace these satellites. In November 1984, the Space Shuttle Discovery retrieved both satellites and returned them to Earth for refurbishment. They were later placed in their correct orbits.

Although the system was flexible and reusable, its complexity, coupled with the ever-present rigors of flying in an aerospace environment, meant that the turnaround time between flights was several months, rather than the several days initially projected. In addition, missions were delayed for a wide range of problems associated with ensuring the safety and performance of such a complex system. The Shuttle was reusable, but it was still expensive to operate.

"For whatever reason, I didn't succumb to the stereotype that science wasn't for girls. I got encouragement from my parents. I never ran into a teacher or a counselor who told me that science was for boys. A lot of my friends did."

SALLY RIDE,
NASA ASTRONAUT AND FIRST
AMERICAN WOMAN IN SPACE

A New Astronaut Corps

1978

NASA embraces diversity, and opens up its astronaut selection process to a wider range of applicants.

Without question, the Shuttle greatly expanded the opportunity for human spaceflight. In 1978, NASA announced plans to train a new group of astronauts, the first selected since 1969. This class was notable for its diversity. While all previous NASA astronauts had been white men with largely military backgrounds, this new intake included the first female astronauts, and the first astronauts of color. The Mercury-era requirement that astronauts should all be experienced pilots was also relaxed, with more scientists and engineers entering the astronaut pool. This led directly to scientist–astronaut Sally Ride becoming the first American woman in space in June 1983 on shuttle mission STS-7. Guion Bluford, another member of the 1978 astronaut intake, became the first black astronaut in orbit on August 1983 on Shuttle mission STS-8.

Astronaut Guion Bluford, a mission specialist on the STS-8 mission, undergoes an in-flight medical test while on a treadmill.

Expanding Science in Space

The launch of Space Shuttle Discovery on mission STS-26 on September 29, 1988, marked both a return to flight and a return to science after the Challenger disaster in 1986.

Throughout its 30-year career, the Space Shuttle served as a versatile platform for scientific inquiry. While science was not a major component of its original design brief as a technology demonstrator and workhorse for space access, the Shuttle eventually served as an exemplary platform for all manner of microgravity and space science experiments. Each of its 135 flights included some kind of science component, ranging from the deployment of space exploration probes to internal microgravity experiments, and even Earth observation missions.

The Shuttle facilitated x-ray astronomy, which can be conducted only from space because Earth's atmosphere blocks cosmic x-rays from reaching the surface of the planet. At more than 45 feet (13 m) in length, and weighing more than 5 tons (4.5 t), the Chandra X-Ray Observatory was carried into orbit by the Space Shuttle Columbia on mission STS-93 on July 22, 1999. The observatory was deployed from the Shuttle's cargo bay at 155 miles (250 km) above the Earth, but used both internal and external rockets to boost itself up to its higher working orbit.

Among many other contributions to science, the Shuttle conducted two Italian Tethered Satellite System missions in 1992 and 1996 that were designed to investigate new ways to study Earth's upper atmosphere and the generation of electricity in space. These missions, which first deployed and then retracted the satellite on tethers, demonstrated that satellites could potentially be dropped into regions of the atmosphere in which it would be difficult for a free-orbiting satellite to maintain an orbit on its own. They also demonstrated that tethering systems could be used to generate electricity by passing electromagnetic materials through the Earth's upper atmosphere to collect static electricity that could be channeled to

BELOW LEFT The Chandra X-Ray Observatory being placed into the Space Shuttle cargo bay ahead of its launch and deployment in 1999.

BELOW RIGHT The Tethered Satellite System deployed from Space Shuttle Atlantis mission STS-46 in 1992.

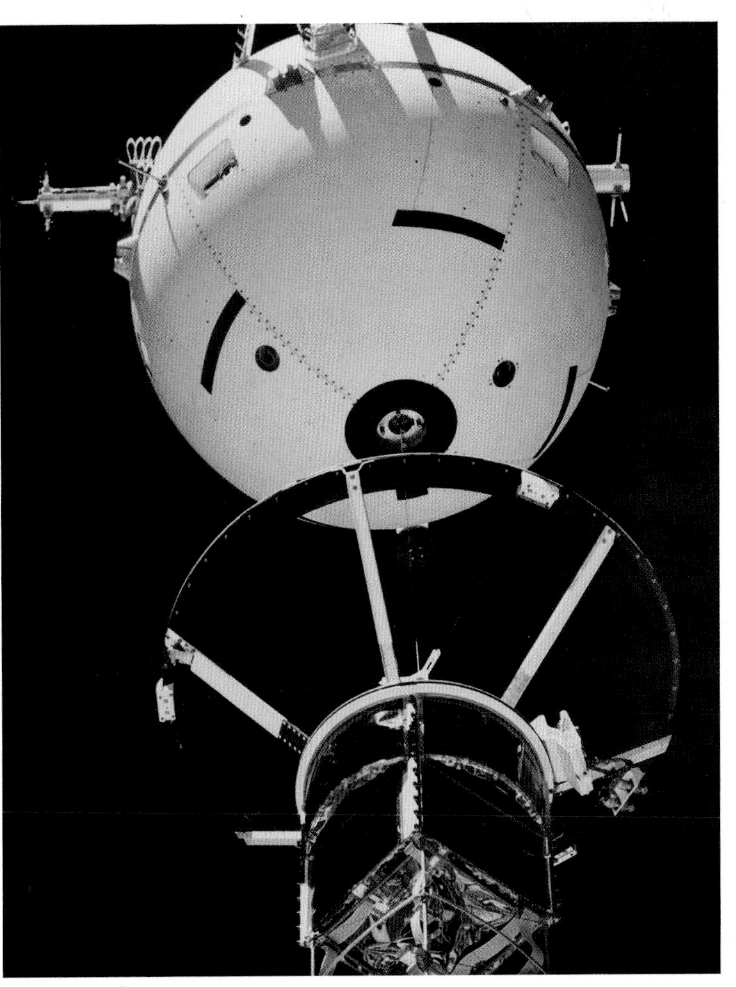

Spacelab

1981–2008

The modular science facility that could turn any Space Shuttle's cargo bay into a fully functional orbital laboratory.

Astronaut Jan Davis carries out life sciences, microgravity, and technology research inside the Spacelab module on Space Shuttle Endeavour mission STS-47 on September 12, 1992.

Working in partnership with NASA, Spacelab was first developed in the late 1970s and early 1980s alongside the Space Shuttle as a means to maximize the spacecraft's potential as a platform for science. The first pressurized Spacelab module flew aboard Space Shuttle Columbia on STS-9 in 1983. This test flight included a total of 73 experiments. The laboratory module flew again on the STS-51-B mission, conducting a series of low-gravity materials processing and fluid experiments. The STS-51-F mission served as a test flight for Spacelab pallets without the main Spacelab module, and concentrated on solar science and measuring the orbital space environment.

Spacelab flights became more dynamic as the program matured, and gradually took on a more international character. For the STS-61-A mission, the Spacelab module was managed by the German space agency, which directed its science activities from its operations center near Munich.

Spacelab's Atmospheric Laboratory for Applications and Science experiments package mapped Earth's atmosphere in order to compare global ozone levels and examine the factors that influenced them. The Spacelab Life Sciences flights studied how people, plants, and animals respond to weightlessness. The United States Microgravity Laboratory missions used the module to advance American expertise in low-gravity research. During the United States Microgravity Payload missions, scientists used Spacelab resources for both crew-intensive life-science research and power-consuming materials experiments. Spacelab J in 1992 featured Japanese materials processing and life-science experiments.

Overall, Spacelab flew on at least 22 major missions between 1983 and 1998, with Spacelab hardware being deployed on 10 further flights as late as 2008, and there is little doubt that it provided the Shuttle fleet with an invaluable resource for space science.

generate power on the spacecraft. The missions further demonstrated that non-electrical tethers could be used to help generate artificial gravity, and to boost payloads to higher orbits.

Another stunning Shuttle-based science experiment was the Shuttle Radar Topography Mission. The experiment itself consisted of a specially modified radar system that flew during the 11-day STS-99 mission in 2000. In the most ambitious Earth mapping mission to that point, the radar captured elevation data on a near-global scale, and thereby generated the most complete, high-resolution, digital topographic map ever, together with an enormously significant dataset for land-use scientists.

Among the various interplanetary exploration probes launched by the Space Shuttle, the most significant were probably the Magellan spacecraft that went to Venus, the Galileo spacecraft that went to Jupiter, and the Ulysses spacecraft that went to study the Sun. The Shuttle also deployed a number of space-based astronomy platforms, including the Compton Gamma Ray Observatory, the Hubble Space Telescope, and the Upper Atmosphere Research Satellite. For many space scientists, the data yielded by these probes and observatories more than justify the cost of the Shuttle program as a whole.

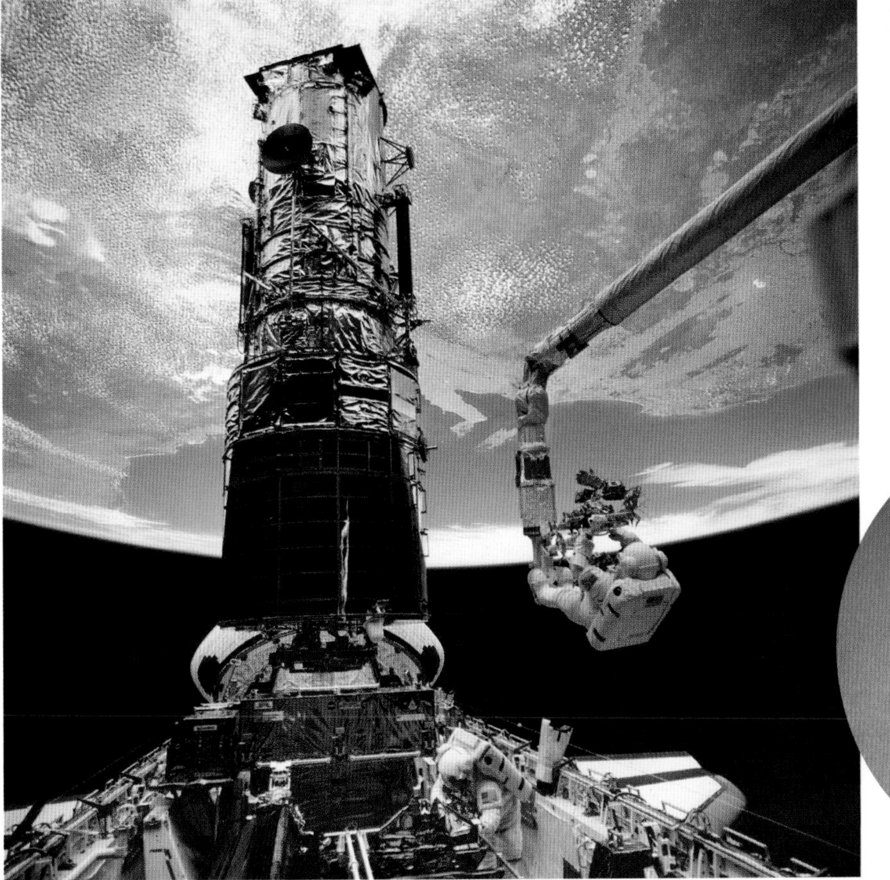

LEFT Astronaut Story Musgrave, anchored to the end of the remote manipulator arm, prepares to be elevated to the top of the towering Hubble Space Telescope to install protective covers on magnetometers on December 9, 1993.

BELOW Astronauts regularly used duct tape for space repairs. This roll returned with the final Hubble Space Telescope servicing crew on Space Shuttle mission STS-125 in 2009.

Expanding the Team of Space Explorers

Space Shuttle mission STS-9 was the first to include payload specialists, a classification that allowed people to train as astronauts and fly on Shuttle missions on which their experience or knowledge would prove valuable. Pictured here (left to right) are: STS-9 Mission Specialist Robert Parker; Payload Specialist Byron Lichtenberg; Mission Specialist Owen Garriott; and Payload Specialist Ulf Merbold.

The Space Shuttle greatly expanded the opportunities for human spaceflight. From crews of three on the Apollo missions, the Shuttle routinely flew crews of seven astronauts. It also expanded the opportunities for non-pilot astronauts, mission specialists, and guest astronauts from other nations and private corporations. Under the payload specialist program, individuals associated with specific spacecraft payloads and experiments were eligible to fly on the Shuttle. Two serving American politicians, Jake Garn (1932–) and Bill Nelson (1942–), both underwent astronaut training to fly as mission specialists in 1985 and 1986, respectively.

NASA also moved to facilitate orbital flight access to those who had no specialist knowledge that would qualify them as payload specialists. The first non-payload specialist civilian selected to go into space by NASA was high school teacher Christa McAuliffe (1948–86), from the Teacher in Space Project.

Even though NASA's attempts to expand access to spaceflight were positively received in many quarters, some critics accused the agency of pandering to the politicians whose support and funding it required, and of offering perquisites to a carefully selected few.

NASA received less criticism over international collaborations that allowed astronauts from other nations to fly aboard the Shuttle, often as mission specialists.

These astronauts helped to broaden human perceptions of spaceflight, and many offered lyrical accounts of their experiences. West German Ulf Merbold, the first ESA astronaut to fly on a Shuttle, commented afterward: "For the first time in my life, I saw the horizon as a curved line. It was accentuated by a thin seam of dark blue light—our atmosphere. Obviously this was not the ocean of air I had been told it was so many times in my life. I was terrified by its fragile appearance." He was not the first space explorer to have the vital importance of environmental protection and conservation impressed upon him by his experiences in orbit. Nor would he be the last.

American astronaut Sally Ride diagnosed what she viewed as the problem and suggested that we must see the Earth anew: "Some of civilization's more unfortunate effects on the environment are also evident from orbit. Oil slicks glisten on the surface of the Persian Gulf, patches of pollution-damaged trees dot the forests of Central Europe. Some cities look out of focus, and their colors muted, when viewed through a pollutant haze. Not surprisingly, the effects are more noticeable now than they were a decade ago." Many have speculated that the more people who can view the Earth from space, the greater the likelihood that humanity will take positive steps to ensure its own long-term survival.

A New Beginning

The 1985 Space Shuttle Discovery's mission STS-51G featured a number of firsts, most achieved by payload specialist Sultan Salman Al-Saud, who was the first Muslim, the first Arab, and the first member of a royal family to fly in space. It was also the first Shuttle mission that did not include any astronauts trained before NASA's 1978 intake at the start of the Space Shuttle era.

Astronaut and fervent space exploration advocate Sally Ride regarding the Earth from the flight deck of Space Shuttle Challenger on June 21, 1983.

The Loss of Challenger

ABOVE LEFT Christa McAuliffe, the teacher from Concord, New Hampshire, whom NASA picked to fly on the Space Shuttle, during microgravity training on the KC-135 reduced-gravity zero-G aircraft, January 8, 1986.

ABOVE RIGHT The crew of Space Shuttle mission Challenger pose for their official portrait on November 15, 1985. In the back row from left to right: Ellison S. Onizuka, Christa McAuliffe, Greg Jarvis, and Judy Resnik. In the front row from left to right: Michael J. Smith, Dick Scobee, and Ron McNair.

The loss of the Space Shuttle Challenger during its launch from the Kennedy Space Center on January 28, 1986, was a tragedy that touched not just everyone in the United States, but also people all around the world. For many amateur space enthusiasts, the STS-51-L mission already had an extra significance, because a member of its crew was a teacher, Christa McAuliffe, who was slated to hold a class and perform several student experiments in orbit. The Teacher in Space Project had been years in the making, and was touted as a major step forward in science education for young people. But the mission ended abruptly and tragically 73 seconds into the flight: a failure in one of the Shuttle's two solid rocket boosters (SRBs) ignited the shuttle's external fuel tank, which exploded in a blazing fireball that consumed Challenger entirely.

But the causes of the Challenger disaster had as much to do with the assumptions of those working at NASA as it did with the failed booster. A thorough investigation revealed that a rubber O-ring used in one of the joints in one of the SRBs had become brittle in the cold weather at the time of the launch. When the rubber gave way, the hot gas from the booster leaked out, damaging the external fuel tank, and causing the explosion. While some specialists understood the problem of rubber in cold environments, the engineers responsible for the O-ring were convinced that the joints were safe, and looked for other possible causes.

After the O-rings had been identified as the cause, NASA made extensive safety improvements, including the installation of a crew escape system, as well as upgrades to the Shuttle's landing systems. NASA also reorganized the Space Shuttle program to ensure that all necessary information would be available to all managers, and that any concerns that a contractor or member of staff may have had about a given piece of equipment could be reported anonymously from any level of the program staff.

Finally, President Ronald Reagan made further policy decisions that had a major effect on the whole Space Shuttle program. Firstly, he directed that NASA would no longer have a monopoly to launch all satellites deployed on behalf of the United States government. Secondly, he took NASA out of the commercial launch business altogether, thereby opening up that market to private sector service providers. The Shuttle itself did not return to flight until the fall of 1988.

Finding Faults

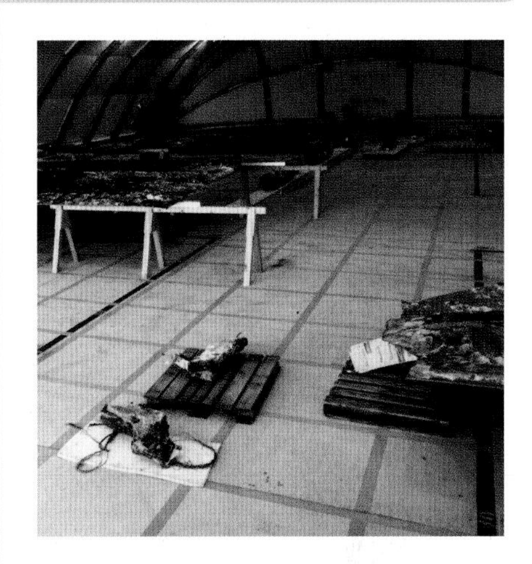

In the wake of the Challenger disaster, the United States Coast Guard and Navy worked to recover as much debris from the spacecraft as possible to help NASA's engineers and scientists work out what had gone wrong, and whether there had been any faults with any part of the Shuttle launch system. During the investigation that followed, the remains of the spacecraft, including parts of the external fuel tank, pictured here, were gathered together in a tent near the Logistics Facility at the Kennedy Space Center.

A solid rocket booster veers away from the massive explosion that engulfed the Space Shuttle Challenger on January 28, 1986.

The Soviet Buran: A Different Space Shuttle

The Soviet Buran shuttle and Energia launch vehicle mounted on a transporter.

In September 1976, NASA publicly unveiled its Space Shuttle for the first time. Its striking design immediately caught the attention of many in the Soviet Space program. Soviet engineers were soon combing over photographs of the new space plane to determine whether or not it was built to give the United States a military advantage in space. What could be lurking behind those large payload doors? Speculation abounded regarding the possibility that the Americans might have developed a new weapon of some kind: could it destroy the Soviet Union's satellites, cripple the nation's communication network, attack it from space?

From the late 1960s, Soviet space scientists had already known that their Cold War rivals were working on a successor to Apollo. They had even learned of its replacement's rough shape and design. But it was only in February 1976 that the Soviet leaders issued decree 132-51, which formally announced the development of their own reusable shuttle, known as Buran.

Like its American counterpart, Buran featured delta-shaped wings that enabled it to glide in to land like an aircraft. It, too, required the aid of a booster to reach orbit, which it received from the massive heavy-lift rocket Energia.

The Soviet Buran was intended to carry out four fundamental missions:

1. Match American capabilities in space.
2. Advance Soviet defenses, technological prowess, and science-gathering capabilities.
3. Justify the creation of a large space complex.
4. Launch and return to Earth cosmonauts and other payloads.

In 1982, Soviet space scientists and engineers began orbital tests of BOR-4, a small, Shuttle-like prototype spacecraft, to gain valuable data on heating and controllability during reentry. The data gathered from these tests aided the development of the thermal protection tiles that would later be used on Buran. By December 1984, the first Buran prototype arrived at the Baikonur Cosmodrome. Almost a year later, it completed its first atmospheric flight test. Utilizing a specially designed transport aircraft, the first flight-ready Buran arrived at Baikonur in 1985. While the final assembly, systems integration, and tests were carried out, the prototype vehicle underwent 24 piloted test flights validating the automated landing systems and computer-controlled flight systems. By October 1987, the final assembly and system checks on the Buran orbiter had been completed.

On November 15, 1988, the Buran-Energia Launch System flew without a crew. It made two orbits, and spent 203 minutes in space before reentering the Earth's atmosphere for an automated landing at the Baikonur Cosmodrome. Although the test was a success, this was Buran's final flight. Its development costs were too great for the Soviet Ministry of Defense to justify, especially when Soyuz already offered a reliable method of transporting cosmonauts and payloads into space. However Buran's demise was not confirmed publicly until 1993, two years after the end of Communism and the breakup of the Soviet Union.

Buran vs. Space Shuttle

Technical and Performance Comparison

	BURAN	SPACE SHUTTLE
Maximum weight	105 tons (95.3 t)	120 tons (109 t)
Payload weight	30 tons (27.2 t)	30 tons (27.2 t)
Crew	2–10 people	2–10 people
Orbit altitude range	150–621 miles (250–1,000 km)	118–596 miles (190–960 km.)
Length	119 feet (36.37 m)	122 feet (37 m)
Height	53 feet, 6 inches (16 m)	56 feet (17 m)
Wing span	Wing span: 78 feet (23 m)	78 feet (23 m)
Payload bay dimensions	19 feet × 50 feet (6 m × 15.5 m)	15 feet × 60 feet (4.6 m × 18 m)

FAR LEFT This poster for the Space Commerce Corporation—the American agents selling space on a variety of launch vehicles—promoted the commercial side of the Soviet space program.

LEFT Like its American counterpart, the Soviet Buran was tested and transported on the back of a large aircraft.

The Mir Space Station

The Soviet Union launched the core of its Mir space station on February 20, 1986. Weighing 22.4 tons (20.4 t) at launch, with 3,000 cubic feet (85 m^3) of habitable space, this initial module was the first piece of a much larger projected orbital space facility. By the early 1990s, its weight had been increased to about 121 tons (110 t)—with 13,000 cubic feet (368 m^3) of habitable space by the addition of further modules. By the time of the breakup of the Soviet Union in 1991, Mir consisted of its original core module, Kvant 1 (launched March 31, 1987); Kvant 2 (launched November 26, 1989); and Kristall (launched May 31, 1990). It was more than 107 feet (32 m) long with its docked Progress-M resupply and Soyuz-TM crew spacecraft, and about 90 feet (27 m) wide across all its modules.

This photo of Mir was taken during STS-86 after Atlantis undocked from the space station on October 20, 1997.

Mir's purposes were to create a permanent Soviet presence in space, and to serve as a research platform to establish the requirements for long-duration human spaceflight. In the latter regard, Mir was a resounding success. Except for two periods—between July 17, 1986, and February 4, 1987, and between April 27 and September 4, 1989—Mir remained permanently occupied. The longest single mission took place between December 21, 1987, and December 21, 1988, a total of 366 days. On that mission, cosmonauts Vladimir Titov and Musa Manarov (1951–) far outdistanced any previous spacefarers up to that point.

Even this long-duration mission paled in comparison to an incredible 439 days in orbit for Valery Polyakov (1942–) between January 8, 1994, and March 22, 1995. To date, Polyakov holds the record for the longest time spent in space on a single trip.

Mir remained in orbit with a crew until 1999. By this time, the ISS (p. 232) was already under construction. With the Russian space program a core participant in this new venture, the outdated Mir was an unnecessary and burdensome expense, and the decision was reluctantly taken to deorbit it. In March 2001, the Mir station made its final descent into Earth's atmosphere, ultimately impacting somewhere in the southern Pacific Ocean.

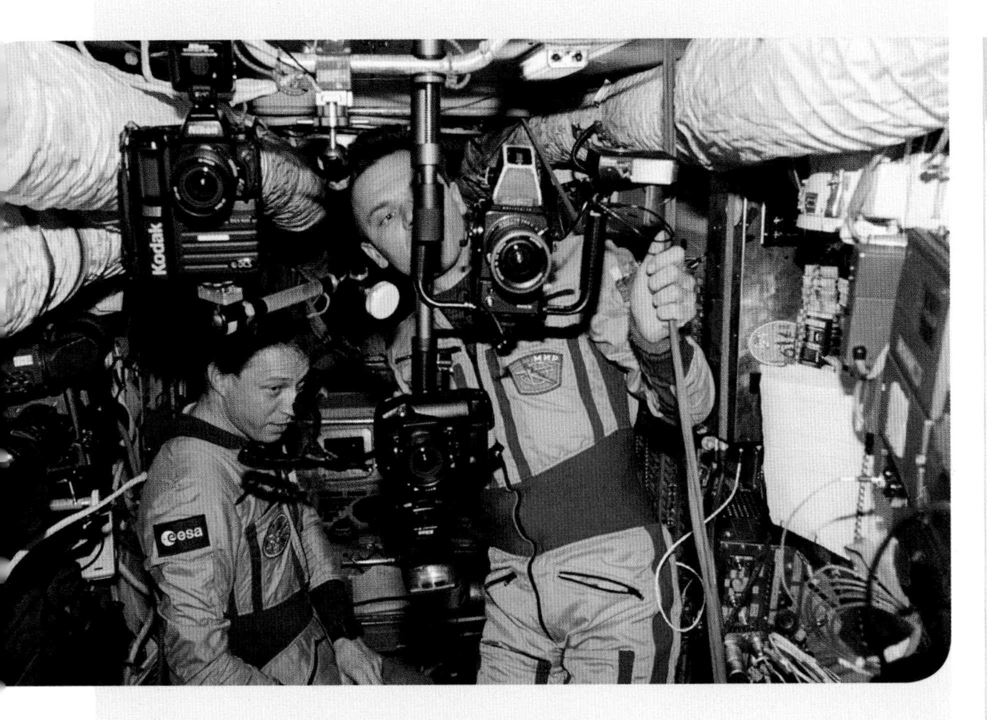

The Shuttle–Mir Program

1995–98

A new chapter in international cooperation in space that helped lead the way to the International Space Station.

After the end of the Cold War, Mir became the focus of a new age of international cooperation in space exploration. In 1995, the United States and Russia began the Shuttle–Mir program, in which the Space Shuttle docked with Mir in a succession of nine missions through 1998, and American astronauts undertook extended stays on the Russian station.

While these joint missions were a triumph of collaboration, and paved the way for Russian participation in the ISS, they also highlighted some of the risks associated with keeping a space station operational over extended periods. On June 25, 1997, a Russian Progress resupply vessel collided with Mir's science module, Spektr. The module decompressed, and its solar arrays were knocked out of service. Although the crew members were uninjured, the accident crippled the space station, and led to a series of crises. A subsequent fire on Mir also almost led to its being abandoned. In a courageous decision, the Russian Space Agency refused to issue an evacuation order. The cosmonaut crew soon brought the fire under control, and restored normal operations. Despite domestic political pressure to remove American astronauts from Mir after these accidents, NASA officials decided to continue their involvement in the cooperative missions.

Crewmembers in Mir on the day of the rendezvous and the docking of STS-74 and Mir 20.

Mir Modules

The Mir core resembled Salyut 7, but had six ports rather than the latter's two. Fore and aft ports were used primarily for docking. The four radial ports in a node at the station's front were for berthing large modules. The core weighed 20.4 tons (18.5 t) at launch in 1986.

Kvant was added to the Mir core's aft port in 1987. This small, 11-ton (10-t) module contained astrophysics instruments and life support and attitude control equipment.

Kvant 2, added in 1989, carried an EVA airlock, solar arrays, and life support equipment. The 19.6-ton (17.7-t) module was based on the transport logistics spacecraft originally intended for the Almaz military space station program of the early 1970s.

Kristall, added in 1990, carried scientific equipment, retractable solar arrays, and a docking node equipped with a special androgynous docking mechanism designed to receive spacecraft.

Spektr was launched on a Russian Proton rocket from the Baikonur launch center in Central Asia on May 20, 1995. Spektr carried four solar arrays and scientific equipment. The module was also used on an Earth observation mission to monitor natural resources and the atmosphere.

Priroda was the last science module to be added to the Mir. Launched from Baikonur on April 23, 1996, it docked with the space station as scheduled on April 26. Its primary purpose was to add Earth remote sensing capability to Mir.

Building the International Space Station

In 1984, a consortium of national space agencies set about a series of negotiations that would lead to the building of a new space station, one that would not ultimately be owned and operated by any single country, but would be a truly international project. Those early negotiations determined the scope and design of the station; who would be responsible for the manufacture of which parts; and how its operation would be organized. The first elements of what was to become the International Space Station (ISS) entered orbit more than a decade later, in 1998, and the installation was not completed until 2011.

The ISS is widely considered one of the signature achievements of space exploration history for three major reasons. Firstly, there is the political feat of bringing a diverse group of nations together for the advancement of a single enterprise. Secondly, there is the technical achievement of successfully completing a large, technically challenging, and costly feat of space engineering with contributions from across the globe. Thirdly, the station internationalized and reinvigorated the human dream of space exploration. ISS operations have energized the imaginations of a new generation, and helped to foster the development of private sector space research, cooperative technological efforts, and scientific investigation around the globe. The ISS also permits long-term research not possible on Earth using microgravity in such areas as materials science, fluid physics, combustion science, biology, and biotechnology.

Mission Specialist James Newman waves back at the camera during the first of three space walks performed during the STS-88 mission, the first Space Shuttle mission to the ISS.

The first two station components—the Zarya module from the Russian space agency Roscosmos, and NASA's Unity module—were launched and joined together in orbit in late 1998. The orbital assembly of the ISS also inaugurated a new era of hands-on labor in space, involving more space walks and a new generation of space robotics, as astronauts worked to join the various parts of the station together. Collectively, the Space Shuttle,

International Space Station Assembly 1998–

#	Element	Launch Date	Launch Vehicle
1	Zarya module (Roscosmos)	1998-11-20	Proton-K
2	Unity module (NASA)	1998-12-04	Endeavour (STS-88)
3	Zvezda service module (Roscosmos)	2000-07-12	Proton-K
4	Z1 Truss	2000-10-11	Discovery (STS-92)
5	P6 truss and solar arrays	2000-11-30	Endeavour (STS-97)
6	Destiny (NASA laboratory)	2001-02-07	Atlantis (STS-98)
7	External stowage platform 1	2001-03-08	Discovery (STS-102)
8	Canadarm2 (CSA)	2001-04-19	Endeavour (STS-100)
9	Quest joint airlock (NASA)	2001-07-12	Atlantis (STS-104)
10	Pirs docking compartment and airlock (Roscosmos)	2001-09-14	Soyuz-U (Progress M-SO1)
11	S0 truss	2002-04-08	Atlantis (STS-110)
12	Mobile base system	2002-06-05	Endeavour (STS-111)
13	S1 truss	2002-10-07	Atlantis (STS-112)
14	P1 truss	2002-11-23	Endeavour (STS-113)
15	External stowage platform 2	2005-07-26	Discovery (STS-114)
16	P3/P4 truss and solar arrays	2006-09-09	Atlantis (STS-115)
17	P5 truss	2006-12-09	Discovery (STS-116)
18	S3/S4 truss and solar arrays	2007-06-08	Atlantis (STS-117)
19	S5 truss and external stowage platform 3	2007-08-08	Endeavour (STS-118)
20	Harmony (ESA) relocation of P6 truss	2007-10-23	Discovery (STS-120)
21	Columbus laboratory (ESA)	2008-02-07	Atlantis (STS-122)
22	Dextre (Canada) Kibo module elements (JAXA)	2008-03-11	Endeavour (STS-123)
23	Kibo module elements and robotic arm (JAXA)	2008-05-31	Discovery (STS-124)
24	S6 truss and solar arrays	2009-03-15	Discovery (STS-119)
25	Exposed facility (Japan)	2009-07-15	Endeavour (STS-127)
26	Poisk module (Russia)	2009-11-10	Soyuz-U
27	ExPRESS logistics carriers 1 and 2 (NASA)	2009-11-16	Atlantis (STS-129)
28	Cupola and Tranquility node (NASA)	2010-02-08	Endeavour (STS-130)
29	Rassvet module (Roscosmos)	2010-05-14	Atlantis (STS-132)
30	Leonardo module (ESA) and ExPRESS logistics carrier 4 (NASA)	2011-02-24	Discovery (STS-133)
31	Alpha magnetic spectrometer, OBSS, & ExPRESS logistics carrier 3 (NASA)	2011-05-16	Endeavour (STS-134)
32	International Docking Adaptor-2	2015-07-16	Falcon 9 (SpaceX-CRS-7)
33	Bigelow expandable activity module (Commercial)	2016-04-08	Falcon 9 (SpaceX CRS-8)
34	Neutron Star Interior Composition ExploreR Mission (NICER)	2017-07-03	Falcon 9 (SpaceX CRX-11)

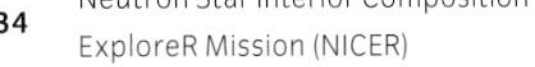

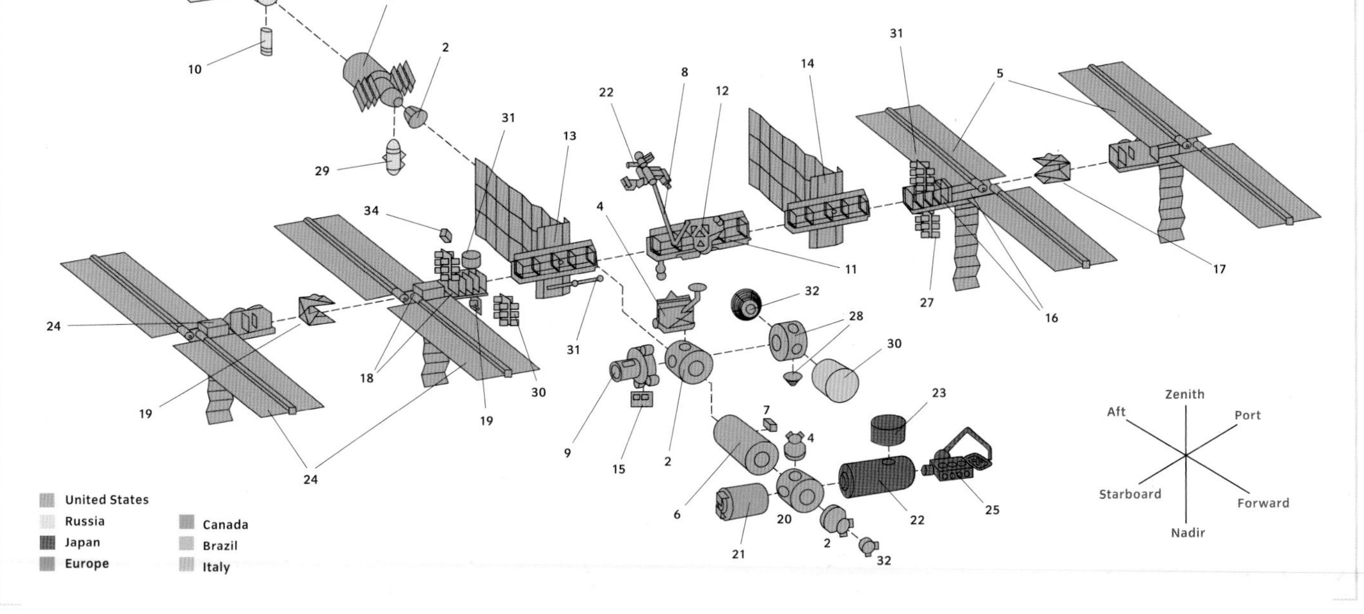

> *"Space station assembly is complete."*
>
> MARK KELLY, ASTRONAUT

the Soyuz crew capsule, and the Progress resupply module flew 46 missions to assemble the ISS.

On October 31, 2000, the station's first crew—American astronaut William Shepherd (1949–), and Russian cosmonauts Yuri Gidzenko (1962–) and Sergei Krikalev (1958–)—lifted off in their Soyuz spacecraft from the Baikonur Cosmodrome en route to their new home. Since that date, not a single day has passed without a human in orbit.

By early 2003, the ISS assembly efforts were hampered by increased costs and technical difficulties. All of this was exacerbated by the tragic loss of the Space Shuttle Columbia and her crew during its reentry descent on February 1, 2003 (p. 238). NASA grounded the Shuttle fleet until the cause of the loss could be established. Without any Space Shuttles to ferry up the larger ISS modules, construction efforts ceased. During this period, access to the station was possible only through the use of Soyuz capsules, and the ISS crew was scaled back to just two members.

Roscosmos flew a total of 14 resupply and crew rotation missions until Shuttles were cleared to return to flight on July 26, 2005. Thereafter, ISS crew members were increased to six. The largest pieces of ISS were finalized in orbit in 2011, just before the retirement of the remaining Shuttle fleet. However, the station has since received further components, such as the Bigelow Expandable Activity Module, as part of various technology demonstration programs. These will continue for the operational lifespan of the ISS, currently projected until at least 2024.

The International Space Station, photographed by an STS-132 crew member on Space Shuttle Atlantis after the station and Shuttle began post-undocking separation on May 23, 2010.

The International ISS Consortium

1984–

The nations behind the single largest international collaboration in human space exploration.

When leaders at NASA started formulating a plan to establish an international consortium to build a space station to achieve a permanent presence in space, they focused their efforts on soliciting contributions from countries and transnational space agencies that were traditionally allies of the United States, such as Brazil, Canada, Japan, and the European Space Agency.

In 1994, Russia's Roscosmos joined the effort, something that would have been almost inconceivable during the Cold War. Thereafter, the partners in this international project contributed the following elements:

Canada A 55-foot- (16.75m-) long robotic arm used for assembly and maintenance tasks.

European Space Agency A pressurized laboratory launched on the Space Shuttle, and logistics and supply services provided by its Automated Transfer Vehicle (ATV), launched on the Ariane 5 rocket.

Japan A laboratory with an attached exposed exterior platform for experiments; logistics transport vehicles.

Russia Two research modules; early living quarters with their own life support and habitation systems; a science power platform of solar arrays that can supply about 20 kilowatts of electrical power; logistics transport vehicles; and Soyuz spacecraft for crew return and transfer.

Both NASA and Roscosmos have full mission control centers for the ISS, but science activities are additionally monitored by whichever partner agency or organization is responsible for conducting the various experiments.

Astronaut William Shepherd enjoying a brief break from installing furnishings inside the ISS's Zvezda module in December 2000.

Science on the ISS

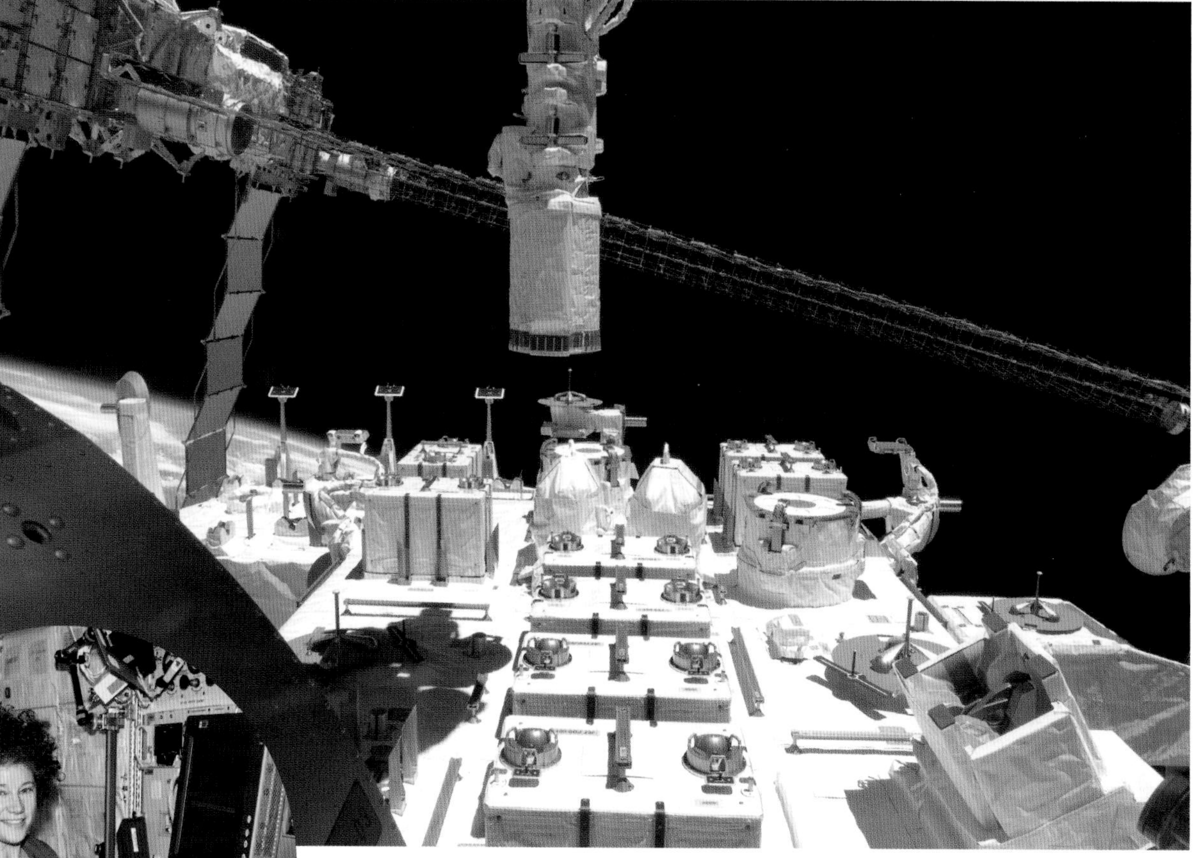

Since its first crew arrived in 2000 the ISS has become the orbital home to some world-class scientific research. This research has covered several key disciplines. Discoveries from its biomedical research have found applications for humans on Earth. Materials science research has expanded human understanding of how the properties of various materials are affected by the microgravity environment. Knowledge about the human body has been boosted by data gathered on the physical changes that the various ISS crews experienced during their time in space. Finally, the ISS has provided a platform for space science research into the origins and evolution of the Sun and the planets.

ABOVE LEFT Astronaut Susan Helms works at the Ultrasound Flat Screen Display and Keyboard Module on the ISS.

ABOVE RIGHT The exterior porch of the ISS Kibo module provides a literal platform for experiments involving the vacuum of space.

Probably the most critical component of research on the ISS has involved learning about how humans react to microgravity over time. It is now known that weightlessness affects almost every aspect of the human body, including the heart, lungs, muscles, bones, immune system, and nerves. Many of the physiological changes in astronauts actually resemble changes in the human body normally associated with aging on Earth. For instance, in addition to losing mass in microgravity, bones and muscles appear to heal less well in space. For astronauts, time spent in microgravity seems to result in dissociation between their physical and chronological ages. By studying the changes in astronauts' bodies, research from the ISS has played a key role in developing a model for the consequences of getting older in space, and the potential responses to it.

Researchers using ISS capabilities have also found that microgravity provides them with new tools to address two fundamental aspects of biotechnology: the growth of larger, high-quality protein crystals, and the growth of three-dimensional tissue samples in laboratory cultures. On Earth, gravity distorts the shape of crystalline structures, limiting their size, while tissue samples grown outside the body do not take on their full three-dimensional structure, making them more difficult to study. Research conducted on the ISS using a bioreactor (a device designed to grow biochemically active organisms) has been conspicuously successful as an advanced cell-culturing technology. It facilitated, for instance, groundbreaking research on lymph tissue function, which has proven pivotal in the study of the Human Immunodeficiency Virus (HIV).

Researchers around the globe have also used the low-gravity environment to enhance their understanding of how gravity influences the production and processing of a wide range of materials. Space research has produced crystals of cadmium zinc telluride 50 times purer than anything grown on Earth. The crystals have a variety of applications in electronics, and the low-gravity experiments have helped to improve the efficiency of cadmium zinc telluride fabrication on Earth.

"The International Space Station is a phenomenal laboratory....Yet I often thought, while silently gazing out the window at Earth, that the actual legacy of humanity's attempts to step into space will be a better understanding of our current planet and how to take care of it."

CHRIS HADFIELD,
CANADIAN ASTRONAUT

NanoRacks and CubeSats 2009–

Automating space science for maximum efficiency.

One of the most significant advances in opening up space science to researchers on the ground has come through the private company NanoRacks LLC. The brainchild of Jeffrey Manber (1958–), NanoRacks facilitates scientific research on the ISS by offering autonomous experiments packages that can be brought to the station by resupply vehicles, loaded into experiment bays, and tended by astronauts. The data from these experiments are sent directly back to researchers on the ground. Since 2009, the company has delivered more than 580 experiments to the ISS.

Another major innovation came on January 9, 2014, when the NanoRacks CubeSat Deployer was added to the ISS. CubeSats typically measure around 4 cubic inches (10 cm^3) in volume and carry a variety of small scientific instruments. With the deployer in place, the ISS has become a major launch point for swarms of CubeSats with a wide variety of functions. Together, these innovations have reduced the costs of space science.

The NanoRack CubeSat Deployer in operation on the ISS on August 4, 2017.

Columbia's Loss and the End of the Space Shuttle Program

NASA personnel planned to celebrate on February 1, 2003, when Columbia returned with its crew after the successful completion of STS-107. This mission had launched from the Kennedy Space Center's Launch Complex 39A on January 16 on a science mission dedicated to research in physical life and space sciences. The Shuttle was carrying a Spacelab module housing approximately 80 separate experiments. The seven astronauts aboard had worked 24 hours a day, in two alternating shifts, to ensure that these experiments were completed on schedule.

Unfortunately, STS-107 never made it home; the vehicle and crew were lost during reentry into Earth's atmosphere. NASA lost communication with Columbia during its descent, and when the Shuttle failed to land at its appointed time it was obvious something had gone wrong. Seven astronauts—Mission Commander Rick Husband; Pilot William McCool; Mission Specialists Kalpana Chawla, David Brown, and Laurel Clark; Payload Commander Michael Anderson; and Payload Specialist Ilan Ramon—all lost their lives in the disaster. A high priority was attached to recovering their remains from the Shuttle debris that was discovered in Texas.

The crew of mission STS-107. Front from left: Rick D. Husband, Kalpana Chawla, and William C. McCool. Behind them, from left: David M. Brown, Laurel B. Clark, Michael P. Anderson, and Ilan Ramon.

NASA immediately formed a Columbia Accident Investigation Board, which soon developed a plausible theory about what had happened. Approximately 81 seconds after launch, insulating foam from the external fuel tank had impacted Columbia's lower left wing, creating a hole through which hot gases could enter the vehicle. While in orbit for 16 days, nothing appeared amiss, but, once the Shuttle reentered the atmosphere, the hot gases entering the wing quickly melted its support structures, causing the whole vehicle to break up.

After the loss of Columbia, NASA grounded the Shuttle fleet until it had ensured that there could be no recurrence of this disaster. The problem was definitively solved in 2005, but by then the remaining days in the fleet's operational lifespan were already numbered.

The "Vision for Space Exploration"

2004

President George W. Bush announced an end to the Space Shuttle program and a new direction for NASA.

The loss of Columbia signaled the beginning of an important debate about the future of American human spaceflight. President George W. Bush announced on January 14, 2004, his "Vision for Space Exploration" that called for humans to reach for the Moon and Mars during the next 30 years. In support of this goal, the Bush administration announced that the United States would:

- Implement a sustained and affordable human and robotic program to explore the solar system and beyond;
- Extend human presence across the solar system, starting with a human return to the Moon by the year 2020;
- Promote international and commercial participation in exploration, to further American scientific, security, and economic interests.

The President also called for the completion of the ISS and the retirement of the Space Shuttle fleet by 2010, a target NASA missed by a year. The funding that had been spent on the Space Shuttle, roughly $3–5 billion per year, was subsequently reinvested in commercial space services from Orbital/ATK, SpaceX, and other carriers, and also put toward the development of a new space launch system capable of traveling to the Moon and beyond.

This impromptu shrine to the Space Shuttle Columbia crewmembers who lost their lives on February 1, 2003, was created at the entrance to the Johnson Space Center in Houston, Texas.

Shuttle Survivors

While the entire crew of Space Shuttle Columbia mission STS-107 perished when the spacecraft broke apart during descent on February 1, 2003, there were some life-forms that survived the tragedy. A locker containing six canisters of *Caenorhabditis elegans* worms from one of the mission's science experiments was recovered intact from the Columbia debris. When the canisters were eventually opened, several months after the accident, investigators were astonished to discover they contained living worms. Since *C. elegans* worms have a life cycle of just seven to ten days, the worms recovered from the wreckage were clearly descendants of the worms that had traveled into space; their survival is a testament to the resilience of life.

An Assessment of the Space Shuttle

The Space Shuttle program operated in Earth orbit between 1981 and 2011, and from the first flight to its final one it was an important symbol of the United States' technological capability, universally recognized as such by both the American people and the international community. Through its 30-year service history, it remained the most highly visible symbol of American technological prowess, and claimed a number of significant achievements. It flew 135 missions, with multiple flights a year, and transported more than 500 people into space. It delivered satellites, telescopes, and planetary probes into orbit. It had experiments on every flight, and full laboratories on many. It made 36 trips to assemble and then to supply the ISS.

There is no question that the Space Shuttle was a magnificent machine. No other nation had the technological capability to build such a sophisticated vehicle during the 1970s. Few nations could do so today. It was a massively complex system, with more than 200,000 separate components that must work in synchronization with each other and to specifications more exacting than any other technological system in human history. The Space Shuttle must be viewed as a triumph of engineering and excellence in technological management.

Despite two catastrophic accidents—the Challenger explosion that killed the crew of seven on January 28, 1986, and the Columbia accident on reentry on February 1, 2003, that led to the deaths of a further seven astronauts—the Space Shuttle was an incredibly resilient spacecraft that was always able to launch on schedule, even when the notice period for urgent flights to the ISS was less than 15 minutes. This is an enormously demanding operational challenge that has never even been considered as a possibility for other launch vehicles, before or since.

Additionally, the Space Shuttle proved itself to be one of the most flexible space vehicles ever flown. With its large payload bay, it was capable of satellite deployment and recapture, as well as in-orbit repair and redeployment. While the program was not explicitly designed as a science platform, it proved to be a well-suited vehicle for all manner of microgravity and space science investigations.

Ultimately, the Space Shuttle may be viewed as both a triumph and a tragedy. As a symbol of American technological excellence, and as a reliable, adaptable system on which stunning scientific experiments were conducted, it receives high marks. But the program still failed to achieve one of its core design objectives: lowering the cost of reaching Earth orbit. Creating a system that significantly lowers launch costs is now one of the principal objectives for the next generation of launch vehicles already under development.

OPPOSITE This image of Space Shuttle Endeavour silhouetted against the edge of the Earth's atmosphere was taken from the ISS on February 9, 2010 during the STS-130 mission.

BELOW Not everything in orbit is about work. Astronaut Ellen Ochoa takes the opportunity to play her flute on the Space Shuttle Discovery's aft flight deck during the STS-56 mission.

The Potential of Space Tourism

> *"Space travel for everyone is the next frontier in the human experience."*
>
> BUZZ ALDRIN,
> NASA ASTRONAUT AND
> SECOND HUMAN ON THE MOON

The notion of space tourism, of traveling into orbit or beyond purely for leisure, dates back at least as far as the invention of the rocket itself, but it received a pretty prominent boost in 1968 when airline Pan American World Airways announced that it would take reservations in anticipation of future space tourism as a promotion for the Stanley Kubrick film, *2001: A Space Odyssey* (p. 365). Reportedly, Pan Am received more than 93,000 reservation requests for a service it did not have the capacity to deliver.

In that same year, hotel magnate Barron Hilton began talking about building some kind of rentable accommodation on the Moon, and proposed a "space shuttle service" that would ferry passengers on round-trips for a price of $1,500 (about $15,000 today), in addition to another $1,000 for a two-week stay at a lunar Hilton. He envisioned his Moon hotel as being big. One account gave it 5,000 rooms, and its own private ocean. Although this was clearly a highly speculative and aspirational plan, as late as 1999 Hilton was reportedly considering a collaborative project with 16 other groups to make a $25 billion space hotel a reality. Under this plan, a two-week trip would initially cost $2 million per person, dropping to $415,000 by the fifth year of operation.

In the classic 1968 film *2001: A Space Odyssey*, Stanley Kubrick envisioned a space station in Earth orbit complete with a Hilton hotel.

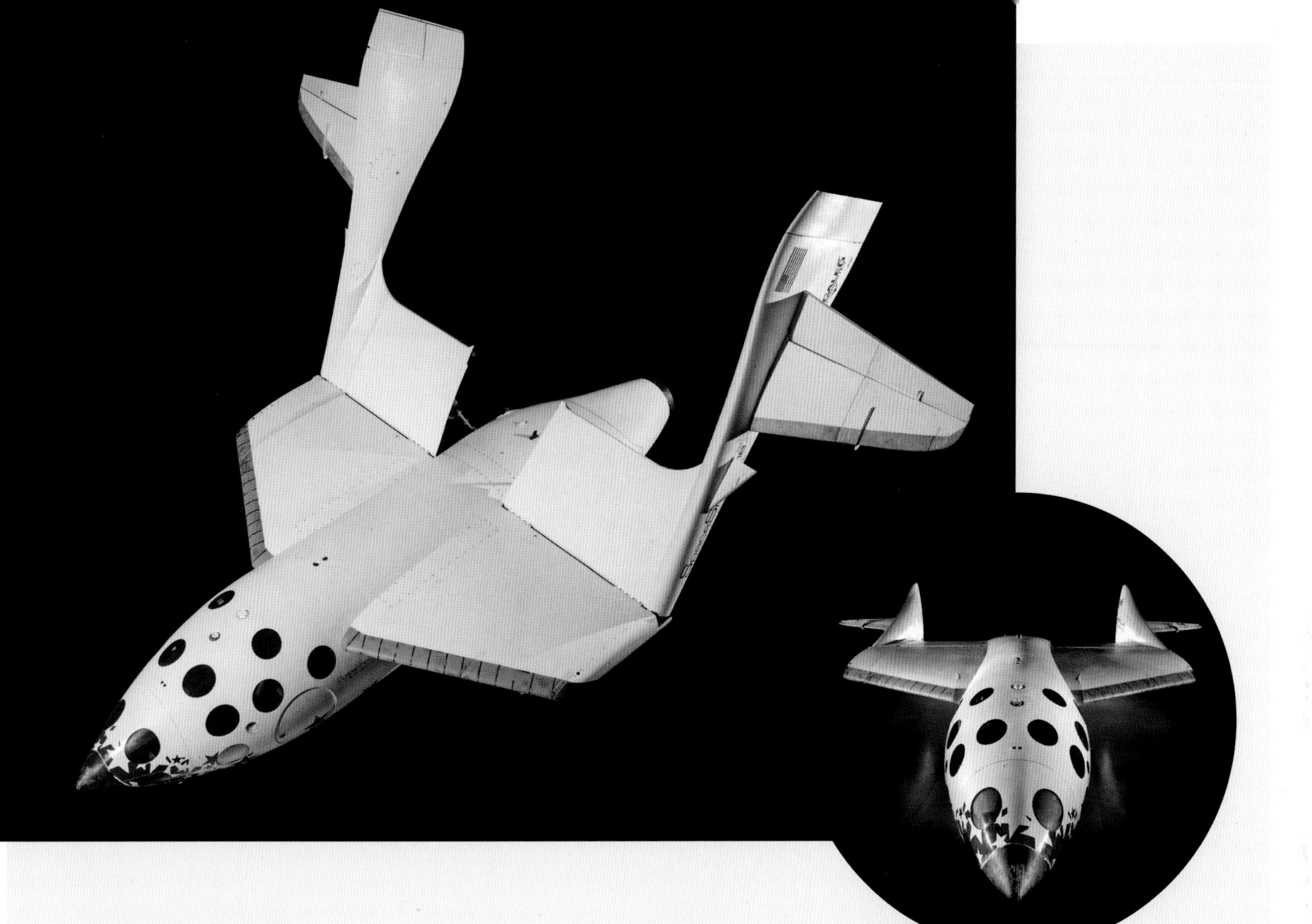

The Saga of SpaceShipOne

2004

The world's first privately financed suborbital spacecraft, which claimed the coveted Ansari X-Prize.

Gifted to the Smithsonian Institution's National Air and Space Museum by investor Paul Allen, SpaceShipOne can be found on display in the Boeing Milestones of Flight Hall.

SpaceShipOne became the first privately developed space vehicle to fly past the Kármán line in 2004. It was designed and built by Scaled Composites, in a joint venture between the aircraft designer Burt Rutan (1943–) and Microsoft cofounder Paul Allen (1953–).

The spacecraft went on to claim the $10 million Ansari X-Prize as the first privately financed, reusable spacecraft to make two suborbital spaceflights within a two-week period. These flights took place on September 29 and October 13, 2004.

Like the Bell X-1 (p. 76), SpaceShipOne launches from the air. To attain the necessary altitude of 50,000 feet (15,000 m) for launch, SpaceShipOne is attached to a long-winged carrier aircraft called White Knight, which carries the spacecraft up to the requisite height for release. The spacecraft's pilot then ignites its hybrid rocket motor, which burns solid rubber fuel with liquid nitrous oxide, to power the vehicle up into suborbital space. To begin its reentry, SpaceShipOne relies on its distinctive swept wings with tail fins. At the highest point in its trajectory, the pilot makes the tail and about a third of the wing tilt upward into what its designer called a "shuttlecock" configuration. This helps the spacecraft to maintain its stability, and to brake for reentry. After deceleration, the pilot lowers the wings and tail back into position for atmospheric flight, before gliding down for a runway landing.

Aside from its prize-winning feats, SpaceShipOne is also the baseline design for a proposed suborbital commercial space tourism vehicle, SpaceShipTwo, currently under construction by Virgin Galactic, Inc.

DENNIS A. TITO
ДЕННИС А. ТИТО

In 1993, a market survey on space tourism conducted in Japan found widespread support for the venture. An international commercial space transportation study in 1994 found similar support: "A new space transportation system would provide tangible and intangible benefits to the general public. The development of new market areas would create new opportunities and capabilities, for example, space tourism."

But it was American engineer and multimillionaire Dennis Tito (1940–) who made space tourism a reality in June 2000 when he signed a deal with the Russian MirCorp to fly aboard a Soyuz rocket to the Russian space station Mir. MirCorp acted as Tito's broker with the Russian space firm Energia, which owned both Mir and the rocket that would get Tito into orbit. While MirCorp had grandiose plans for operating a tourist service to the space station, it failed to obtain the venture capital necessary to make it a reality.

This forced Tito to look elsewhere for a trip into space. Eventually, he negotiated a $20 million deal that would enable him to fly to the ISS on a Soyuz spacecraft in 2001. South African software engineer Mark Shuttleworth (1973–) paid a similar sum, and also flew to the ISS later the same year. A number of other multimillionaires have made the excursion since, and more will doubtless follow in the future, either paying their own way, or obtaining corporate sponsorships. But while the current price tag places space tourism out of reach for all but a small circle of the ultra-wealthy, Tito's trip set a precedent that opened the way for spaceflight outside of the dominion of various national space agencies. It will almost certainly require the development of more cost-effective launch technology to drive down prices to an affordable level. Space tourism for all is now closer to reality than ever before in the history of space exploration.

The First Space Tourist

On April 28, 2001, American engineer and multimillionaire Dennis Tito became the world's first space tourist when he joined the crew of Soyuz TM-32 on a journey to the ISS. While NASA's leadership was initially resistant to commercial space tourism, and specifically to Tito's plan to visit the ISS, Russian space organizations were more open to the notion, and worked to facilitate Tito's trip. Cosmonauts Talgat Musabayev and Yuri Baturin, Tito's crewmates on TM-32, supervised their paying passenger during his week-long space excursion. They also helped the tourist to adjust to the microgravity environment before accompanying him back to Earth.

OPPOSITE The Russian-made Sokol KV-2 Pressure Suit Dennis Tito wore on his trip to the International Space Station in 2001.

LEFT An artist's impression of the Kankoh Maru, a proposed civilian space transport aimed at the tourist market.

8

The Lure of the Red Planet

Mars has always held a special fascination for humanity. Its reddish color in the night sky has strong associations with war and destruction, while astronomers have long speculated on the possibility that life might exist there, either now, or at some time in the past. Viewed from Earth-based telescopes, its features have prompted all manner of speculation about the planet and what humans might encounter there.

More than a century of fictional depictions of Mars served only to further excite humanity's desire to visit the planet. Astronomers Giovanni Schiaparelli (1835–1910) and Percival Lowell (1855–1916) became interested in Mars during the late nineteenth century. Based on their observations, both men advocated that Mars had once been a watery planet, speculating that topographical features had been built by intelligent beings. From their time until the present, the idea of intelligent life on Mars has remained a potent theme in popular culture.

Both the Soviet Union and the United States made Mars an early target for exploration in the Cold War space race of the 1960s. While the Soviets reached Mars first, in June 1963, they gained little scientific return for their achievement. The first American probe to flyby Mars took photographs that showed a cratered, Moon-like surface, dashing the hopes of many that intelligent life might be present there. Subsequent missions focused on finding water, a key requirement of life on Earth, as a possible indicator that simple life might have evolved on the planet.

Various nations have now declared their intention to pursue a human expedition to Mars in the future. Although it may yet be many years before we finally see space explorers walking on the Red Planet, few scientists now doubt that such an expedition will eventually come to pass.

OPPOSITE NASA assembled this mosaic image of Mars from images taken by Viking Orbiter in 1998. The north polar cap is visible at the top, the great equatorial canyon system is below center, and four huge Tharsis volcanoes are on the left.

Expectations about Life on Mars

In the nineteenth century, many people around the globe embraced the idea of life on the Red Planet. The Italian astronomer Giovanni Schiaparelli created a map of Mars based on observations made during the Great Opposition of 1877, when Mars and the Earth were approaching their closest proximity in their respective orbital cycles. Schiaparelli's map showed a network of linear structures on the surface of Mars, which he called *canali*, the Italian word for "channels." Many English-speakers misinterpreted *canali* as canals: purpose-built artificial waterways, as opposed to channels, which could be caused by various kinds of natural processes. This misunderstanding gave credence to a theory about the evolution of the solar system, which persisted for many years, that the Sun had been gradually cooling since its origin and, as it did so, each planet had a period in which it was habitable. Mars had once been in that category, but its civilizations were dying out or had already expired, while Earth was currently flourishing, and Venus would become habitable in the future. Thus, the expectation grew that the first probe to reach Mars would uncover signs that the planet has once been inhabited by some form of civilization.

Astronomer Giovanni Schiaparelli observed dark lines crisscrossing the surface of Mars.

Based on this theory, those who had misinterpreted Schiaparelli's observations began to formulate all manner of speculations about life on the Red Planet. For example, in his book *Uranie* (1889), French author Camille Flammarion described what life on Mars might be like: "They have straightened and enlarged the watercourses and made them like canals, and have constructed a network of immense canals all over the continents. The continents themselves are not bristling all over with Alpine or Himalayan upheavals like those of the terrestrial globe, but are immense plains, crossed in all directions by canals, which connect all the seas with one another, and by streams made to resemble canals."

American astronomer Percival Lowell became interested in Mars during the latter part of the nineteenth century, and built what became the Lowell Observatory near Flagstaff, Arizona, to study the Red Planet. Accepting the idea of Martian canals, he advanced the argument that Mars had once been a watery planet, and that the topographical features subsequently identified as canals had been

"That Mars is inhabited by beings of some sort or other we may consider as certain as it is uncertain what these beings may be."

PERCIVAL LOWELL, AMERICAN ASTRONOMER

built by intelligent beings. Over the course of the first 40 years of the twentieth century, others used Lowell's observations as a foundation for their arguments supporting the idea of life on Mars.

Over his lifetime, Lowell published three books relating to the subject: *Mars* (1895), *Mars and Its Canals* (1906), and *Mars As the Abode of Life* (1908). These writings, more even than Schiaparelli's maps, served to spread the notion that canals existed on Mars, and that they were evidence of intelligent life. Lowell offered a compelling portrait of a dying planet, whose inhabitants had constructed a vast irrigation system of canals to distribute water to the population centers nearer the Martian equator. Despite its popular appeal, few astronomers accepted this theory. Subsequent telescope-based astronomical observations failed to confirm Lowell's findings, and many, including Italian astronomer Vincenzo Cerulli (1859–1927), concluded that these channels were merely optical illusions.

Nonetheless, Lowell's observations added further fuel to the popular Martian myth. People genuinely expected explorers to find life on Mars. Magazine illustrations commonly portrayed the planet with a network of canals. Editors at

ABOVE LEFT This computer-generated image depicts a mythical Martian canal of the kind Percival Lowell imagined.

ABOVE RIGHT Lowell, an astronomer who founded the Lowell Observatory in Flagstaff, Arizona, and propagated the notion that there were canals on Mars.

Going Global

Numerous maps and globes of Mars were created across the world in the wake of the Great Opposition of 1877. This small example was manufactured by J. Lebèque & Co., in Brussels, Belgium, around 1892. The map, drawn by Belgian astronomer Louis Niesten (1844–1920), incorporates the dark areas described by Schiaparelli and the ideas of the British astronomer Nathaniel Everett Green (1823–99), who believed the lines to be some form of optical illusion.

The map of the surface of Mars created by American astronomer Percival Lowell in 1907.

Life magazine informed readers in a 1944 issue of their periodical that the canals served to irrigate patches of vegetation "that change from green to brown in seasonal cycles." Willy Ley (1906–69), one of the leading writers of popular science in the 1950s, assured readers of a 1952 issue of *Collier's* magazine that primitive plant life "like lichens and algae" surely existed on Mars. Where plants exist, Ley added, animals must have followed. Walt Disney, in a widely viewed 1957 television broadcast, showed animated drawings of flying saucers skimming over fields of Martian plants and animal life.

It was only with the dawn of the space age and the arrival of the first probes to Mars in 1965 that these misconceptions were challenged with hard data. But even this new evidence was not enough to banish the belief of life on Mars entirely. After all, even in the absence of animals, vegetation, and any relics of a sophisticated but departed civilization, it was still possible that some less complex life had once existed on the Red Planet. The search for evidence to prove, or disprove, the existence of such simple life has been a core feature of many of the following missions to explore the Red Planet.

Early Maps of Mars

1887

The source of most of the early theories supporting the notion that Mars once hosted complex life.

Both Giovanni Schiaparelli and Percival Lowell prepared maps of the Martian surface. Schiaparelli's was the first detailed cartographic depiction of the Red Planet, and he was largely responsible for naming its various promontories. Schiaparelli was aware that people had misinterpreted his initial descriptions of Mars, and wrote in his 1893 book, *Life on Mars*: "Rather than true channels in a form familiar to us, we must imagine depressions in the soil that are not very deep, extended in a straight direction for thousands of miles, over a width of 100, 200 kilometers and maybe more. I have already pointed out that, in the absence of rain on Mars, these channels are probably the main mechanism by which the water (and with it organic life) can spread on the dry surface of the planet."

This corrective did not persuade true believers, such as Lowell. He continued to argue for the existence of a Martian civilization until his death in 1916. His maps of the Red Planet always depicted a world that had been shaped by intelligent action. Lowell even made a globe of Mars, based on his own observations of the planet in 1901. Lowell's globe depicted the "non-natural features" of the planet's surface, including canals, oases, and intersections. The Lowell globe of Mars, like his writings on the subject, carried to their logical conclusion wishful thinking about the prospect of finding life on the Red Planet.

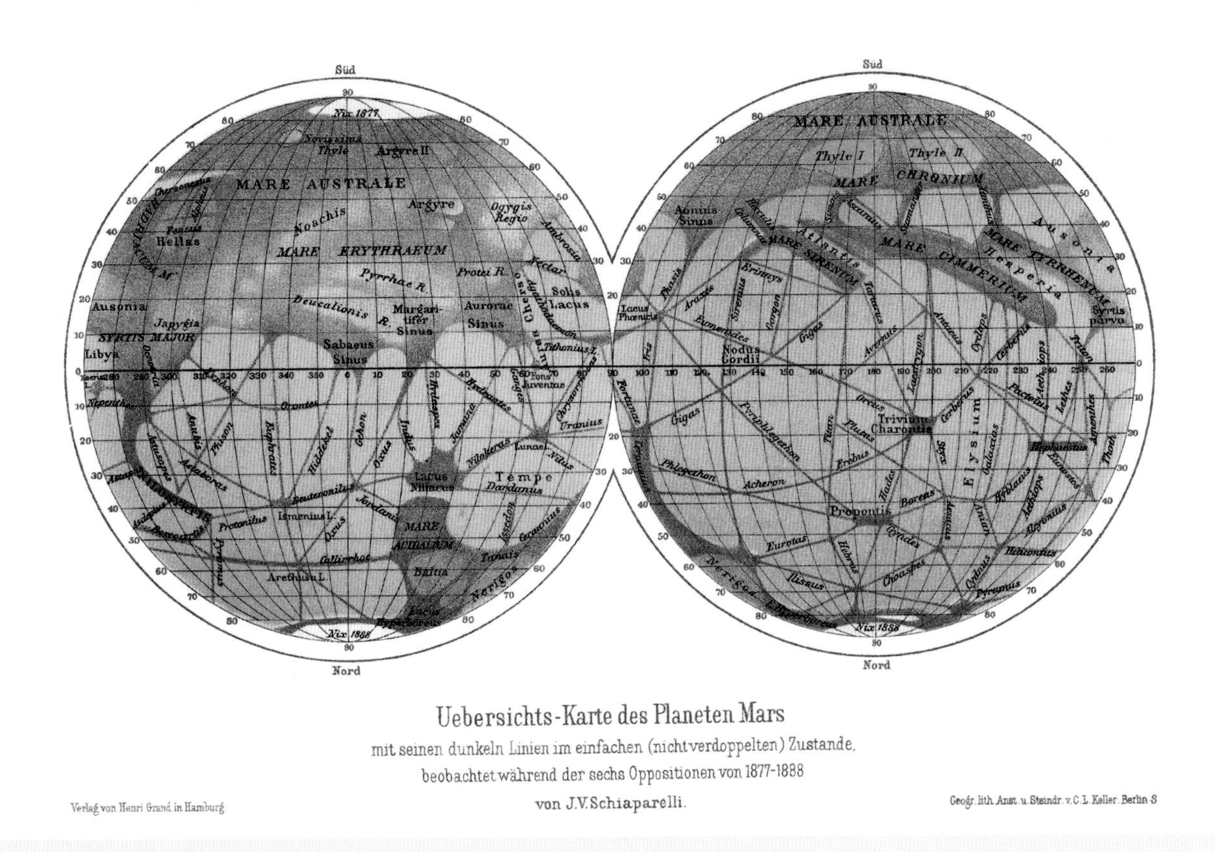

This drawing of the two hemispheres of Mars was made by Giovanni Schiaparelli between 1877 and 1888. It gave rise to the belief that intelligent beings had built canals to bring water from the poles to an arid equatorial region, and the assertion that the patterns on the surface changed with the Martian seasons. Scientists have subsequently identified powerful dust storms as the cause of the planet's periodically changing appearance.

Mars in Twentieth-Century Fiction and Popular Culture

> *"In one respect at least, the Martians are a happy people, they have no lawyers."*
>
> EDGAR RICE BURROUGHS, AMERICAN AUTHOR

Ray Bradbury's *The Martian Chronicles* was adapted into a popular television miniseries in 1979, starring Rock Hudson.

No planet in the solar system has more expectation, excitement, misapprehension, and trepidation associated with it than Mars. Humans have long speculated on the nature of the planet, and the life that might exist there. Martian visitations remain a persistent feature in fiction and popular culture, with the aliens sometimes being portrayed as friendly, but more often as dangerous, and potentially threatening to the whole human species.

One of the earliest mass-market novels about Mars was an 11-volume series by American writer Edgar Rice Burroughs (1875–1950) between 1912 and 1948. Starting with *A Princess of Mars*, in which American Civil War soldier John Carter is magically transported to Mars for a series of adventures, Burroughs presented a somewhat quixotic vision of a dying Mars, which was called Barsoom by its factionalized and warlike inhabitants. The series still has its fans to this day, and it was adapted into a big-budget Hollywood movie in 2012.

Another American science fiction writer, Ray Bradbury (1920–2012), penned a wildly popular series of stories set on and around Mars in *The Martian Chronicles*. Bradbury's book presented powerful images of Mars in 1950, and emphasized the adventures of humans fleeing a troubled and eventually nuclear war-riven Earth. The humans interact and enter into conflict with some natives of Mars who have powers of empathy and telepathy. Since Bradbury first published most of his works in the series as short stories, and only later wove them together into this novel, it sometimes requires effort from the reader to make the leaps between the various characters and stories, but the book remains an important reflection of Mars in popular fiction in the mid-twentieth century. On the flip side of this is the cult classic British television series and feature film *Quatermass and the Pit* (also known as *Five Million Years to Earth*), a science-fiction thriller, crafted by British writer Nigel Kneale (1922–2006), and set on an Earth in which humans discover a buried alien ship containing long-dead Martians. In a terrifying twist, it is revealed that humanity's tendency toward war is part of a Martian inheritance from the dead aliens.

A further American author, Robert Heinlein, took a different approach to the Red Planet in the 1961 novel *Stranger in a Strange Land*. This book tells the story of Valentine Michael Smith, a human who comes to Earth in early adulthood after being born on Mars and raised by Martians. This experience gives Smith an alternative insight into Earth culture and how it might be transformed. He eventually becomes a messianic figure, leading millions on a journey of self-discovery. Not surprisingly, not everyone accepted Smith's teachings, and eventually he becomes a martyr to his cause as his followers continue to perpetuate his message. *Stranger in a Strange Land* became a touchstone for the 1960s' counterculture, and remains an important text for social reformers seeking to alter the landscape of world politics.

Mars was also a popular subject for movies throughout the twentieth century, although a large number of these films proved far from memorable. *Just Imagine* (1930) tells a love story between a human and a Martian. *Rocketship X-M* (1950) tells the story of humans visiting Mars and encountering

The poster for the 1964 movie *Robinson Crusoe on Mars* claimed it was a "scientifically authentic" film. The following year, Mariner 4 would completely revolutionize the scientific community's understanding of the Red Planet.

ABOVE The film poster for the 1953 film version of *The War of the Worlds*. The movie gave H. G. Wells's novel a vivid visual dimension, but did not provoke the same audience reactions as the earlier radio version.

LEFT Orson Welles speaks with newsmen in the wake of the sensational broadcast of his dramatic radio adaptation of *The War of the Worlds*.

The War of the Worlds Radio Broadcast

1938

The radio adaptation that had some Americans believing Martians were real, and invading New Jersey.

On October 30, 1938, Orson Welles (1915–85) and his Mercury Theatre of the Air broadcast their very special radio adaptation of H. G. Wells's classic science fiction novel *The War of the Worlds* over the NBC radio network in the United States.

But instead of simply following the text of the original story, Welles's version was deliberately written to ape the style of a contemporary radio news program, as if Martians had actually landed in New Jersey and an alien invasion were underway. Although the broadcast was peppered with intermittent announcements that it was a drama rather than actual news, many listeners were apparently completely fooled. The broadcast was credited with sparking widespread panic about an unfolding alien invasion, although there is some dispute over the exact level of fear it actually engendered in the American public.

Welles's broadcast is still widely cited as a striking example of the power of drama to create havoc among an unsuspecting public. There have been various movie and comic book adaptations, video games, a television series, and a number of sequels based on *The War of the Worlds* over the years, but none of these packed the wallop of Welles's radio version.

survivors of a nuclear war. *Robinson Crusoe on Mars* (1964) draws on contemporary science to describe a survival story of astronauts stranded on Mars. *Capricorn One* (1978) concerns a fictional government conspiracy about a faked landing on Mars. The astronauts involved eventually expose the hoax, and embarrass the government. It is no coincidence that, soon after this film hit the theaters, the idea that NASA had faked the Moon landings in the 1960s and 1970s began to take off. It has remained a persistent belief held by a small percentage of the public worldwide ever since, despite the lack of evidence to support the notion.

Beyond these, *Total Recall* (1990) poses the question of whether or not we live a real existence, and if so, how do we know? Of course, not all Mars movies focus on conspiracies and war: there are actually a number of comedies involving the inhabitants of the red planet. While some of these, such as *Mars Attacks!* (1996), Tim Burton's film adaptation of the trading card series of the same name, focuses on slapstick and gruesomeness, a surprising number are actually aimed at younger viewers. *Santa Claus Conquers the Martians* (1965) presents an outrageous story of Santa Claus and two children defeating laughably inept Martians.

Beyond film and literature, Mars has also inspired a wide variety of toys, games, and memorabilia. The Red Planet has spawned lunch boxes, astronaut toys, action figures, and a host of other items. A number of toy companies, including Lego and Hot Wheels, went one stage further, creating toys that resemble actual Martian probes, landers, rovers, and research stations, enabling a generation of children to explore the Mars of their own imaginations.

"Santa, you will never return to Earth, you belong to Mars now."

KIMAR THE MARTIAN,
SANTA CLAUS CONQUERS THE MARTIANS

BELOW LEFT Hollywood's finest took on the Martian menace in Tim Burton's sci-fi comedy *Mars Attacks!*

BELOW RIGHT The American government attempts to fake a Mars landing on the set of the 1978 film *Capricorn One*.

Wernher von Braun and the Exploration of Mars

ABOVE LEFT Wernher von Braun and fellow rocket scientist Ernst Stuhlinger discuss the practicalities of a mission to Mars in Disney's *Tomorrowland* television series.

ABOVE RIGHT A concept model for an electric propulsion vehicle for a manned mission to Mars, thought to have been designed by Wernher von Braun's rocket group. It was transferred to the Smithsonian Institution's National Air and Space Museum in 1974.

A human mission to Mars has long been considered a grand, and as yet unclaimed, prize in space exploration. Like a mountain waiting to be climbed, it remains a tantalizing goal, and one that, in the early years of the space age, seemed to be slowly moving within humanity's reach.

One of the many explorers who coveted the chance to explore Mars was the rocket pioneer Wernher von Braun. He tried to sell the idea in a novel, *The Mars Project*, which he wrote in 1948. In his book, von Braun made detailed engineering diagrams and calculations to support the concept, and presented in it the first "technically comprehensive design" of both an interplanetary spacecraft and a full mission. Unfortunately, the novel was not a particularly engaging read, and it did not immediately find a publisher. Instead, von Braun used it as grist for his presentation at the First Symposium on Spaceflight held at the Hayden Planetarium in New York City in 1951. He published the technical information on a Mars mission in West Germany under the same title in 1952, with the University of Illinois Press publishing an English-language version in 1953.

Von Braun's concept for this Mars mission bore considerable resemblance to the large Antarctic expeditions of the late 1940s, especially Operation Highjump, carried out by the United States Navy from 1946 to 1947. Since Antarctica is so remote, explorers were cut off from the rest of the world, and had to take all the necessary equipment and supplies required for any eventuality with them. Mars is even more remote, and von Braun expected any expedition to the Red Planet to require a large, well-equipped crew to be successful.

Accordingly, von Braun proposed an "enormous scientific expedition," involving an armada of ten spacecraft, seven of which would be reserved for passengers and three for cargo. In total, he predicted a total of 70 crew members would be necessary. The armada would assemble in Earth orbit around a giant space station before departing on the nine-month flight to Mars. Upon arrival, von Braun predicted that the expedition would send advance gliders through the Martian atmosphere to land on the ice at one of the poles. The crew would then travel 4,000 miles (6,500 km) overland on crawlers toward Mars's equatorial region. They would then build a landing strip for others to come to the surface. Von Braun estimated that such an expedition would spend 443 days on the surface of Mars before returning to Earth.

As temptingly plausible as this concept may sound at first, most of von Braun's contemporaries recognized that it was impractical, and, in some ways, naïve. Von Braun was greatly underestimating the complexity of mounting an interplanetary exploration mission, conveniently overlooking such aspects as radiation beyond Earth's magnetosphere, and greatly underestimating the costs involved. Realizing the overambitious nature of his vision, in 1956 von Braun revised his plan downward to only two spacecraft and 12 crewmembers. This was essentially the approach he advocated in the 1957 television show he worked on for Walt Disney. While this broadcast contained many ideas that were fantastical, including some of von Braun's speculations on what bizarre flora and fauna might be found on the planet, his enthusiasm for an expedition to Mars was validated by its inclusion as a central objective for NASA's long-range plan for space exploration during the 1970s, announced in 1959.

> *"Our sun is one of 100 billion stars in our galaxy. Our galaxy is one of billions of galaxies populating the universe. It would be the height of presumption to think that we are the only living thing in that enormous immensity."*
>
> WERNHER VON BRAUN

BELOW LEFT This concept art from 1989 depicts a possible scene that the first human travelers to Mars might face when they walk out onto the surface of Mars.

BELOW RIGHT Werher von Braun's 1952 German-language publication, *Das Marsprojekt*, detailed the technical requirements of a mission to Mars.

The First Soviet Attempts to Reach Mars

"20 seconds, then silence."

TARIQ MALIK,
SPACE.COM MANAGING EDITOR,
ASSESSING THE SOVIET LANDING
OF MARS 2 IN 1971

The desire to explore Mars was not just an American dream in the 1950s. Even before its scientists and engineers had developed rockets powerful enough to escape Earth's orbit, many in the Soviet Union had already expressed a fervent desire to reach the Red Planet. In his 1923 novel *Aelita*, which was adapted into a movie of the same name in 1924 (p. 32), author Aleksey Tolstoy (1883–1945) depicted Mars as the site of a workers' rebellion in the mode of the Bolshevik Revolution. Around this time, Soviet rocketeer Friedrich Tsander (1887–1933) coined the slogan "Forward to Mars!" as a rallying cry for space exploration.

In 1959, the Dovzhenko Film Studio produced a movie entitled *The Sky Calls* about a fictional race to Mars between two superpowers, although the identity of the nations involved is not explicitly revealed. The film's narrative opens in the 1950s, with a reporter talking to scientists about the future of space exploration. Later, the reporter has a dream in which the Mars race unfolds, with the two rival teams of explorers eventually joining forces to reach their goal. Upon arrival on Mars, the explorers encounter monsters and survive various other hazards.

While the Soviet leadership didn't give its support to any human-led trip to Mars, it did approve robotic efforts to reach the Red Planet. But it was in undertaking this

ABOVE LEFT Friedrich Tsander was an early rocket experimenter who was more effective as an advocate for space travel than as an engineer.

ABOVE RIGHT The Rodina Motherland rocket from the 1959 movie *The Sky Calls*, taking cosmonauts from a space station on toward Mars.

challenge that Soviet spaceflight officials learned firsthand how difficult it could be to fly a mission to Mars. Between October 1960 and November 1964, Soviet engineers launched six missions to the Red Planet, but not a single one of them was successful.

The project fared better in the 1970s, as shown in the table below, both orbiting and then landing on the Red Planet. The first significant Soviet success on Mars came in 1971 with the Mars 2 orbiter, which launched on May 19, 1971; unfortunately, its lander crashed on the planet's surface in December of that year. Mars 3, launched on May 28, 1971, arrived at Mars on December 3, although it failed to enter the desired orbit. Regardless, the two orbiting probes sent back a total of 60 images, although unusually fierce dust storms on the surface meant that few details were visible.

On August 22, 1972, the Mars 3 orbiter was commanded to dispatch its lander to the surface. Although successful, data transmission from Mars 3 was lost 14 seconds after touchdown on the surface, leaving scientists frustrated with just one grainy and incomplete image of the planet.

This high-definition image of Mars was constructed from data returned in 1972 by the Soviet probes, Mars 2 and Mars 3.

Early Soviet Missions to Mars

#	Launch Date	Spacecraft	Launcher	Mission Goal	Results
1	October 10, 1960	1M No. 1 (Marsnik 1)	8K78/L1-4 (Molniya)	Mars flyby	Third-stage failure at T+300 seconds
2	October 14, 1960	1M No. 2 (Marsnik 2)	8K78/L1-5 (Molniya)	Mars flyby	Third-stage failure at T+290 seconds
3	October 24, 1962	2MV-4 No. 3 (Sputnik 22)	8K78/T103-15 (Molniya)	Mars flyby	Fourth-stage failure in Earth orbit
4	November 1, 1962	2MV-4 No. 4 (Mars 1)	8K78/T-103-16 (Molniya)	Mars flyby	Failed on its way to Mars
5	November 4, 1962	2MV-3 No. 1 (Sputnik 24)	8K78/T-103-17 (Molniya)	Mars landing	Stranded in Earth orbit
6	November 30, 1964	3MV-4 No. 2 (Zond 2)	8K78 (Molniya)	Mars impact	Failed on its way to Mars
7	March 27, 1969	M-69 No. 240, 521	UR-500	Mars orbit	Exploded at T+438 seconds
8	April 2, 1969	M-69, No. 522	UR-500	Mars orbit	Failed at T+0.02 seconds; destroyed
9	May 10, 1971	M-71 No. 170 (Kosmos 419)	UR-500	Mars orbit	Failed to leave Earth orbit
10	May 19, 1971	M-71 No. 171 (Mars 2)	UR-500	Mars orbit	Orbited Mars
11	May 28, 1971	M-71 No. 172 (Mars 3)	UR-500	Mars orbit, landing	Orbited Mars; ; lander dispatched August 22, 1972 but failed 14 seconds after touchdown and transmitted only one incomplete image
12	July 21, 1973	M-73 No. 52 (Mars 4)	UR-500	Mars orbit	Failed to orbit Mars
13	July 25, 1973	M-73 No. 53 (Mars 5)	UR-500	Mars orbit	Entered Mars orbit
14	August 5, 1973	M-73 No. 50 (Mars 6)	UR-500	Mars flyby, landing	Flew by Mars, landed capsule
15	August 9, 1973	M-73 No. 51 (Mars 7)	UR-500	Mars flyby, landing	Flew by Mars, capsule missed the planet

The Soviet Mars 3 space probe seen before launch on May 28, 1971.

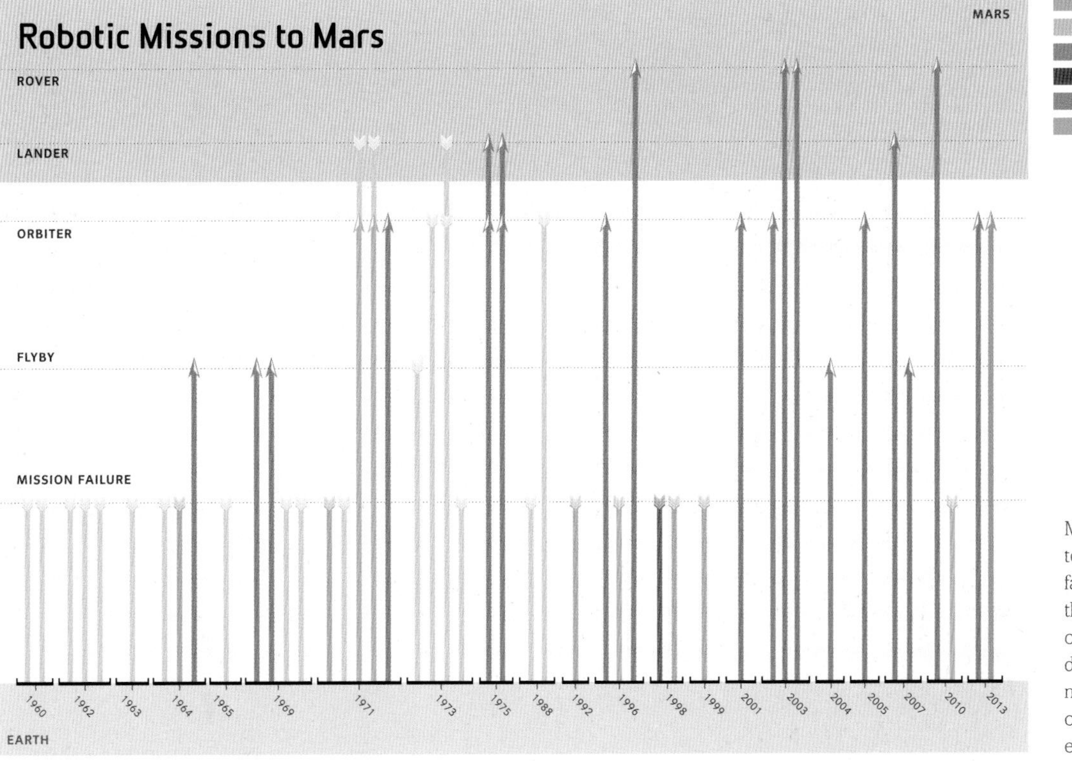

Mars's reputation as being difficult to explore has its roots in the failed Soviet attempts to reach the Red Planet in the 1960s, but other nations have faced similar difficulties when attempting Mars missions. This chart shows the overall success rate of all Mars exploration attempts to date.

Sergei Korolev's Plans for a Human Mission to Mars

1959–66

Could the Soviet space program have scored a further propaganda coup with a human mission to Mars?

In 1959, Sergei Korolev (1907–66), fresh from the successes of the Sputnik (p. 90) and Luna probe (p. 130) launches, persuaded Soviet leaders to support the development of robotic missions to Mars. Korolev's team began developing a heavy interplanetary spacecraft, designated TMK, to be flown on what became the failed Soviet N-1 Moon rocket (p. 138). Measuring over 403 feet (123 m) tall, 64 feet (19.6 m) in diameter, and weighing 75 tons (76 t), the TMK would carry a crew of three on a three-year journey to Mars. On reaching their destination, the crew would drop a robotic probe, the N-II, down to the surface.

On April 12, 1960, Korolev unveiled a modified version of this plan to the Kremlin, adding the prospect of a human landing mission with three to four piloted spacecraft flying in formation to the Red Planet. Although this initiative did not gain Kremlin approval, it was not the end of Soviet efforts to send humans to Mars. Planetary exploration further evolved with studies of a human mission to Mars gaining support within the Soviet spaceflight community. This led to engineering studies of a mission using a nuclear-powered spacecraft with a crew of six that would be assembled in Earth orbit out of elements launched by individual rockets. Once it reached Mars, the spacecraft would deploy a "train" of five structures, including a crew habitat with drilling equipment, a pod for exploration aircraft, two pods carrying main and backup return rockets, and a nuclear generator. The crew would spend a year on Mars, traveling the terrain, collecting samples, and relaying data to an orbiting spacecraft overhead.

Despite all these plans, the Kremlin refused to approve any human mission to Mars, instead giving the go-ahead to the attempt to beat the Americans in their already-announced Moon landing mission (p. 148).

Early Mariner Missions to Mars

ABOVE Engineers assemble a mock-up of Mariner 2 by in the Arts and Industries Building of the Smithsonian in 1962.

OPPOSITE Too anxious to wait for the first image of Mars taken by Mariner 4 to download, staff at NASA's Jet Propulsion Laboratory converted the raw data from the probe into this "paint by numbers" artwork, which they hand-colored while the computer processed the digital version of the image.

Like the early Soviet missions to Mars, NASA also had its share of problems in its initial efforts to reach the Red Planet. In 1964, the organization launched two probes to Mars, but only one, Mariner 4, successfully reached its destination, arriving in the vicinity of the planet on July 15, 1965, flying within 6,118 miles (9,846 km) of its surface.

By April 1965, Mariner 4 had broken both the record of continuous space probe operation (129 days, previously set by Mariner 2 around Venus), and the record for long-distance radio communications of 62 million miles (100 million km), previously set by Zond-1 around Venus. En route to Mars, Mariner 4 also detected the effects of solar flares, and measured cosmic radiation.

By July 7, the probe had entered the gravitational field of Mars, and four days later it was only one million miles (1.6 million km) away. By July 14, NASA's JPL had sent the signal to aim the cameras at the Martian surface for the rapid flyby. Mariner 4 began capturing images on magnetic tape for later transmission to Earth. It was not until July 15 that JPL received the first image from the probe, transmitted at only eight bits per second. Since each image had a size of 250,000 bits, it took nine days before the full set of 21 images arrived. These photographs, showing a cratered, Moon-like surface, dashed the hopes of many that life might be present on the planet. The images depicted a planet without structures or canals, and nothing that even remotely resembled evidence of intelligent life.

After capturing its photographs, Mariner 4 entered cruise mode, and activated other science instruments. One of the greatest achievements of the mission was the ability of flight controllers to modify their plans while the spacecraft was en route to Mars, something NASA would end up doing regularly on subsequent missions. For example, after launch, the probe's flight path was redesigned to fly behind Mars, allowing NASA's scientists to measure the effects of the Martian ionosphere and atmosphere on the spacecraft's radio signals. This experiment indicated surface pressures on Mars of somewhere between 4.1 and 7.0 millibars. This suggested that the Martian atmosphere was much thinner than previously believed. It also enabled improvements in the calculation of the planet's mass. Mariner 4's magnetometer also revealed that Mars had no magnetic field equivalent to that of the Earth.

Based on this information, a weekly magazine, *U.S. News and World Report*, announced definitively in its August 9, 1965 issue that "Mars is dead." This led NASA scientists to look for evidence that simple life had once existed on Mars.

"The Mars we had found was just a big moon with a thin atmosphere and no life. There were no Martians, no canals, no water, no plants, no surface characteristics that even faintly resembled Earth's."

BRUCE MURRAY,
PLANETARY SCIENTIST AND
DIRECTOR OF NASA'S JPL

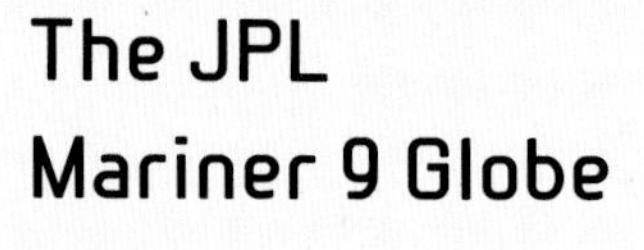

The JPL Mariner 9 Globe

1971–72

Providing a global view of the Martian surface that guided the planning of NASA's future Mars missions.

ABOVE LEFT One of the photomosaic Mars Globes based on the photographs produced by Mariner 9 at the Smithsonian Institution's National Air and Space Museum.

ABOVE RIGHT This reprocessed Mariner 9 image shows Olympus Mons, an extinct shield volcano 374 miles (624 km) in diameter, and 16 miles (25 km) high.

Mariner 9's extended Mars mapping mission revolutionized scientists' understanding of the planet. Arriving in November 1971, the probe immediately encountered a huge surprise: the entire planet was engulfed in a dust storm. What's more, some features appeared to be protruding above the plumes.

When the dust settled on the surface, scientists identified that the features the probe had initially spotted were the tops of dormant volcanoes. Mariner 9 also discovered a huge rift across the surface of the planet that was later named the Mariner Valley after the spacecraft that discovered it.

The probe spent nearly a year orbiting Mars, and during that time returned a total of 7,329 photos. Those images were pieced together by JPL scientists, processed through a computer, and assembled into high-resolution photomosaic globes that showed subtle details of the planet's surface, such as the relics of ancient riverbeds, and gullies in the now-desolate landscape.

JPL's Elmer Christensen supervised the construction of five of these globes. In the process, he enhanced surface details to simulate the ridges and craters, and the shading on the planet's dark regions and its bright polar caps.

Completed in September 1973, these globes showed Mars's large-scale geological features with a perspective that could not be obtained through the flat maps previously produced. The globes went on to become a primary source for planning on the various NASA Mars missions that were to follow.

The next time that Mars's orbit brought it close to Earth, in 1969, NASA launched two more probes, Mariner 6 and Mariner 7, within a month of each other. Each achieved its objectives, reaching Mars in August 1969. The pair sent back data that revealed further information about the Martian atmosphere and surface, but their pictures confirmed the planet's Moon-like appearance, and provided no hint of evidence that Mars had ever been able to support life. However, these probes did lay the groundwork for a future NASA mission to land a spacecraft on the planet.

Before that lander mission. NASA launched a further two orbital probes to Mars. While Mariner 8's launch vehicle failed, Mariner 9 reached its destination, and provided significant data as it mapped the planet's surface. Mariner 9's photos identified that volcanoes had once been active on the planet, that the frost observed seasonally on the poles was made of carbon dioxide, and that considerable geological activity had occurred at some point in the planet's past. Suddenly, Mars fascinated scientists, reporters, and the public once again, largely because these new findings hinted that some kind of life really might have once existed there.

Dual Television Camera System

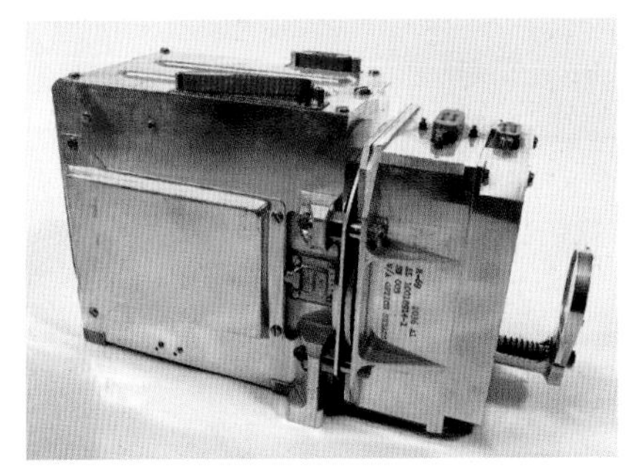

This dual television camera system is identical to those flown to Mars on Mariner 6 and Mariner 7 in 1969. The device was capable of capturing both wide-angle images in medium resolution and narrow-angle images in high resolution, and was integral to the scientific instrumentation package of both probes.

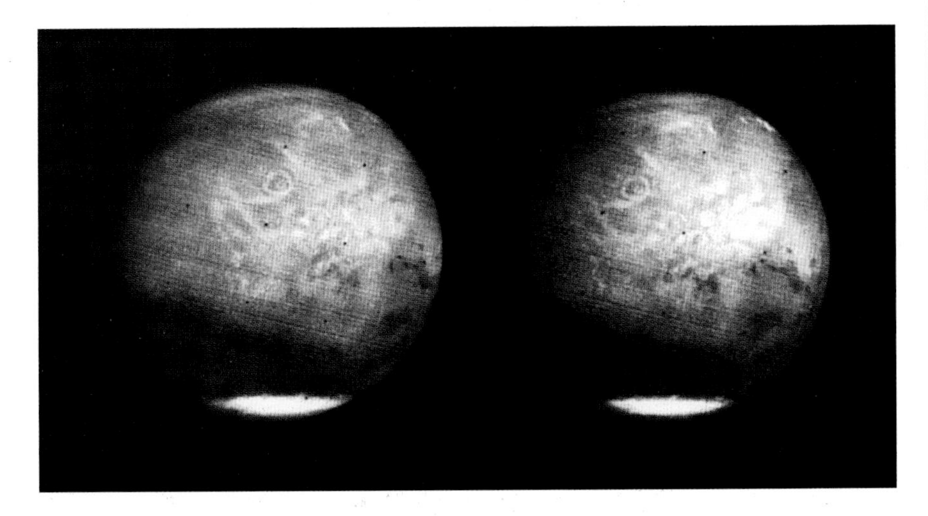

Views of Mars from Mariner 7, showing Olympus Mons and the southern polar cap in 1969.

Early NASA Missions to Mars

#	Launch Date	Spacecraft	Launcher	Mission Goal	Results
1	November 5, 1964	Mariner 3	Atlas Agena D	Mars flyby	Failure of the payload fairing to separate
2	November 28, 1964	Mariner 4	Atlas Agena D	Mars flyby	First flyby of Mars and transmission of images
3	February 25, 1969	Mariner 6	Atlas Centaur	Mars flyby	Transmitted 75 photos of Mars
4	March 27, 1969	Mariner 7	Atlas Centaur	Mars flyby	Transmitted 126 photos of Mars
5	May 8, 1971	Mariner 8	Atlas Centaur	Mars orbit	Launch failure
6	May 30, 1971	Mariner 9	Atlas Centaur	Mars orbit	Orbited Mars; transmitted numerous images

Vikings 1 and 2 Reach Mars

ABOVE LEFT The boulder-strewn field of red rocks reaches to the horizon nearly 2 miles (3.2 km) from Viking 2 in the Nowhere Plains region of Mars.

ABOVE RIGHT A duplicate of the two Viking landers, used by NASA scientists and engineers to model how the craft would respond to various radio commands.

Perhaps the most audacious and exciting mission to Mars was Project Viking, conducted by NASA between 1975 and 1982. Consisting of twin orbiters, each with its own lander, the mission was designed to build the experience and knowledge gained from the previous Mariner missions, and to answer definitively whether or not some kind of simple life might exist on the Red Planet. Launched on August 20, 1975, from the Kennedy Space Center, Florida, Viking 1 spent nearly a year cruising to Mars before reaching orbit. The Viking 1 lander touched down on July 20, 1976, on the Golden Plain region of Mars. On landing, the probe immediately took its first photographic image of its surrounds, and began transmitting it back to Earth. In contrast to the Mariner 4, the first black-and-white images took just 4 minutes to arrive at NASA's control center, with the first panoramic image taking a further 7 minutes. The Viking 2 lander followed in September 1976, landing about 120 miles (200 km) west of the crater Mie in the Nowhere Plains. Although earlier spacecraft had already landed on Mars, the Viking landers were the first probes to

"The Viking lander on Mars is a toy grown to large size, a metaphor of a dream; that dream of extending our will, our hand, our seeing eye to another world. It is not a machine, it is us."

RAY BRADBURY,
AMERICAN SCIENCE FICTION WRITER

The rocket carrying the *Viking 1* orbiter and lander lifts off from Florida on an 11-month journey to Mars on August 20, 1975.

The Fate of the Face

Mars Global Surveyor (MGS) took this image of the so-called "Face on Mars" on April 8, 2001. Using its higher resolution camera, the MGS demonstrated that the "face" was merely an optical illusion created by nothing more than light and shadow on a Martian hilltop.

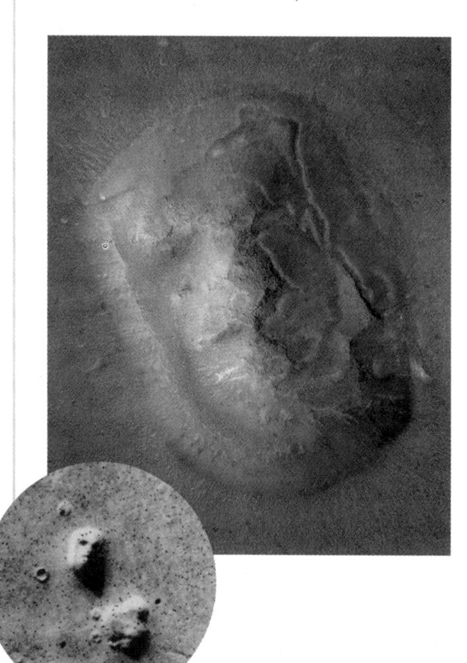

BELOW The first panoramic view of the surface of Mars taken by Viking 1. The horizon features are approximately 1.8 miles (3 km) away.

operate on another planet for an extended period. The two landers continuously monitored the weather around them, and found both exciting, cyclical variations, and an exceptionally harsh climate that seemed to preclude any possibility of life. The atmospheric temperature at the more southern Viking 1 landing site, for instance, was only as high as +7°F (-13°C) at midday, but the predawn summer temperature was -161°F (-107°C). The lowest predawn temperature recorded was -184°F (-120°C), about the point at which carbon dioxide begins to frost.

Three biology experiments contained on the landers discovered unexpected and enigmatic chemical activity in the Martian soil, but they provided no clear evidence for the presence of living microorganisms, at least not near the landing sites. This led some NASA scientists to conclude that Mars could be self-sterilizing, with the combination of strong ultraviolet radiation from the Sun that saturates the surface, the extreme dryness of the soil, and the oxidizing nature of its chemistry preventing the formation of even the simplest microorganisms.

In part because of this null result, NASA did not send another probe to Mars for two decades. The Viking program's chief scientist, Gerald Soffen (1926–2000), commented in 1992: "If somebody back then had given me 100 to 1 odds that we wouldn't go back to Mars for 17 years, I would've said, 'You're crazy'." But NASA and other national space agencies would eventually return.

One image in particular from the Viking mission captured the imagination of the public. On July 25, 1976, the Viking 1 orbiter took a photo of the Cydonia region of Mars that looked like a human face. Even at the time it was first released to the public, all evidence suggested that the effect was the result of shadows on a hilltop, as Soffen himself explained at a press conference, but that did not convince some members of the public.

Belief in the Martian "face" persisted for decades. As a NASA official stated in 2001: "The 'Face on Mars' has since become a pop icon. It has starred in a Hollywood film, appeared in books, magazines, radio talk shows—even haunted grocery store checkout lines for 25 years! Some people think the Face is bona fide evidence of life on Mars—evidence that NASA would rather hide, say conspiracy theorists. Meanwhile, defenders of the NASA budget wish there was an ancient civilization on Mars."

Carl Sagan and Mars Exploration

The NASA scientist who became the public face of space exploration for a generation.

Carl Sagan (1934–96) played a leading role in Mars exploration from its earliest years. A trained scientist, Sagan worked with NASA from its very beginnings in the late 1950s, taught Apollo astronauts about lunar science, and labored as a scientist on the Mariner, Viking, Voyager, and Galileo missions to the planets. He helped to solve the mysteries of the high temperature of Venus, which he suggested were caused by the insulating effect of its thick and cloudy atmosphere, and the seasonal changes on Mars, caused by windblown dust. Fittingly, Asteroid 2709 Sagan was named for him. As an activist, he was a cofounder of the Planetary Society, an important advocacy group for space exploration.

As a Pulitzer Prize-winning author, Sagan wrote many commercially successful works, including *Cosmos*, which became the best-selling science book published in English up to that point. This work accompanied his award-winning 13-part television documentary series, *Cosmos: A Personal Voyage*, seen by 500 million people in 60 countries. He also coproduced and cowrote the script for the film *Contact*, which was adapted from his 1985 science fiction novel of the same name. Both the book and the film dealt with humanity's encounter with extraterrestrial life.

Sagan fully believed that the Viking program would find evidence of life on Mars. He trumpeted this possibility in his statements before the landings in 1976 to such an extent that some of his NASA colleagues believed he was raising public expectations too high. Planetary scientist and JPL director Bruce Murray complained that Sagan's over-optimism could dash the hopes of the public if Viking found no evidence of life. While Murray's assessment arguably proved correct, no one doubted the powerful intellect, striking charm, and enormous charisma that enabled Sagan to become the public face of space science and exploration.

Carl Sagan with a Viking lander mock-up in Death Valley, California, on October 26, 1980.

Mars Hiatus

"I believe the biggest impediment we have right now with going to Mars is public commitment. More people need to see themselves as a part of space travel; we need to see more inclusiveness."

MAE JEMISON,
NASA ASTRONAUT

It was not a hiatus that was planned but, after the remarkably successful Viking program, no further probes reached Mars for more than two decades. And it was not for lack of trying. One NASA wit offered the explanation that, after the landing by the two Viking probes in 1976, the Martians had developed a very effective planetary defense system, and it was now knocking out everything that came near the Red Planet.

The Soviet Union made two attempts to reach Phobos, one of the moons of Mars, in 1988, just a year before the collapse of the Soviet Union and the end of the Cold War. Both missions failed. Phobos 1, launched on July 7, 1988, was a Mars orbiter with a small lander intended to touch down on Phobos, but was lost in August 1988 en route to Mars. Phobos 2 was launched on July 12, 1988 and was lost in March 1989, near Phobos. These two failures set back later Russian exploration of Mars. The Russian space program attempted only one further flight in 1996, Mars 96, but this failed even to get out of Earth orbit. As the Russian economy struggled in the 1990s, its space program focused on continuing to fund the Mir space station, and Russia's human spaceflight capability.

NASA fared little better. Its Mars Observer, launched on September 25, 1992, was intended to orbit Mars and collect the most detailed data since the Viking missions of the mid-1970s. The mission progressed smoothly until August 21, 1993, three days before the spacecraft was expected to enter orbit around Mars. Suddenly, and without warning, controllers lost contact with the over $800-million spacecraft. Mars Observer was never heard from again. The best estimate for what happened, based on the mission's flight telemetry and analysis, was that when mission controllers contacted the spacecraft three days before its Mars encounter and directed it to correct course, a failure in the fuel line caused an explosion. Sadly, we will never know for sure.

The loss of Mars Observer led NASA to restructure its Mars exploration program. NASA Administrator Daniel Goldin swore that he would not countenance another mission in which scientists tried to do too much at one time. He recognized that attempting to pack a single probe with a greater number of instruments was akin to putting all the organization's science eggs into one complex basket in which there are too many potential points of failure. His more cost-effective and risk-efficient alternative was to build smaller, less-expensive probes with more limited objectives. The phrase "faster, better, cheaper" emerged from this restructuring, and the result was a series of missions that limited the costs of its failures as it explored the Red Planet. Some were successful, such as Mars Pathfinder in 1997, but Mars Climate Orbiter and Mars Polar Lander, launched in 1998 and 1999 respectively, failed, and proved an embarrassment to NASA.

Overall, the success rate for Mars missions is surprisingly low, with roughly twice as many failures or partial successes as fully successful missions flown. Fortunately, it appears that space scientists have continued to learn from these failures, and the number of successful expeditions to the Red Planet has continued to increase over time. It seems that the Martian planetary defense system, once suggested in jest, is finally beginning to fall to science.

Mars Missions Successes, 1960–2017

Nation/Agency	Mission Goal	Successful Missions	Partially Successful Missions	Unsuccessful Missions
United States	Mars flyby	14	0	5
Japan	Mars flyby	0	0	1
India	Mars flyby	1	0	0
ESA	Mars flyby	0	2	0
Soviet Union/ Russia	Mars orbit	1	5	15
Total	Mars orbit	16	7	21

The Failed Observer

Seen here being prepared for integration with its launch vehicle at NASA's Payload Hazardous Servicing Facility, Mars Observer was designed to study the Martian surface, atmosphere, and climate, as well as the planet's magnetic field. In addition to these extensive mission goals, there was even provision for the probe to aid data recovery from Russia's planned Mars 96 mission.

OPPOSITE TOP A model of the Soviet Union's twin probes, Phobos 1 and Phobos 2. They were launched in July 1988 to study Mars and its moons, Phobos and Deimos, but failed en route.

OPPOSITE BOTTOM The Titan III rocket launches Mars Observer from the Cape Canaveral launch facility, Florida, on September 25, 1992.

Mars Pathfinder and the Renaissance in Mars Exploration

Advocates of Mars exploration rejoiced in 1997, when Mars Pathfinder, and its tiny rover, Sojourner, became the first NASA probes to set down on the Red Planet's surface since the Viking 2 lander, more than 20 years previously. Spurred by interest in life that can tolerate and thrive in extremely harsh conditions, such as those found in the depths of the oceans on Earth, and by previous data suggesting that life might have once existed on Mars, the Pathfinder probe became an urgent priority for NASA. Launched in December 1996, it reached the Martian surface on July 4, 1997.

This view of Mars Pathfinder's landing site reveals traces of a warmer, wetter past, showing a floodplain covered with a variety of rocks showing signs of erosion.

Nearly everything about Pathfinder generated public excitement. As the craft descended through the Martian atmosphere, it did not use retro-rockets and parachutes, as the Viking mission had done, but inflatable airbags that allowed the probe to bounce along the surface until coming to rest. As the airbags deflated, three solar panels unfolded, revealing the probe's most innovative component: the small, 23-pound (10-kg) solar-powered robotic rover, Sojourner.

Pathfinder's solar panels provided an impromptu ramp for Sojourner to venture out onto the surface of Mars. The rover then proceeded to take close-up images of the surface, using its two front-mounted color cameras and single rear-mounted black-and-white camera. The rover also contained a rear-mounted alpha-particle x-ray spectrometer that provided basic data on the chemical composition of the soil and rocks it traveled over. Sojourner recorded important new data about rocks washed down into the Mars Valley floodplain, an ancient outflow channel in Mars's northern hemisphere. Projected to operate for 30 days, the rover worked for nearly three months, capturing far more data on the atmosphere, weather, and geology of Mars than anyone at NASA had expected. In all, Pathfinder returned more than 1.2 gigabits of data, and more than 10,000 pictures of the Martian surface.

With its semi-autonomous behaviors, the plucky Sojourner rover soon became a favorite of the public, who followed the mission with great interest on the newly emerging World Wide Web. Websites detailing the events of the mission recorded 565 million hits worldwide between July 1 and August 4, 1997. At one point, 47 million people were logged attempting to access data on the mission.

Martian Meteorite ALH84001

1996

Does this lump of Mars rock provide evidence that life existed on Mars long ago?

In August 1996, a team of scientists from NASA and Stanford University announced that a Mars meteorite found in Antarctica contained possible evidence of ancient Martian life. The scientists hypothesized that the 4-pound (1.8-kg) potato-sized rock, to which they gave the less-than-catchy name ALH84001, formed as an igneous rock about 4.5 billion years ago, when Mars was much warmer, and probably possessed oceans hospitable to life. Then, around 15 million years ago, a large asteroid hit the Red Planet, forcing a sizable amount of debris, including the rock, out into space, where it remained until crashing into Antarctica sometime around 11,000 BCE.

In the summer of 1996, a NASA-funded research team announced in the journal *Science* that it had identified evidence of organic molecules in this Martian meteorite found in Antarctica, which suggested that primitive life may have existed on Mars in its early history. This discovery stimulated enthusiasm at NASA and across the world for a new Mars exploration mission, which had already been in the planning stages prior to the meteorite study in *Science*. NASA went on to form a multidisciplinary group of scientists to develop strategies for searching for signs of life on future Mars missions.

While some scientists disputed the significance of the ALH84001 findings, the paper nevertheless succeeded in reviving scientific interest in the question of life on Mars, and galvanized NASA's efforts to pursue further missions to the Red Planet.

TOP The 4.5 billion-year-old rock dubbed meteorite ALH84001.

ABOVE A high-resolution image from an electron microscope showing an unusual, carbonite, tube-like structure in meteorite ALH8400. This is believed by some scientists to be evidence that Mars once harbored life.

This image is a composite that creatively combines two images from the Mars Pathfinder mission to create a sunset scene in Mars Valley in July 1997.

Follow the Water

"Mars once was wet and fertile. It's now bone dry. Something bad happened on Mars. I want to know what happened on Mars so that we may prevent it from happening here on Earth."

NEIL DEGRASSE TYSON

Mars Probe Specifications

Mars Global Surveyor

Launch Date November 7, 1996

Arrival at Mars September 12, 1997

Launch Mass 2,272 pounds (1,030 kg)

Dry Mass 2,337 pounds (1,060 kg)

Payload Mass 171 pounds (78 kg)

Power 980 watts

Status Ceased operating November 2, 2006

Mars Reconnaissance Orbiter

Launch Date August 12, 2005

Arrival at Mars March 10, 2006

Launch Mass 4,810 pounds (2,180 kg)

Dry Mass 2,273 pounds (1,031 kg)

Payload Mass 306 pounds (39 kg)

Power 2,000 watts

Status Still operating

Following on from the success of its Mars Pathfinder probe in identifying evidence of water, as well as of geological activity in Mars's ancient past, scientists at NASA began to shift the focus of their Mars exploration efforts.

Working on the notion that all life on Earth was first formed in liquid water, and that any similar life elsewhere would probably have similar chemistry requirements, NASA's scientists surmised that in order to find evidence of life on Mars, past or present, they would need to follow the water. There was already some evidence that water had been present on the planet in the past; the next question was if any remained there currently.

The next NASA probe to reach the Red Planet was the Mars Global Surveyor (MGS), which entered a Martian orbit in 1998, and began to map the planet's surface in greater detail. Using MGS images, scientists identified more than 150 geographic features from across the planet's surface that were probably created by fast-flowing water. These results led Michael Malin (1950–), the space scientist who designed the probe's main camera, to suggest that there might still be some water on Mars, hidden beneath the Martian soil, deep within the planet's substrata.

Further images from MGS and NASA's subsequent probe, Mars Reconnaissance Orbiter (MRO), showed further evidence of dry riverbeds, floodplains, and gullies on Martian cliffs and crater walls, as well as sedimentary deposits that suggested the presence of water flowing on the surface of Mars, which may even have supported simple microbial life at some earlier point in its history when it was wetter, and possibly warmer.

But perhaps more intriguing is the possibility that simple life forms might still be living on Mars today, lying beneath the planet's polar caps or in subterranean hot springs warmed by vents from the Martian core. Discoveries from Mars Pathfinder and other spacecraft at the Red Planet suggested as much. These might be Martian equivalents of the single-celled microbes that dwell in Earth's bedrock. Most Mars space scientists would be quick to add, however, that these are unproven theories that currently lack any evidential support. The only way for scientists to find out for sure will be to continue the strategy of "follow the water." If the evidence exists of life on Mars, finding it is just a matter of time and continued exploration.

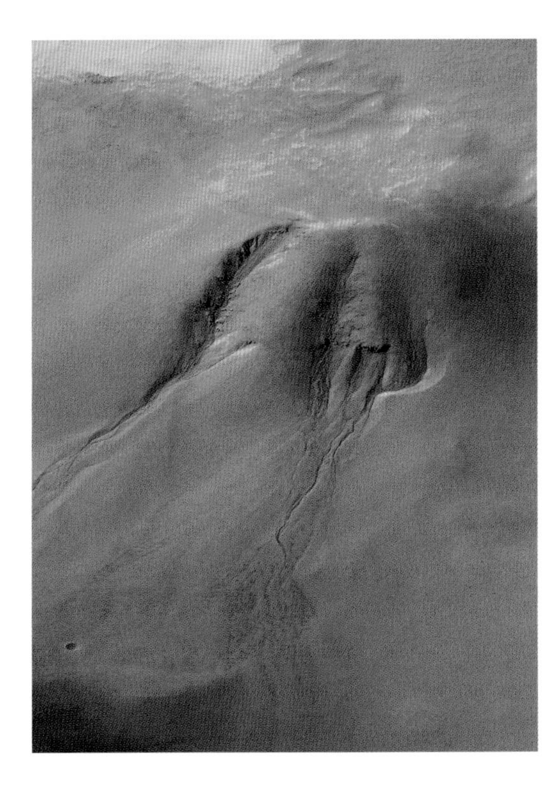

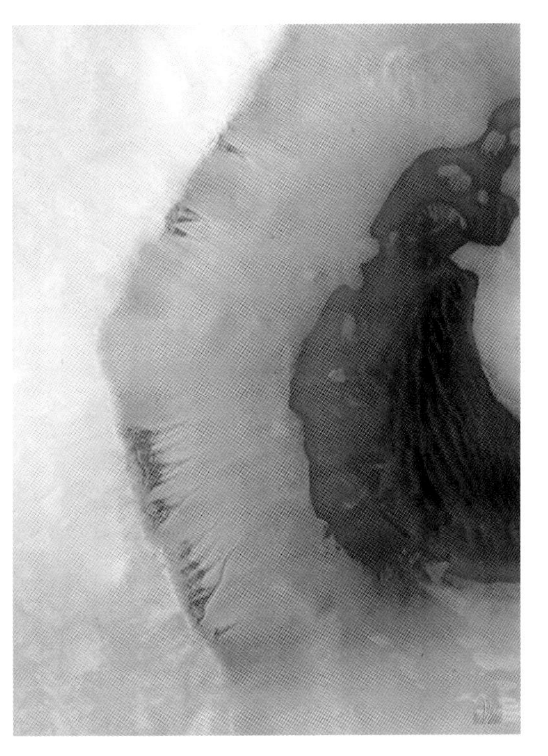

FAR LEFT The channeled aprons on this small feature within Newton Crater on Mars suggest the action of liquid water in the planet's geological history.

LEFT Taken on June 15, 1998 by the Mars Global Surveyor, this image appears to show groundwater seepage and surface runoff on Mars. Some scientists have theorized that the location and young age of the gullies depicted suggest the possibility that subsurface liquid water still exists on the Red Planet.

Surveying and Reconnoitering the Red Planet 1996–

The orbiting probes that demonstrated NASA's commitment to finding evidence of water on Mars.

NASA's Mars Global Surveyor (MGS) and Mars Reconnaissance Orbiter (MRO) probes both made significant contributions to the search for evidence of possible life on Mars. Launched in November 1996, MGS's primary goal was its global mapping mission, but, after this was completed in January 2001, it was assigned a variety of other tasks, including working as a relay platform for communication between Earth and the various NASA lander missions that made it to Mars during its long service life. Its extended mission lasted until November 2006, when NASA lost contact with the spacecraft.

Launched in August 2005 with a more sophisticated instrument package than its predecessor, MRO started its operations on Mars on March 10, 2006. Upon arrival, it joined five other probes exploring the planet, either in orbit or on the planet's surface. At the time, this set a record for the most operational spacecraft at Mars. As of November 2017, MRO was fully operational and continuing in its search for water.

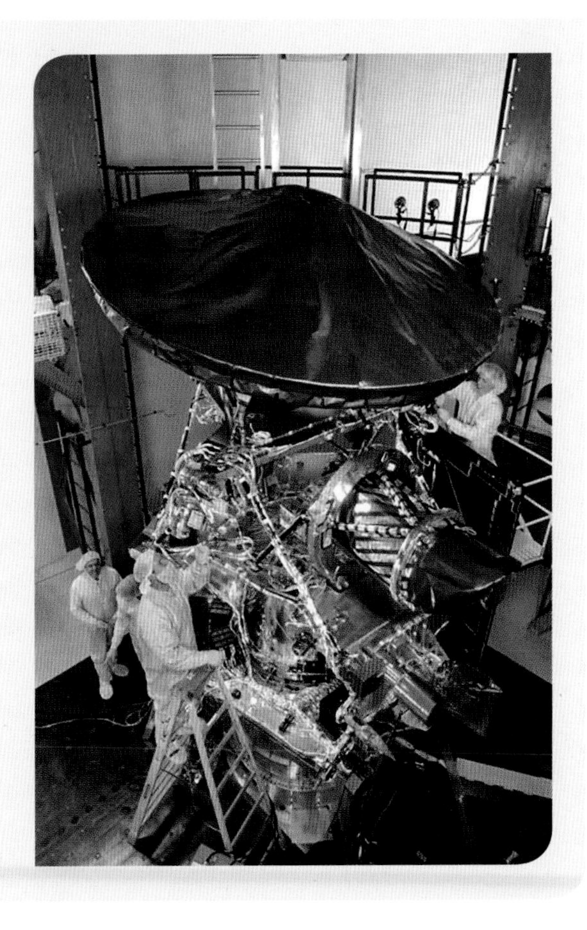

Mars Reconnaissance Orbiter, pictured here in early January 2005 undergoing environmental testing to ensure its performance in its ongoing mission.

Spirit and Opportunity

An early piece of concept art showing NASA's Mars exploration rovers, Spirit and Opportunity, on the surface of Mars.

Even before they reached their destination in January 2004, the Mars Exploration Rovers (MERs), Spirit and Opportunity, captured the adoration of people across the globe. Originally designated simply as MER-1 and MER-2 when they were designed and built, the rovers were renamed as part of a student essay competition. The winning entry was penned by Sofi Collis, a third-grade Russian-American student from Arizona, who wrote, "I used to live in an orphanage. It was dark and cold and lonely. At night, I looked up at the sparkly sky and felt better. I dreamed I could fly there. In America, I can make all my dreams come true…Thank you for the 'Spirit' and the 'Opportunity.'"

The initial mission for the two rovers was designed to last just 90 days, during which time they would search for evidence of the effects of water in shaping Martian geology. In the end, the two hardy rovers wound up massively exceeding this service life, remaining operational for years after their arrival.

Despite their different names, Spirit and Opportunity were completely identical in design. Each weighed around 396 pounds (180 kg), and was armed with the same suite of five scientific instruments: a panoramic camera, a miniature thermal emission spectrometer, a Mössbauer spectrometer, an alpha-particle X-ray spectrometer, and a microscopic imager. Each also carried a rock abrasion tool for grinding weathered rock surfaces to expose the interior, and a dexterous robotic arm with its own microscopic camera that enabled the rovers to manipulate and view the rock samples.

Where they differed was in their points of departure on Mars. Spirit landed in the middle of the massive Gusev Crater on January 4, 2004, while Opportunity landed on the other side of the planet in its Southern Plains region three weeks later. After their airbag-protected landing, each rover settled onto the Martian surface, rolling out to take panoramic images of the surroundings. These first images gave mission scientists the information they needed to select promising geological features for the rovers to examine. Once the targets had been chosen, the rovers then drove to the designated destination to investigate the geology.

Spirit explored the Gusev Crater and its basalt base that revealed little evidence of past water on Mars. From there, it traveled to a feature named Columbia Hills in honor of the astronauts who died in the 2003 disintegration of Space Shuttle Columbia. There, the rover found a variety of rocks indicating that early Mars was

Mars Exploration Rover Specifications

Operator
NASA Jet Propulsion Laboratory

Launched
Spirit: June 10, 2003, on Delta II rocket from Cape Canaveral, Florida

Opportunity: July 7, 2003, on Delta II rocket from Cape Canaveral, Florida

Landed
Spirit: Gusev Crater, January 3, 2004

Opportunity: Southern Plains, January 24, 2004

Rover Dimensions
5 feet (1.6 m) long, 5 feet (1.5 m) tall

Total Launch Mass
2,343 pounds (1,063 kg)

Rover Mass
408 pounds (185 kg)

Rover Payload
Panoramic camera, miniature thermal emission spectrometer, Mössbauer spectrometer, alpha-particle X-ray spectrometer, magnets, microscopic imager, rock abrasion tool

Rover Propulsion
Cleated wheels each driven by individual electric motors

Rover Electrical Power
Max 140 W from solar array

Heating Element
Radioisotope Heating Units

Prime Mission
90 days/0.75 miles (1 km)

Status
Spirit: operated for 2,208 sols (Martian days), lost contact March 22, 2010

Opportunity: Still operational

NASA engineers complete the assembly and testing of Spirit and Opportunity at JPL on February 10, 2003.

Between October 26 and 29, 2007, Spirit took the images to assemble this rover selfie. As is evident in the composite image, Spirit's solar panels were covered in so much dust that the rover almost blended into the Martian surface. The dust on the solar panels reduced the amount of electrical power the rover could generate from the Sun, which impaired its functionality until NASA eventually lost contact with the vehicle.

> *"We have concluded that the rocks here were once soaked in liquid water. It changed their texture, and it changed their chemistry. We've been able to read the telltale clues the water left behind, giving us confidence in that conclusion."*
>
> STEVEN SQUYRES, SCIENCE TEAM LEADER FOR THE MARS EXPLORATION ROVERS, MARCH 2, 2004

characterized by meteorite impacts, explosive volcanism, and abundant subsurface water. Unusual-looking bright patches of soil turned out to be extremely salty and affected by past water. At a nearby circular feature dubbed Home Plate, Spirit discovered finely layered rock, and even took the opportunity to gaze upward and observe the Martian moons, Phobos and Deimos.

By contrast, Opportunity landed close to a thin outcrop of rock that lent itself to an analysis confirming that a body of salty water once flowed gently over the area. The rover also tested sands that had become petrified by water and wind action, and soaked by groundwater. It continued to examine more sedimentary bedrock exposures along its route, where an even broader, deeper section of layered rock revealed new aspects of Martian geological history. Scientific analysis on Earth pointed to a past environment that could have been hospitable to life, and also fossil preservation, if life capable of such preservation ever existed on the planet.

Over the course of their operations, the two rovers drove more than seven times farther than originally planned. Spirit covered 3.4 miles (5.5 km) and, by February 6, 2007, Opportunity had almost doubled that distance, having traveled almost 6.2 miles (10 km). Along the way, both contended with hills and craters, escaped sand traps, and overcame numerous technical problems to continue their mission.

Setting a Roving Record

2004–

In addition to exploring the planet, one NASA rover literally broke new ground on the Martian surface.

While the Spirit rover ceased operations in 2010, as of 2017 Opportunity remained fully mobile, as it continued its scientific endeavors. As Mars Exploration Rover Project Manager John Callas notes: "Opportunity has driven farther than any other wheeled vehicle on another world... This is so remarkable considering Opportunity was intended to drive about 1 km, and was never designed for distance. But what is really important is not how many miles the rover has racked up, but how much exploration and discovery we have accomplished over that distance."

A drive of 157 feet (48 m) on July 27, 2014, put Opportunity's total distance traveled at 25.01 miles (40.25 km). Before Opportunity, the record for distance traveled by a space exploration rover was held by the Soviet Lunokhod 2 (p. 161), which landed on the Moon on January 15, 1973, and drove about 24 miles (39 km) in less than five months. With the massive head start it has already amassed, and the fact that it is still going strong to this day, it seems that the indefatigable little Mars rover has established a record that will prove difficult to beat.

Opportunity made this curving 52-foot (15.8-m) drive while testing a new piece of navigational software named Field D-star. The software, which was part of an upgrade package sent to the rovers in 2006, enabled them to plan the optimal long-range route around any obstacles to their designated destination.

International Probes to Mars

ABOVE LEFT This image, taken by the High Resolution Stereo Camera (HRSC) on ESA's Mars Express probe, shows the main channel of Kasei Valley, which is thought to have been formed by a series of gigantic flood events.

ABOVE RIGHT Engineers preparing India's Mangalyaan probe for launch in October 2013. It reached Mars in 2014, and is projected to be in service until 2024.

Why is it so hard to conduct a successful mission to Mars? That is a question with many possible answers, none of which really explains why so many probes bound for the Red Planet have failed before they were able to relay any meaningful data back to Earth. Problems and mishaps have afflicted Mars missions undertaken by a wide variety of space agencies.

In 2003, the Japanese probe Nozomi suffered a fatal failure in its electrical systems. The ESA successfully reached Mars in 2004 with its Mars Express probe, but its lander component, the British Beagle 2 spacecraft, failed to make contact with mission controllers from the Martian surface.

Mars Express was the first probe to reach the Red Planet that was not launched by either the United States or the Soviet Union/Russia. This signaled an important transformation in Mars exploration. Since then, six additional international missions have sent spacecraft to Mars, and of those, only two have ended in complete failure, with the other four either completely or partially successful in their missions. As of 2017, ESA's Mars Express and the ExoMars Trace Gas Orbiter, as well as the Mangalyaan orbiter operated by the Indian Space Research Organisation (ISRO), are all still actively exploring Mars. Further international cooperative missions are in the planning stage.

The Saga of Mars Express

2003–

ESA's first mission to the Red Planet has met with its own share of successes and failures.

Mars Express represented one of the most stunning Mars missions of the last 20 years. The mission had two major components. The first was an orbiter that reached the Red Planet in 2003 and explored Mars using its high-resolution cameras and powerful radar capable of sounding the subsurface structure of Mars down to its subterranean permafrost deposits. The spacecraft also completed an analysis of the planet's atmospheric composition and global circulation, as well as the relationship between Mars, solar radiation, and the interplanetary medium.

The second part of the probe was a lander, Beagle 2, intended to carry out the long-standing task of seeking knowledge about the possibility of life on Mars. Its astrobiology and geochemistry instruments were the most sophisticated ever sent to the planet, and it even carried an instrument calibration chart designed by artist Damien Hirst. Unfortunately, something went wrong during Beagle 2's descent through the Martian atmosphere on the morning of December 25, 2003. It failed to make contact with either Mars Express or NASA's Mars Odyssey orbiter upon reaching the surface. ESA formally declared Beagle 2 lost on February 6, 2004.

Despite the failure of its lander, Mars Express continues to function well, and was fully operational as of 2017. Its longevity has made it a uniquely valuable platform for the acquisition of scientific data about the Red Planet, and has honed the skills of ESA scientists and engineers in operating a long-duration, long-distance planetary mission.

ABOVE Concept artwork for ESA's Mars Express spacecraft depicting the probe in orbit around the Red Planet.

RIGHT Concept artwork showing the British Beagle 2 probe on the surface of Mars.

Curiosity self-portrait at the Big Sky drilling site in Gale Crater on Mars, October 6, 2015.

Curiosity

While NASA's first few Mars rovers were all relatively small, its most recent Mars exploration vehicle towers over them all, both in terms of scale and ambition. Weighing 1,982 pounds (899 kg), and the size of a small car, Curiosity is the first full-scale astrobiology mission to Mars since the Viking landers of 1976.

Launched on November 26, 2011, Curiosity entered the thin Martian atmosphere on August 6, 2012. As the Rover's large size precluded the use of airbags as the sole means of cushioning its landing, NASA scientists instead opted to employ an innovative automated "sky crane" system to lower Curiosity gently onto the surface. Approximately three minutes prior to touchdown, the descent module deployed a parachute, then switched to retrorockets before, in the final phase of the landing, it gently lowered the rover down to the surface on a tether.

The excitement that greeted the landing was palpable, and extended far beyond NASA's mission control. In New York's Times Square, thousands of people gathered to watch the landing on the big screen in a state of euphoria, chanting "sci-ence, sci-ence, sci-ence." The first data returned from the mission demonstrated that the rover had reached the surface of the Red Planet safely, and its first images showed the view from its landing site within the Gale Crater. After orientating itself, Curiosity's first goal was to explore its new crater home, and to drive up the mound at its center, known as Mount Sharp.

Subsequently, Curiosity has made some stunning discoveries; it measured the radiation levels on the Martian surface, and found them to be comparable to those experienced by astronauts aboard the ISS. This enhances the possibility that a human expedition to Mars might be feasible. Additionally, Curiosity found evidence of an ancient streambed, where water flowed roughly knee-deep for thousands of years. In drilling into the soil, Curiosity also identified some of the key chemical ingredients for life as we know it, including sulfur, nitrogen, hydrogen, oxygen, phosphorus, and carbon.

Having followed the water, and found evidence of it, the next question Curiosity will face is whether or not Mars could have supported, or might still support, simple life. The quest to ascertain whether there is life on Mars continues.

Key Science Objectives

NASA's Curiosity rover has a total of eight key science objectives for its mission to Mars:

- Determine the nature and inventory of organic carbon compounds on Mars
- Inventory the chemical building blocks of life present on Mars
- Identify features that may represent the effects of biological processes
- Investigate the chemical, isotopic, and mineralogical composition of the Martian surface and near-surface geological materials
- Interpret the processes that have formed and modified rocks and soils
- Assess four-billion-years' worth of atmospheric evolution processes
- Determine the present state, distribution, and cycling of water and carbon dioxide
- Characterize the broad spectrum of surface radiation, including galactic cosmic radiation and solar proton events

Curiosity's view of an alluring Martian panorama, looking south toward Mount Sharp.

A Mars Sample Return?

ABOVE LEFT Concept art from 1995 imagining what a Mars sample return mission might look like.

ABOVE RIGHT Concept art showing a human space explorer examining a sample of Mars rock on the planet's surface.

For as long as humanity has been actively engaged in Mars exploration, scientists have wanted to bring samples from the Martian surface back to Earth. Today, scientists across the globe generally accept that a sample return mission would be an important step forward in the exploration of Mars. Conceivably, such samples could further the search for life on Mars, enabling Earth-bound scientists to employ the full range geochemical studies and age-dating techniques that cannot easily be carried out remotely by rovers and landers. It will likely require exhaustive analysis of a range of samples from a range of locations across the Martian surface in laboratories on Earth to give scientists a chance of determining conclusively whether or not life ever existed on Mars.

In planning for Mars exploration during the early 1970s, several scientists at NASA suggested the use of small, automated Mars sample return probes that could collect Martian dust and rock fragments from the planet's surface for return to Earth for tests. The scientists estimated that such a mission could be mounted by the 1980s, but nothing came of their proposal. In 1989, President George H. W. Bush announced a Space Exploration Initiative in which America would "return to the Moon, and go on to Mars." A Mars sample return mission was identified as a necessary step in preparation for human landings on the Red Planet. Again, NASA scientists predicted that such a mission could be accomplished within ten years, but the project was quietly terminated in the early 1990s on grounds of cost. Due in part to the excitement generated by the Martian meteorite ALH84001 in 1996 (p. 273), a sample return mission once again rose to the top of NASA's agenda. As earlier, it was estimated that such a mission could be undertaken within ten years. It remained on the scientific wish list, but no funding was made available. In the first decade of the twenty-first century, NASA proposed a decade-long program of strategic Mars missions, including a sample return component. This drive ultimately resulted in the Mars Reconnaissance Orbiter (p. 275) and the Curiosity rover (p. 283). But, by the time Curiosity had reached the Martian surface in 2012, the idea of a sample return mission had once again been dropped.

A Mars sample return mission seems perpetually always a decade away, for three interrelated reasons. Firstly, although offering tremendous scientific potential, sample return would be an incredibly expensive enterprise, requiring difficult choices to be made to balance the scientific yield and cost. Secondly, it is difficult to justify this massive expenditure politically when the weight of material that would likely be yielded would be so small. Finally, contamination of samples being brought to Earth has emerged as a major concern. Ensuring both that no Earth biological material contaminates the Mars sample, and that whatever might exist on Mars is thoroughly isolated, so that it cannot enter into the Earth's biosphere, would be necessary elements of any sample return mission. Assuring the sterility of the samples would be difficult to guarantee. While there may yet be a Mars sample return mission in the future, there are serious obstacles to carrying one out. It may yet prove more cost-effective to send humans to Mars to collect the samples manually, just as NASA astronauts did on the Apollo program.

BELOW LEFT Based on feasibility studies of Mars sample return missions, NASA commissioned a series of black-and-white drawings to show what a Martian sample return probe might look like.

BELOW RIGHT This Lunar Sample Return Container was used during the Apollo Moon landings. A similar case might eventually be employed on a future human mission to Mars.

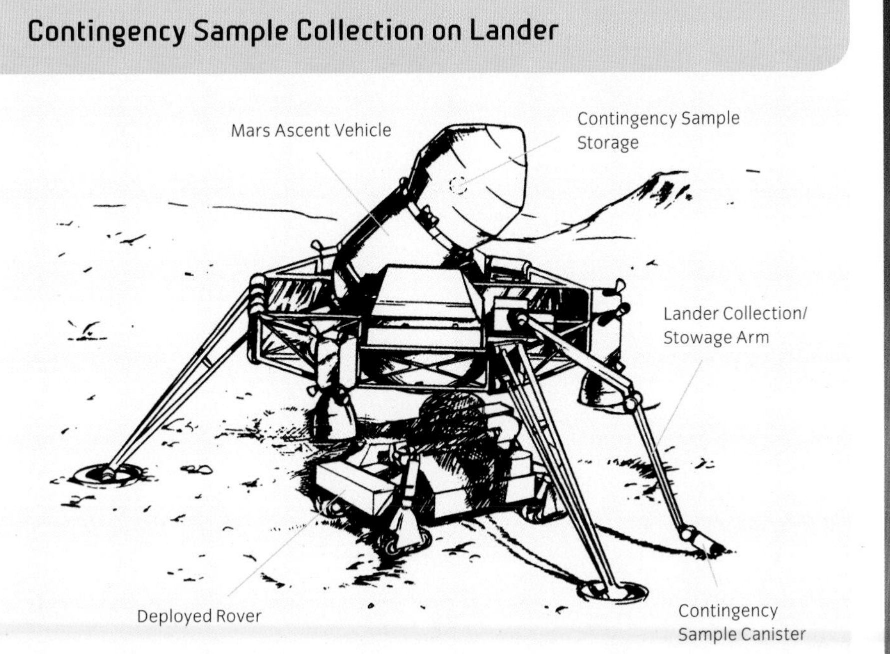

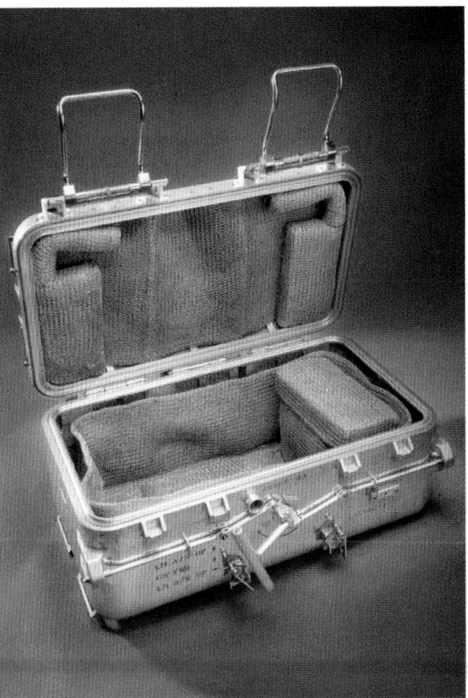

Mars in Twenty-First-Century Popular Culture

BELOW The Mars habitat for the 2000 Brian De Palma movie *Mission to Mars* was based on NASA plans.

BOTTOM Actors Kim Delaney and Gary Sinise experience weightlessness in a scene from *Mission to Mars*.

Mars has been the setting for several significant feature films in the twenty-first century. Some of these are entertaining, a few are scientifically embarrassing, and many more are entirely forgettable, but two hold up remarkably well under scientific scrutiny. Among the entertaining romps are such titles as *Red Planet* (2000), *Ghosts of Mars* (2001), and *Stranded* (2001), and the ridiculous, big-budget, special effects-laden, actioner *John Carter* (2012).

In addition to the 2015 Ridley Scott film *The Martian*, the other most significant Mars-related film in recent years was *Mission to Mars* (2000), directed by Brian De Palma. The film's script was adapted from an original screenplay by Jim Thomas, John Thomas, and Graham Yost, all writers with a great pedigree in space-related feature films and documentaries. Starring Gary Sinise, Tim Robbins, Don Cheadle, Connie Nielsen, Jerry O'Connell, and Kim Delaney as a team of astronauts, the film tells the story of a disastrous mission to Mars, of which Don Cheadle's character is the sole survivor. A second mission sets out to rescue him, and eventually returns to Earth after an encounter with the "Face on Mars" (p. 268), which reveals that life was first seeded from the Red Planet.

Aside from references to the famed face, *Mission to Mars* is scientifically significant for two reasons. Firstly, like only a small number of filmmakers before them, the producers of the film secured support from NASA for both filming in its facilities and the portrayal of its Mars mission planning concepts in the film.

Secondly, the movie's main premise—that Mars seeded Earth with protoplasm—was the first broad examination in a modern Hollywood film of the scientific theory of panspermia, which holds that microscopic life forms embedded in rock from a planet that supports such life can survive the effects of space travel, and subsequently deposit that life on any world it happens to land on. This method of distributing life throughout the universe gained widespread currency at NASA and beyond after the excitement surrounding Martian meteorite ALH84001 in 1996 (p. 273). The film cites this theory as the explanation for the origins of life on Earth, as revealed to the astronauts by a Martian hologram.

Major Feature Films Relating to Mars 2000–2017

Mission to Mars • 2000
NASA-supported film about the first human mission to Mars and a sole survivor's rescue.

Red Planet • 2000
Terraforming Mars proves more difficult than anticipated, and a mission sent to help crash-lands on the planet.

Ghosts of Mars • 2001
A Mars-based police team fights zombies.

Stranded (Náufragos) • 2001
A human mission crash-lands on Mars.

John Carter • 2012
Based on the Edgar Rice Burroughs's 1912 novel, *A Princess of Mars*, former American Civil War soldier, John Carter, is transported to Mars and engages in intrigue and adventure.

The Martian • 2015
A single astronaut is stranded after an accident during a human mission to Mars, and has to survive alone until he is rescued.

The Space Between Us • 2017
Mostly taking place on Earth, this film tells the story of a boy born in a Mars colony who visits Earth.

Live Like a Martian

The science fiction tale of survival on a hostile planet that is based on science fact.

Few films about Mars exploration have struck the balance between entertainment and scientific accuracy as well as director Ridley Scott's 2015 film *The Martian*, based on the book of the same name by Andy Weir.

Its plot is relatively simple. In 2035, an expedition to Mars is forced to leave early after a dust storm threatens to overturn its ascent vehicle. Everyone escapes except astronaut Mark Watney (Matt Damon), who is presumed dead. However, Watney survives and, despite sustaining an injury, goes on to thrive through the use of his engineering and scientific skills to repair equipment, grow potatoes (using his own excrement), make oxygen, and signal Earth. Spoiler: in the end, he is rescued by his fellow astronauts.

In writing *The Martian*, Weir heavily researched NASA's planning for survival during a Mars mission, and depicted, in both the novel and the film, scientifically legitimate episodes, and explanations for various problems a stranded astronaut might face. The film benefited from the assistance of NASA experts, who provided technical expertise, images, and advice. It was also a major critical and box-office success that served as an evocative rallying cry for a future human mission to Mars.

ABOVE LEFT (Left to right) NASA astronaut Drew Feustel, actor Matt Damon, director Ridley Scott, author Andy Weir, and Director of the Planetary Science Division at NASA Headquarters Jim Green participate in a Q&A session on *The Martian*.

ABOVE RIGHT Matt Damon stranded on Mars in *The Martian*.

Boots on the Ground: By the 2030s?

ABOVE Concept art of a future Mars mission visiting the site of a Viking lander.

OPPOSITE TOP Concept art depicting the ascent stage of a two-stage lander rising to Mars orbit on the first leg of a long trek home.

OPPOSITE BOTTOM Concept art for a potential Mars research station being assembled by future Mars explorers.

Sending humans to Mars presents a significant challenge, but it remains a potentially very rewarding accomplishment. All that is required is a political decision by a spacefaring nation, or coalition of nations, to expend the resources necessary to accomplish the task. Most plans formulated to this point have been too large, too complex, and too expensive to be feasible. However, some studies have recommended a leaner operation, and may be possible within a budget of approximately $250 billion, which is roughly what the ISS cost to build and maintain. Such a plan could be actioned as soon as the 2030s.

For example, a proposal to "live off the land," using resources on the Red Planet, might dramatically simplify exploration plans. The first humans to Mars may well extract fuel and consumables from the Martian environment. Such a mission would require a two-year-plus timetable to fly to Mars, work on the surface, and then return to Earth. It would also require a vehicle for getting to Mars, a lander with a scientific laboratory and habitat, a power plant for generating electricity on the surface, rovers, human transports on the surface, food, a manufacturing plant capable of producing its propellant and, most critically, an ascent vehicle for leaving Mars for the journey home.

Fuel could be manufactured on Mars from the local atmosphere, which consists mainly of carbon dioxide. This gas would be pumped into a reaction chamber in the manufacturing plant, where it would be mixed with liquid hydrogen and heated. The resulting process, discovered in the nineteenth century by French chemist Paul Sabatier (1854–1941), produces methane and water. The methane would be pumped through a cryogenic cooler, which would reduce it to a liquid state that could be stored for use as rocket fuel. The resulting water could be pumped into an electrolysis unit, where electrodes separate it into hydrogen and oxygen.

Upon arrival, humans would need to deploy an inflatable greenhouse to grow food. Using automated rovers, the crew could then begin explorations of the surrounding terrain. They would collect rock samples for analysis in a small laboratory set up in their habitat module. They could also drill into the Martian substrata in search of water and any subterranean life that may exist. They could even search for fossils, and seek to confirm the existence of further natural resources that have been detected by satellites orbiting Mars. Once their time on the planet came to an end, the crew would undertake a 110-day trip back to Earth.

The technical problems of such a mission are considerable. The crew would be exposed to two types of radiation: cosmic radiation invading the solar system from the galaxy beyond, and solar flares of radiation running the whole electromagnetic spectrum from the Sun. A fast transit time is the best protection against galactic radiation, as is the local atmosphere on Mars. Solar flares, on the other hand, can be lethal, especially in the unprotected vacuum of space. Engineers may opt to shield the crew from such flares with water, using a donut-shaped water tank into which the explorers could retreat until the solar storm subsides. It may also be necessary to maintain some artificial gravity on the spacecraft carrying the crew to Mars, to help minimize biomedical problems associated with prolonged exposure to low-gravity environments. This could be accomplished by using rotating sections to create artificial gravity.

Most scientific and technical challenges can be overcome with sufficient funding. The major obstacle for a human Mars mission remains cost. President Donald Trump announced on December 11, 2017, that he intended to re-vector NASA toward a return to the Moon and the establishment of a Moon base prior to a human mission to Mars. This may push a Mars landing a decade or more into the future, or it might energize other nations to take leadership for a national or international Mars mission. Getting humans to Mars in the 2030s can be done, but only if we are willing to spend enough money to overcome all the obstacles.

"People will visit Mars, they will settle Mars, and we should because it's cool."

JEFF BEZOS,
FOUNDER AND CEO OF AMAZON.COM

9

Beyond Mars

Since the early 1970s, humans have sent probes to explore every planet in the outer solar system, as well as several of its asteroids (p. 294) and comets (p. 306). We have placed spacecraft in orbit around Jupiter (p. 304) and Saturn (p. 316), and landed on Saturn's moon Titan (p. 313). The New Horizons probe has visited Pluto and the farthest dwarf planets of our solar system in the Kuiper Belt. These stunning missions have yielded a treasure trove of knowledge about our solar system, how it originated, and how it works. They have also helped to teach us how our solar system falls into three distinct zones, only one of which contains areas conducive to the development of life.

To learn about the wider universe, scientists have created a variety of Earth-orbiting observatories (p. 326), through which we can study the cosmos. These have also revolutionized humanity's understanding of everything from star formation to the very origins of the universe itself.

But perhaps the most enduring aspect of humanity's exploration of space beyond the orbit of Mars is the hope and expectation that we might find some form of intelligent life out there. It is a hope that inspired us to attach descriptive plaques and records to the probes we have sent out of our solar system.

Through our studies of our own solar system, we have identified many key ingredients that would make a planet habitable, and scientists have even come up with an equation that helps us to quantify what we are looking for. We have also identified thousands of worlds orbiting distant stars, at least some of which have some resemblance to our own world.

Humanity may never be able to answer the question of whether life exists on other worlds definitively, but it won't be for a lack of searching.

OPPOSITE Captured by the Hubble Space Telescope, this photo from 2017 shows a small portion of the Carina Nebula. Towers of cool hydrogen laced with dust rise from the wall of the nebula.

Three Zones of the Solar System

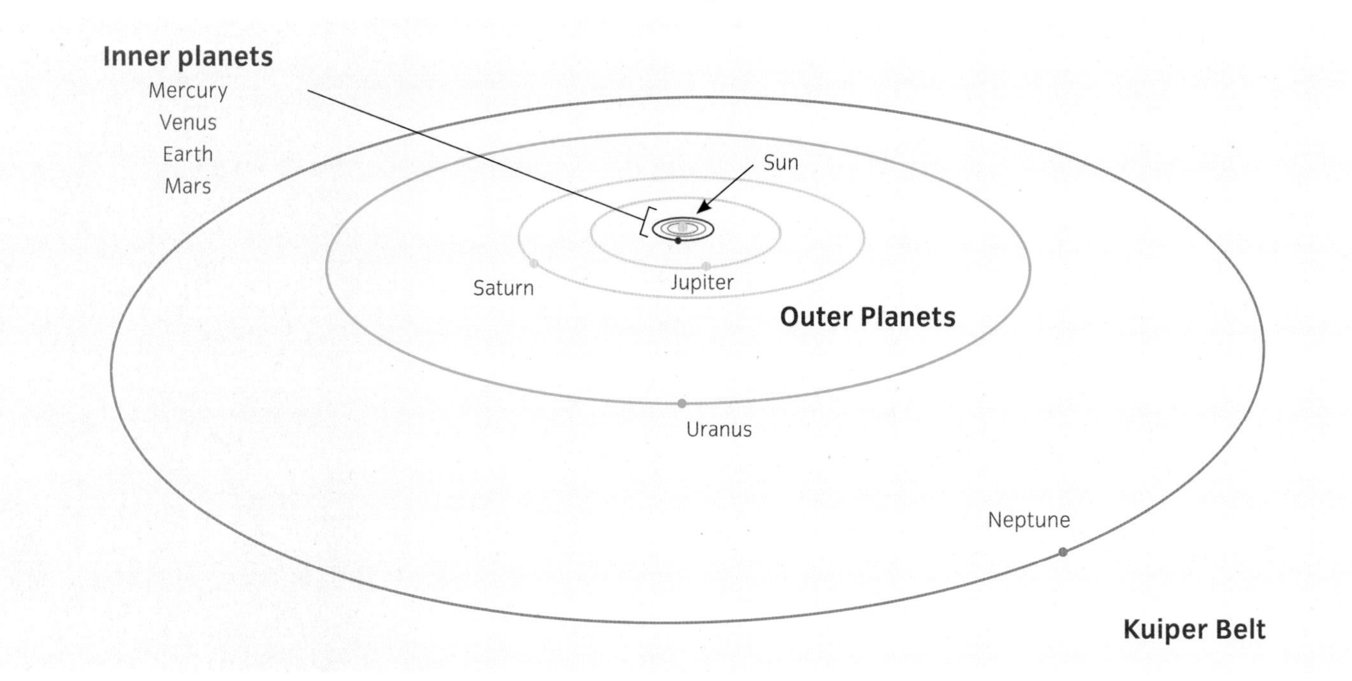

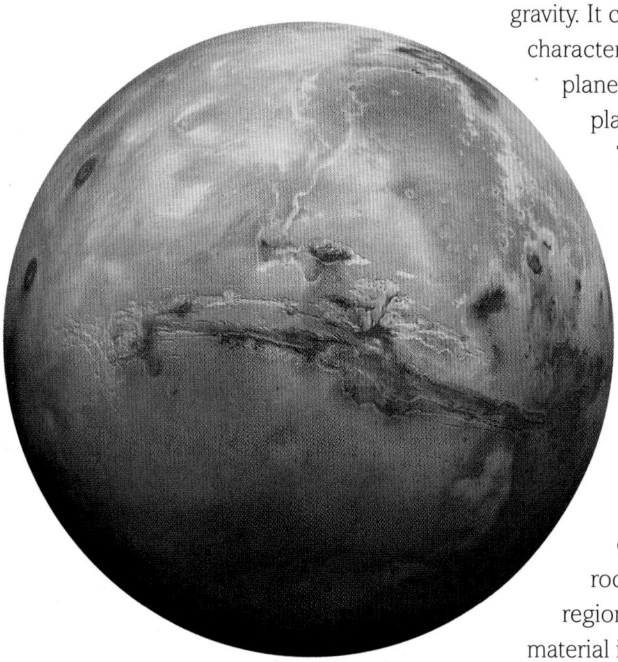

The solar system consists of the Sun and all the objects bound to it through gravity. It can be conveniently divided into three distinct zones, each with its own characteristics. The inner zone closest to the Sun contains the terrestrial, or rocky, planets: Mercury, Venus, Earth, and Mars. The second zone of the outer planets contains the four gas giants: Jupiter, Saturn, Uranus, and Neptune. The third zone includes the Kuiper Belt, of which Pluto is the most prominent constituent, and the Oort Cloud of distant comets, asteroids, and other bodies.

Each of these three zones has its own challenges and opportunities for exploration. The inner planets occupy the solar system's habitable zone, the region around a star that has conditions most generally propitious to life, although relatively mild temperatures that support liquid water, another key requirement of life, have thus far been found only on Earth. Humans have sent probes to all of the planets in the inner zone, from which we have learned much about planetary science. Also in this zone, beyond the orbit of Mars, lies the main asteroid belt, consisting of thousands of rocky and mineral-rich bodies. Few probes from Earth have visited this region, but scientists have been able to determine that the total mass of the material in the asteroid belt is a small fraction of that contained in Earth's Moon.

The second zone of the outer planets contains the gas giants that have captivated astronomers since Galileo Galilei first turned his early telescope on them. It is little wonder that these worlds, also termed the Jovian planets, were early targets for exploration by probes. The most stunning of these probes were the Voyager missions to the outer solar system. Launched in 1977, the Voyager 1 and Voyager 2 spacecraft visited all the giant outer planets. Between them, the two spacecraft took well over 100,000 images of these planets and their associated features, including their rings and satellites, as well as millions of magnetic, chemical spectra, and radiation measurements in their vicinity.

The Kuiper Belt comprises the third zone, and is the most mysterious region of the solar system. While it was known to host numerous icy dwarf planets, the area remained relatively unknown until 2016 when the New Horizons probe reached the dwarf planet Pluto and the other bodies nearby. Much about these Kuiper Belt objects remains mysterious. Do they represent the remains of planetary embryos whose growth was stunted? Are they relics of the origin of the solar system? The current consensus among planetary scientists is that they are both. Efforts to explore this zone of the solar system have just begun, and are hampered by the time it takes for spacecraft to reach so far from the Earth, but further exploration will push humanity closer to completing the first stage of its reconnaissance of the solar system. As with previous exploration attempts, what is learned from these efforts will inform the direction of future research.

OPPOSITE TOP The three zones of the solar system go by a variety of names, but can be divided into (1) the inner, or terrestrial, planets, (2) the outer planets, also known as the Jovian planets, or gas giants, and (3) the Kuiper Belt and other icy bodies.

OPPOSITE BOTTOM Mars, the outermost of the terrestrial planets.

BELOW The dwarf planet Pluto, viewed from the New Horizons spacecraft in July 2015, is representative of the Kuiper Belt. The mountains have been accentuated to emphasize Pluto's features.

The First Deep Space Reconnaissance Missions: Pioneers 10 and 11

Concept art showing Pioneer 10 encountering Jupiter.

As America moved toward placing the first humans on the Moon in the 1960s, NASA began to consider sending probes to the outer solar system. At around the same time, two NASA personnel made a remarkable discovery that would greatly facilitate trips to Jupiter and the other gas giants. Gary Flandro (1934–), an engineer working on flight trajectories for NASA probes, and mathematician Michael Minovich (1936–) found that, once every 176 years, the Earth and all of the gas giants gather on one side of the Sun in a geometric sequence that makes it possible to visit all of those planets in a single, highly economical flight trajectory that soon became known as the "Grand Tour" flight path.

The gravity of each planet on the route would bend the spacecraft's trajectory, and increase its velocity enough to deliver it to the next destination. This would create a slingshot-like effect that reduced the flight time to reach all four of the gas giants from 30 years to 12 years. The realization that this optimal flight path was due to occur in the late 1970s led to one of NASA's most focused and significant space exploration efforts.

To help maximize the science yield of the "Grand Tour," NASA commissioned two probes that would allow the agency trial the technologies needed for long-term, long-distance space exploration. Pioneer 10 and Pioneer 11 were both to be sent to Jupiter, with the latter probe traveling on to an encounter with Saturn. Both were small, nuclear-powered, spin-stabilized spacecraft that rotated around the axis of their dish-like, high-gain antenna to maintain stability. Pioneer 10 was launched on an Atlas-Centaur rocket on March 3, 1972, and arrived at Jupiter on December 3, 1973. Although many were concerned that the spacecraft might be damaged by the intense radiation it discovered in Jupiter's orbital plane, the probe was still able to transmit data about the planet back to Earth before it continued on its way out of the solar system. By May 1991, it was about 52 Astronomical Units (AU) away, or 52 times the distance from the Earth to the Sun, and still transmitting data.

The Deep Space Network, a collection of large radio telescopes around the globe, continued to track Pioneer 10, until it was 82 AUs from Earth, at which point, on January 23, 2003, the network received its last very weak signal from the probe.

Specifications for the Pioneer Probes

Width
9 feet (2.7 m)

Length
9 feet, 6 inches (2.9 m)

Diameter
9 feet (2.74 m)

Weight, Empty
568 pounds (257 kg)

Power Source
Plutonium-238 radioisotope thermoelectric generator

Launch Vehicle
Atlas-Centaur

Launch
Pioneer 10—March 3, 1972
Pioneer 11—April 6, 1973

End of Mission
Pioneer 10—January 23, 2003
Pioneer 11—September 30, 1995

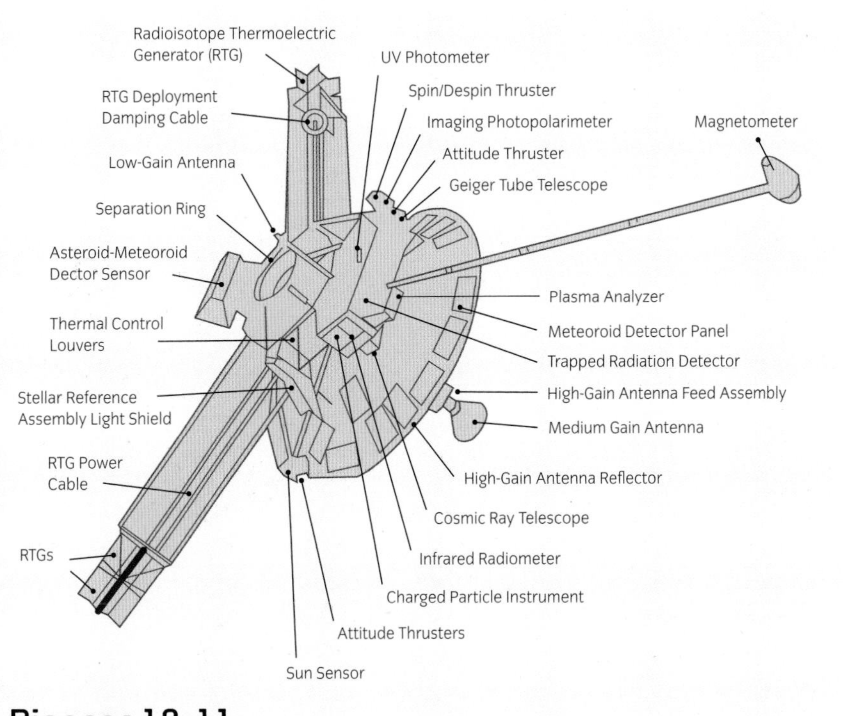

Schematic of the Pioneer probes.

It has not been heard from since, but it continues to travel beyond the solar system, coasting silently on a trajectory that is taking it into interstellar space toward the red star Aldebaran, which it will reach in a little over two million years' time.

Launched on April 6, 1973, Pioneer 11 reached Jupiter in 1974, and in December 1974 provided scientists with their closest view of the largest planet in the solar system, from 26,600 miles (42,809 km) above its cloud tops. This close approach to Jupiter gave the spacecraft a gravitational boost to speeds of 107,373 mph (172,800 km/h) over the 1.5 billion miles (2.4 billion km) distance toward Saturn, which it reached in 1979.

> *"The heritage of mankind is not the Earth but the entire universe."*
>
> PIERRE BOULLE, FRENCH AUTHOR

Closing to within 13,000 miles (20,900 km) of the planet, Pioneer 11 discovered two new moonlets and a new Saturnine ring. It also charted the planet's magnetic field, its climate, and the general structure of its gaseous interior before continuing on toward the center of our own galaxy, the Milky Way. Contact was lost with Pioneer 11 on September 30, 1995, but, like Pioneer 10, it will continue to travel silently on its current trajectory.

Both Pioneer 10 and Pioneer 11 were remarkable space probes. They were built for missions spanning a mere 30 months, but returned important data not only about Jupiter and Saturn, but also, over the course of their extended secondary missions, which lasted more than 30 years, they revealed more details about the Sun's influence in the outer solar system more generally.

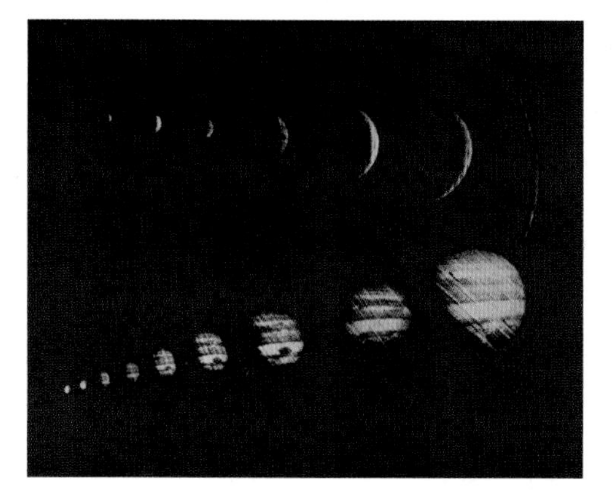

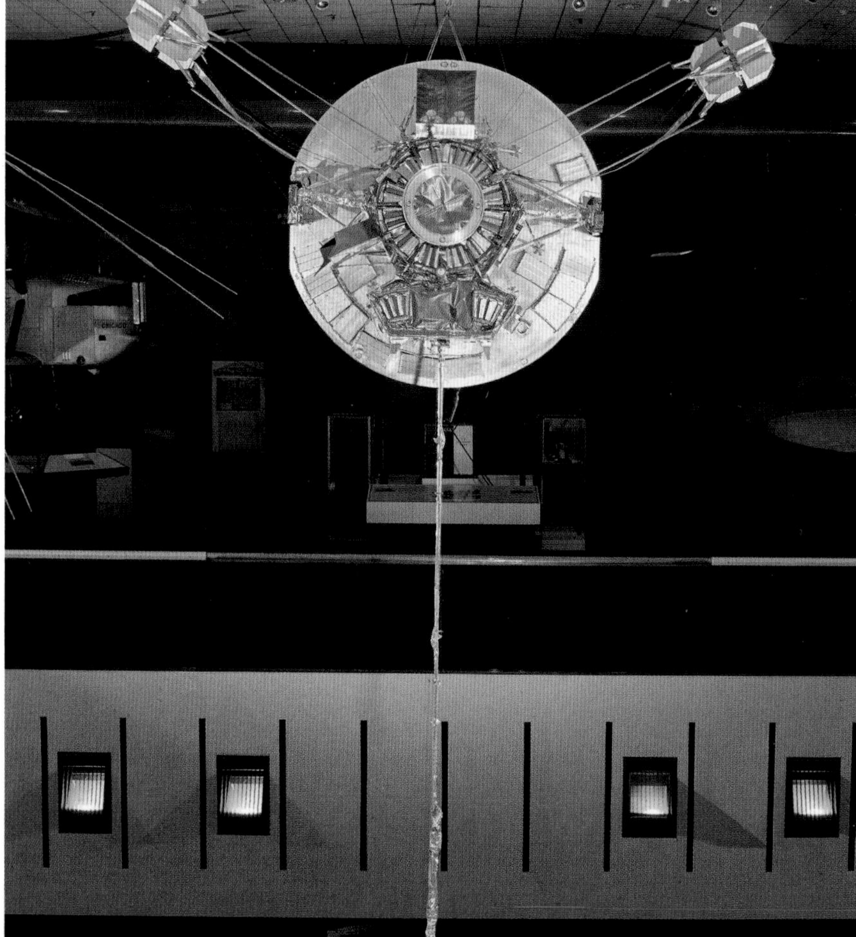

ABOVE As NASA's Pioneer 10 spacecraft approached Jupiter it captured increasingly large images of the gas giant.

RIGHT Reconstructed mock-up of the Pioneer probes on display at the Smithsonian Institution's National Air and Space Museum.

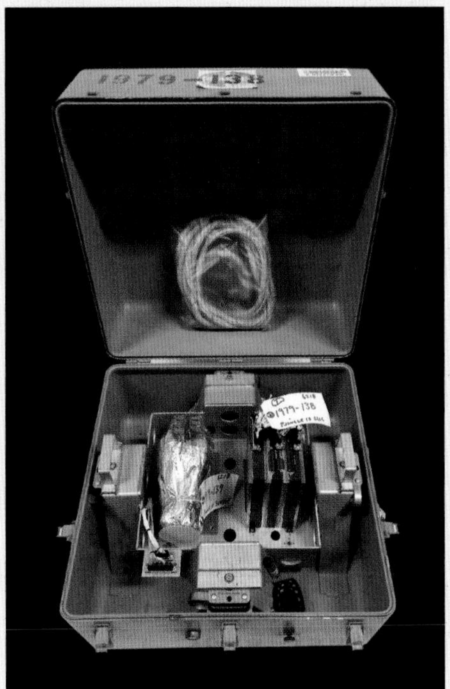

Radioisotope Thermoelectric Generators and Spacecraft Power

To keep going over the vast distences to the outer planets and beyond, spacecraft need nuclear power sources.

The long-duration, deep-space missions undertaken by the Pioneer probes were feasible only because of the nuclear radioisotope thermoelectric generator (RTG) that powered the spacecraft, providing them with electricity. Pioneer 10, Pioneer 11, and all subsequent missions to the Jovian planets and beyond used RTGs. Solar energy is sufficient to power spacecraft only as far from the Sun as Mars. When dealing with distances as great as those between the Sun and Jupiter and farther, the Sun is too weak to generate the energy required to power multiple scientific instruments. Other forms of power storage—batteries and fuel cells—do not retain their charge long enough. Nuclear power is the only reliable alternative.

RTGs operate in the simplest manner possible: radioactive plutonium-238 (Pu-238) decays, producing heat that flows through thermocouples, which generate electricity. The thermocouples are connected through a closed loop that feeds electrical current to the power system of the spacecraft. The strength of the current provided declines over time, in line with the half-life of the isotope used; in the case of plutonium-238, that is around 87 years.

RTGs have evolved over more than 40 years since their first use in space exploration. The current system has flown on a wide range of spacecraft. The Galileo probe that went to Jupiter contained two RTGs, while New Horizons, which visited the Kuiper Belt, had one RTG to power its systems. These RTGs had a thermal power of 4.4 kilowatts, and each carried a little less than 21 pounds (9.4 kg) of plutonium. Because of the low wattage of the systems on board these probes, all missions flown to the outer planets and beyond thus far have been constrained by the limited power available for their instruments. To help resolve this issue, a twin-pronged development effort is now underway to enhance the wattage of RTGs, and to improve the efficiency of scientific instruments developed for such long missions.

Despite the risks of nuclear contamination in the event of an explosive failure in a launch vehicle carrying an RTG power source in its payload, no deep-space mission is currently possible without one.

ABOVE LEFT Lockheed Martin employees Joe Collingwood (right) and Ken Dickinson retract pins in the storage base to release a radioisotope thermoelectric generator in preparation for hoisting the unit into place on the Cassini spacecraft on its mission to Saturn.

ABOVE RIGHT The electronics package for Pioneer 10's magnetometer requires an RTG to keep running. This back-up version was passed to the Smithsonian Institution's National Air and Space Museum from NASA in 1979.

Voyagers 1 and 2

Voyager chief scientist Edward Stone (1936–) once called Voyager 1 and Voyager 2 "the little spacecraft that could." That may have been an understatement. Since their launch in 1977, they have visited all of the four gas giants—Jupiter, Saturn, Uranus, and Neptune—in a single broad sweeping mission.

To keep the probes as cost-effective as possible, the mission planners focused on flyby studies of Jupiter and Saturn, but engineers designed the two probes to conduct as much science and to last as long as possible. Voyager 2 was launched first, lifting off on August 20, 1977, from the Kennedy Space Center. Voyager 1 followed on September 5, 1977, taking a faster trajectory. It soon overtook its sibling to reach Jupiter and then Saturn.

The twin spacecraft carried a complex array of scientific instruments to capture measurements and imagery of the outer solar system. Two long antennas allowed for radio astronomy and plasma-wave studies, while magnetometers on a 42-foot (13-m) boom offered a wide spectrum of measurements of magnetic fields. Instruments to detect and measure cosmic rays, charged particles of various frequencies, and spectrometers operating in the infrared and ultraviolet zones, collected data across a broad range of the electromagnetic spectrum. A suite of visible light cameras of various types and sensitivities completed the instrumentation. The cameras included two television-type cameras, one operating at a low resolution, and the other at a high resolution. These instruments proved exceptionally sturdy through the two crafts' extended missions.

Voyager 1 arrived at Jupiter in 1979, and at Saturn in 1980, where it took pictures of, among other things, the planet and its moon Iapetus. Voyager 2 came within 63,000 miles (101,000 km) of Saturn on August 26, 1981. It captured stunning images of the planet's complex rings system, including the F-ring, which had been discovered only a few years previously by Pioneer 11. The probe also captured images of the moons Hyperion, Enceladus, Tethys, and Phoebe.

Voyager 1 came within 2,500 miles (4,000 km) of Saturn during its flyby of the planet. Saturn's moon Titan was of particular interest to the Voyager 1 mission planners. They directed the probe to examine the moon, which was revealed to have an atmosphere, thick clouds, and water ice. Voyager 1 also found that Titan's atmosphere was composed of 90 percent nitrogen, and that the pressure and temperature at the surface were 1.6 atmospheres and -350°F (-180°C). Consequently, Titan instantly became one of the most enticing prospects for life beyond Earth anywhere in the solar system, and a key target for the Cassini/Huygens mission that followed.

To perform its close pass of Titan, Voyager 1 had to fly out of the ecliptic plane shared by the rest of the planets in the solar system, rendering it unable to continue to Uranus and Neptune. However, on its way out of the solar system, the probe did capture one final portrait of Earth and six other planets, even though they appeared as little more than dots illuminated by the distant Sun.

Specifications for the Voyager Probes

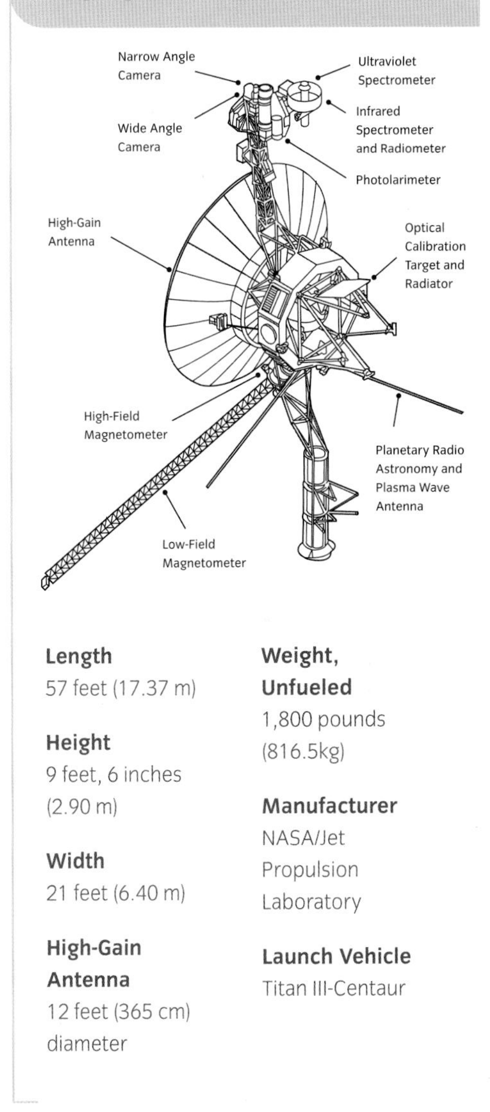

Length
57 feet (17.37 m)

Height
9 feet, 6 inches (2.90 m)

Width
21 feet (6.40 m)

High-Gain Antenna
12 feet (365 cm) diameter

Weight, Unfueled
1,800 pounds (816.5kg)

Manufacturer
NASA/Jet Propulsion Laboratory

Launch Vehicle
Titan III-Centaur

OPPOSITE Jupiter and its four largest moons, called the Galilean satellites, were photographed in early March 1979 by Voyager 1 and assembled into this collage which does not show the Moons to scale but does show their relative positions.

> *"Billions of years from now our Sun, then a distended red giant star, will have reduced Earth to a charred cinder. But the Voyager record will still be largely intact, in some other remote region of the Milky Way galaxy, preserving a murmur of an ancient civilization."*
>
> CARL SAGAN,
> AMERICAN ASTROPHYSICIST

A montage of the Saturn system, assembled from images captured on the Voyager mission.

Flybys of the two outermost gas giants fell to Voyager 2, which arrived in the vicinity of Uranus in 1986 and Neptune in 1989, and captured the first close glimpses of these far-off worlds.

Collectively, the two Voyager probes sent back pictures of Saturn's rings, found active volcanoes on Jupiter's moon, Io, discovered several new satellites across all four planets, and observed unexpected behavior in the particles that make up Saturn's rings. They returned information that revolutionized planetary science, helping to resolve some key uncertainties about the origins and evolution of the solar system, while posing intriguing new questions. Easily the greatest unexpected discovery was at Jupiter, where scientists found active volcanism on the satellite Io, the first time that active volcanoes had been seen on another body in the solar system. Together, Voyager 1 and Voyager 2 observed the eruption of nine volcanoes on Io, and there is evidence that other eruptions occurred between the two probes' encounters. They also demonstrated that plumes from Io's volcanoes extended more than 190 miles (300 km) above the moon's surface. Io's volcanoes are thought to result from the interaction between Jupiter's gravity and that of its other moons. This interaction heats Io's interior. But all of this planetary exploration was just the beginning. Soon after Voyager 2 completed its flyby of Neptune, the probes were handed a new task in the form of the Voyager Interstellar Mission (VIM), which would see them attempt to chart the limits and extent of the Sun's influence.

The Voyager Golden Record

1977

A time capsule of human civilization sent from the late twentieth century out into the vastness of space.

Voyager 1 and Voyager 2 were the third and fourth human artifacts to leave the solar system and encounter interstellar space, after Pioneer 10 and Pioneer 11. For the Voyager missions, NASA affixed time capsules to the exterior of the two spacecraft to leave a message that might be found by life beyond Earth.

The result was the Voyager "Sounds of Earth" record, a 1970s-era analog musical disc. The brainchild of the NASA astrophysicist Carl Sagan (p. 269), the record contained sounds and images selected to portray the diversity of life and culture on Earth.

The two gold-plated copper Voyager discs were each 12 inches (30 cm) in diameter, and contained 122 images, as well as human greetings in 55 languages, various sounds of Earth, and a selection of music. Sagan sought to represent the broad range of human experience. The spoken greetings included Akkadian (spoken in the Sumer civilization of Mesopotamia about 6,000 years ago) and Wu, a modern Chinese dialect, as well as many other current languages. The music included an eclectic 90-minute musical program, which included Chuck Berry's "Johnny B Goode" and three classical pieces by Johann Sebastian Bach.

Each of the two Voyager spacecraft carries a copy of the Voyager Golden Record inside a cover, shown here, bearing pictograms explaining where the disc came from.

Voyager Discoveries

A list of major scientific discoveries from the Voyager missions to the outer planets included the following:

- Discovered the Uranian and Neptunian magnetospheres, both of them highly inclined and offset from the planets' rotational axes, suggesting their sources are significantly different from other planet's magnetospheres.
- Found 22 new satellites: three at Jupiter, three at Saturn, ten at Uranus, and six at Neptune.
- Observed that Io exhibited active volcanism, the only solar system body other than the Earth to be so confirmed. Found Triton possessed active geyser-like structures and an atmosphere.
- Discovered auroral zones on Jupiter, Saturn, and Neptune.
- Found rings around Jupiter. Observed spokes in Saturn's B-ring and a braided structure in its F-ring. Discovered new rings around Uranus and Neptune.
- Observed large-scale storms (notably the Great Dark Spot on Neptune), which was originally thought to be too cold to support such atmospheric disturbances.

Studying Asteroids and Other Cold and Icy Bodies

Several hundred thousand asteroids have been discovered and given names since humans first began charting the heavens, and all available evidence points to there being many more million yet be discovered. They are the cold and icy bodies of the solar system, that are most abundant in the asteroid belt between Mars and Jupiter and the Kuiper Belt on the outskirts of the Sun's gravitational pull, but many also travel outside these areas.

A few of these wandering asteroids end up on a direct collision course with the Earth. Such bodies range in size, from tiny particles that burn up in the Earth's atmosphere, to larger objects as much as 164 miles (200 km) in diameter. Earth is bombarded every day by more than 100 tons of dust and sand-sized particles. Usually about once a year, an object the size and weight of an automobile streaks into the Earth's atmosphere, and creates an impressive fireball as it is incinerated into ash during its descent, long before it can reach the Earth's surface. Every few hundred years, a meteorite weighing a few tons hits the Earth and causes significant damage across a wide area.

In 2017, there were more than 50 asteroids identified as being larger than 100 miles (162 km) in diameter, mostly in the asteroid belt between Mars and Jupiter. Scientists believe that they have identified 99 percent of asteroids larger than 62 miles (100 km) in diameter, but of the smaller asteroids in the 6.2–62 mile (10–100 km) diameter range, only about half have been catalogued. Most of these are also located in the asteroid belt between Mars and Jupiter, and pose no threat to Earth. However, the concern remains that eventually a body that is on an impact course for Earth will be discovered. It is important to track such objects, because a sufficiently large meteorite strike could prove catastrophic to life on this planet. A constant international effort to survey the sky, identifying and tracking these objects, is already underway, but much further work remains to be done.

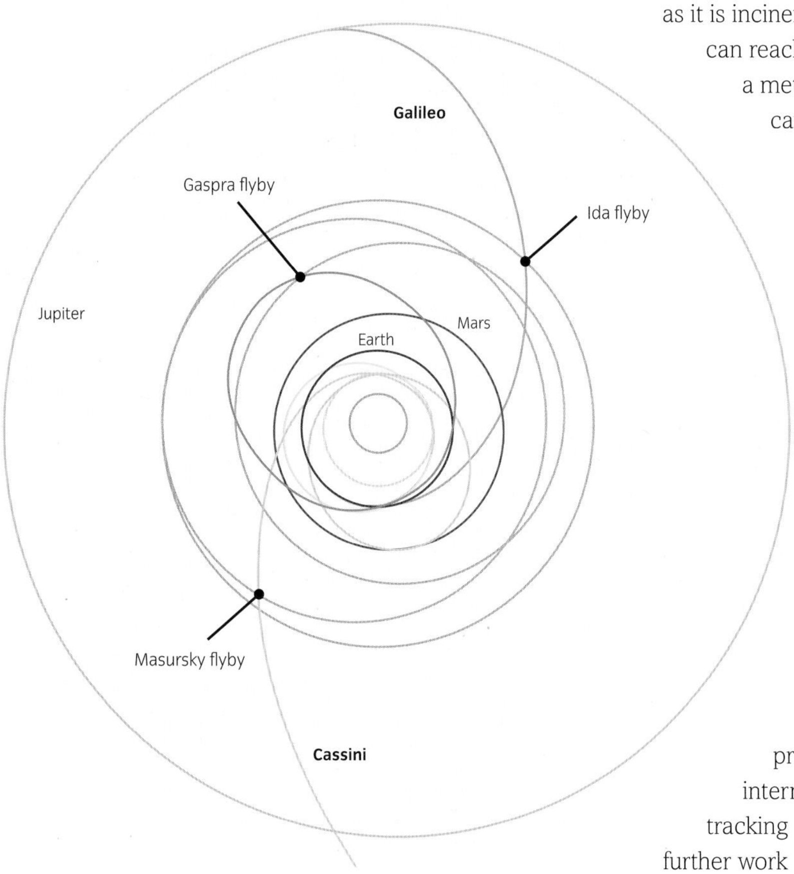

The largest object ever to be identified as an asteroid in the solar system is Ceres. Approaching 587 miles (945 km) in diameter, Ceres has a mass that is equivalent to about 25 percent of that of all the other known asteroids combined, and is now considered a dwarf planet. The next largest are Vesta, Pallas, and Hygiea, which are between 248 miles (400 km) and 326 miles (525 km) in diameter. All other known asteroids are less than 211 miles (340 km) across.

Several asteroids have been visited by various spacecraft, even if sometimes only in a cursory fashion. The Galileo space probe encountered two significant specimens on its way to Jupiter. It passed asteroid 951 Gaspra at a distance of 1,000 miles (1,600 km) in 1991. Galileo's science team noted that Gaspra is an irregularly shaped object that lacks the kind of large craters common on many planetary satellites. Galileo later encountered 243 Ida, identifying it as the first known example of an asteroid with a moon of its own. Subsequently named Dactyl, this mini-moon measures about 1 mile (1.5 km) in diameter and orbits about 62 miles (100 km) from Ida's center.

OPPOSITE This diagram shows the looping trajectories the space probes Galileo and Cassini performed through the asteroid belt on their way to Jupiter and Saturn, respectively. During its flight, Galileo encountered asteroids 951 Gaspra and 243 Ida. Cassini encountered the asteroid Masursky during its passage through the asteroid belt.

ABOVE This computer-generated image shows NASA's Dawn spacecraft heading toward the dwarf planet Ceres.

Since that time, there have been several additional probes that have had encounters with asteroids, some on missions specifically designed to investigate them. Among the highlights of these are the following:

- **Near-Earth Asteroid Rendezvous Shoemaker, also known as NEAR Shoemaker:** launched by NASA in 1996, NEAR Shoemaker flew within 753 miles (1,212 km) of Asteroid Mathilde in June 1997. It made rendezvous with, and landed on, the asteroid Eros on February 12, 2001.

- **Deep Space 1:** launched in 1998, Deep Space 1 flew past asteroid 9969 Braille in July 1999. During a flyby at a range of only 16 miles (26 km), the probe's instruments found intriguing similarities between Braille and asteroid Vesta, one of the largest asteroids in the solar system.

- **Stardust:** built for a comet rendezvous and sample return, this probe encountered asteroid Annefrank in 2002, passing within 2,050 miles (3,300 km) of it, finding it irregularly shaped, cratered, and about 5 miles (8 km) in diameter. It then encountered comet Wild 2, captured cometary particles, and returned to Earth with samples on January 15, 2006.

- **Hayabusa:** launched on May 9, 2003, by JAXA, Hayabusa was a sample return probe sent to the asteroid Itokawa (p. 199). It landed there successfully on November 25, 2005. In April 2007, Hayabusa started its return to Earth, arriving on June 13, 2010. Mission controllers attempted a parachute landing in the South Australian outback, but the spacecraft broke up on reentry, and was incinerated before reaching the ground.

- **Rosetta:** launched in 2004, this ESA probe flew by 2867 Šteins in 2008 and 21 Lutetia in 2010.

- **Dawn:** launched in 2007, Dawn encountered the asteroid, Vesta, in 2011, and the dwarf planet, Ceres, in 2015.

- **Chang'e 2:** launched by the CNSA on December 13, 2012, this probe flew within 2 miles (3.2 km) of the asteroid 4179 Toutatis on its extended post-lunar mission.

- **Hayabusa 2:** JAXA launched this probe in December 2014, and plans to have it return samples from 162173 Ryugu in 2020.

- **OSIRIS-REx:** launched by NASA on September 8, 2016, this probe is part of a sample return mission to asteroid 101955 Bennu.

BELOW The ESA's asteroid lander, Philae, took this parting shot of a solar array on its mothership, Rosetta, shortly after the lander began its descent toward its asteroid target on November 12, 2014.

BOTTOM NEAR Shoemaker took these images of Eros on October 16, 2000, while orbiting 34 miles (54 km) above the asteroid.

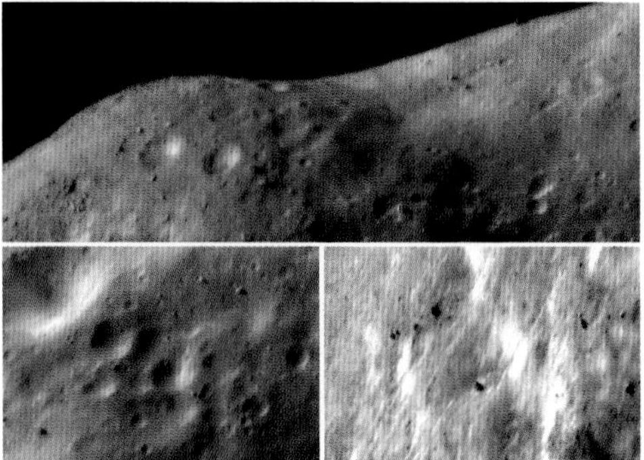

The Stardust Sample Return Mission

1999–2006

The first sample return mission to bring cometary material back for study by scientists on Earth.

ABOVE LEFT This is a computer-generated rendering of Stardust's encounter with comet Wild 2.

ABOVE RIGHT The Stardust sample return capsule successfully landed at the United States Air Force Utah Test and Training Range on January 15, 2006, with its cometary and interstellar samples intact.

NASA's Stardust Sample Return Mission was a significant milestone in the annals of space exploration. Launched on February 7, 1999, its primary goal was to collect carbon-based samples during its encounter with comet Wild 2. Its journey ended with its return capsule streaking across the sky to a successful parachute landing on January 15, 2006. It was the first NASA space mission dedicated solely to the exploration of a comet, and obtaining extraterrestrial material from outside the Earth-Moon orbit.

The return element from the Stardust probe had five major components: a heat shield, a back shell, a sample canister, a parachute system, and avionics. Ablative materials protected the samples stowed in its interior. To capture the cometary samples and interstellar dust, scientists used a specially engineered silicon-based material with a porous, sponge-like structure called aerogel, which was deployed on a gridded tennis racket-like structure at the end of the spacecraft's arm to collect ejecta from the comet.

The samples of interstellar comet dust are believed to consist of ancient material left over from the formation of the solar system. Since their collection, the samples have been distributed to space scientists around the world for continued study, and are expected to enhance our understanding of comets, and possibly even of the beginnings of the universe.

When Asteroids and Comets Come Crashing to Earth

Several thousand interplanetary objects are on collision courses with Earth, and many pieces that are mostly small enter the atmosphere every day. Asteroids and other cold and icy bodies may not seem important, but there is a statistically greater chance—small though it might be—of being killed by some object from space hitting us than of being killed in an airplane crash. According to asteroid tracking experts William Burrows and Don Yeomans, every person alive today faces one chance in 500,000 that he or she will be killed by some type of extraterrestrial impact from a comet or meteorite. Since several thousand meteorites, comets, and asteroids cross Earth's orbit every day, this risk can feel difficult to quantify, but it is only a matter of time before a big object hits the Earth.

One need only look at the craters on the Moon, and on Earth, such as Meteor Crater in Arizona, to verify that planets make fine targets for comets and asteroids. Around 48,000 BCE, a 50,000-ton (45,500 t) iron meteorite hit Earth in what is now northern Arizona, leaving a crater about 4,000 feet (1,200 m) wide, and 590 feet (180 m) deep. It is Earth's largest existing impact crater with associated meteorite fragments. Modeling suggests that the asteroid was moving at around 28,000 mph (46,000 km/h) when it made impact. The meteor was mostly vaporized upon impact, leaving little material from the asteroid in the crater. The impact also resulted in countless animal deaths caused by the impact itself, the destructive air blast it caused across the surrounding landscape, and the changes it created in the immediate local climate.

Among previous asteroid impacts for which there are contemporary historical accounts, the most devastating incident was probably the Chiling-Yang incident. In 1490 CE, an asteroid making its way to Earth burst apart in the upper atmosphere. This provoked mass panic among the large number of people in and around the city of Chiling-Yang, China, who witnessed it. Ming Dynasty records of the time estimate that the death toll relating to the incident was as high as 10,000 people.

OPPOSITE This artist's impression of the Tunguska meteorite event, approaching Siberia, Russia, in 1908, shows how a huge fireball of rock might have looked as it hit the Earth's atmosphere.

BELOW Meteor Crater in Arizona is one of the youngest, and best-preserved, impact craters on the Earth today.

On the morning of June 30, 1908, thousands of Russians witnessed the fiery ball of light in the sky that heralded the Tunguska asteroid impact in central Siberia. Thousands of trees were flattened in the incident. It burned an area of 620 square miles (1,000 km^2), throwing ash and soil into the air, and incinerating countless reindeer, as well as a few people. Eyewitnesses spoke of a "split in the sky," a "second sun," and of being violently thrown to the ground.

Most recently, just after dawn on February 15, 2013, a meteor descended at more than 34,000 mph (55,000 km/h) over the Ural Mountains in Russia. It exploded above the ground, and created a momentary flash that generated a shockwave. Fragments of the object fell in and around the city of Chelyabinsk, roughly 130 miles (210 km) south of Yekaterinburg. Russian Interior Ministry spokesman Vadim Kolesnikov said that 1,100 people called for medical assistance following the incident, mostly for treatment of injuries from glass. The asteroid's size was estimated at 55 feet (17 m) in diameter, with a mass of some 5 tons (4.5 t).

An impact large enough to produce a hole the size of Arizona's Meteor Crater statistically occurs on average once every 1,000 to 2,000 years somewhere on Earth. Taking 1,500 years as the mean of these estimates, the chances of such an impact occurring in any given year can be expressed as 1 in 1,500.

This concept art depicting an asteroid apparently heading for Earth was used as an illustration in a joint NASA/FEMA (Federal Emergency Management Agency) planning exercise held in October 2016. The exercise explored possible responses to a large comet or meteorite strike on the planet.

The K-T Extinction

66,000,000 BCE

The event responsible for the death of the dinosaurs may have been a comet or meteorite strike.

The reality of a massive threat to life on Earth from space first really entered the public consciousness in the 1980s, when geologists began to reach a consensus over the hypothesis that dinosaurs became extinct after an asteroid or comet measuring around 9 miles (14 km) wide came crashing to Earth, and left a crater measuring 186 miles (300 km) just off the Yucatán Peninsula in the Gulf of Mexico. This impact is thought to have thrown enough dust into the atmosphere to cause prolonged global climate change.

Evidence for this mass extinction event is largely based on the fossil record, which marks the boundary between the Cretaceous period (K), in which dinosaur fossils are present, and the Tertiary period (T), in which they are not. The fossil record also shows evidence of several other similar extinction events over the course of Earth's history. These findings have prompted scientists to speculate whether humanity might face a similar situation, should another asteroid or comet of sufficient scale impact the Earth.

This image of an asteroid streaking through the sky above Chelyabinsk, Russia, on February 15, 2013, was captured by local resident M. Ahmetvaleev.

However, even when such an event strikes, the chances of any one person being killed by such an impact are tiny, especially when taking into account how much of the Earth's total surface area is taken up with oceans and seas. The area devastated around Meteor Crater was 300–580 square miles (800–1,500 km^2). The surface area of Earth is roughly 200 million square miles (510 million km^2). Accordingly, the chances of the land on which you happen to be standing being destroyed by a meteorite in the event of a meteor strike is only about 1 in 500,000, and much less if you factor in the rarity of such events.

Nevertheless, this is a genuine threat, and taking notice of all of the objects orbiting the Sun is becoming a major aspect of space exploration moving forward. Adequate preparation for the threat of such strikes will involve learning more about the asteroids and comets that cause them. NASA Astronaut John Young said it best when he paraphrased the words of the title character from the American newspaper strip *Pogo* by cartoonist Walt Kelly: "I have met an endangered species, and it is us." At some point in the future, a comet or asteroid will again hit Earth with disastrous consequences.

Meanwhile, humanity must do everything in its power to minimize the risks: cataloging all Earth-crossing asteroids, tracking their trajectories, and even developing countermeasures to destroy or deflect those objects identified to be on a collision course with Earth—all that is just the start. The only way to secure the long-term survival of our species is to ensure that it is not based exclusively on one planet on which all life could be extinguished with a single asteroid impact of sufficient scale. Humanity needs to find viable alternative worlds to inhabit, and the only way to do that is to start building outposts of humanity elsewhere in our solar system (p. 384).

"If the Earth gets hit by an asteroid, it's game over. It's Control-Alt-Delete for civilization."

BILL NYE,
AMERICAN SCIENCE COMMUNICATOR

Mission to Halley's Comet

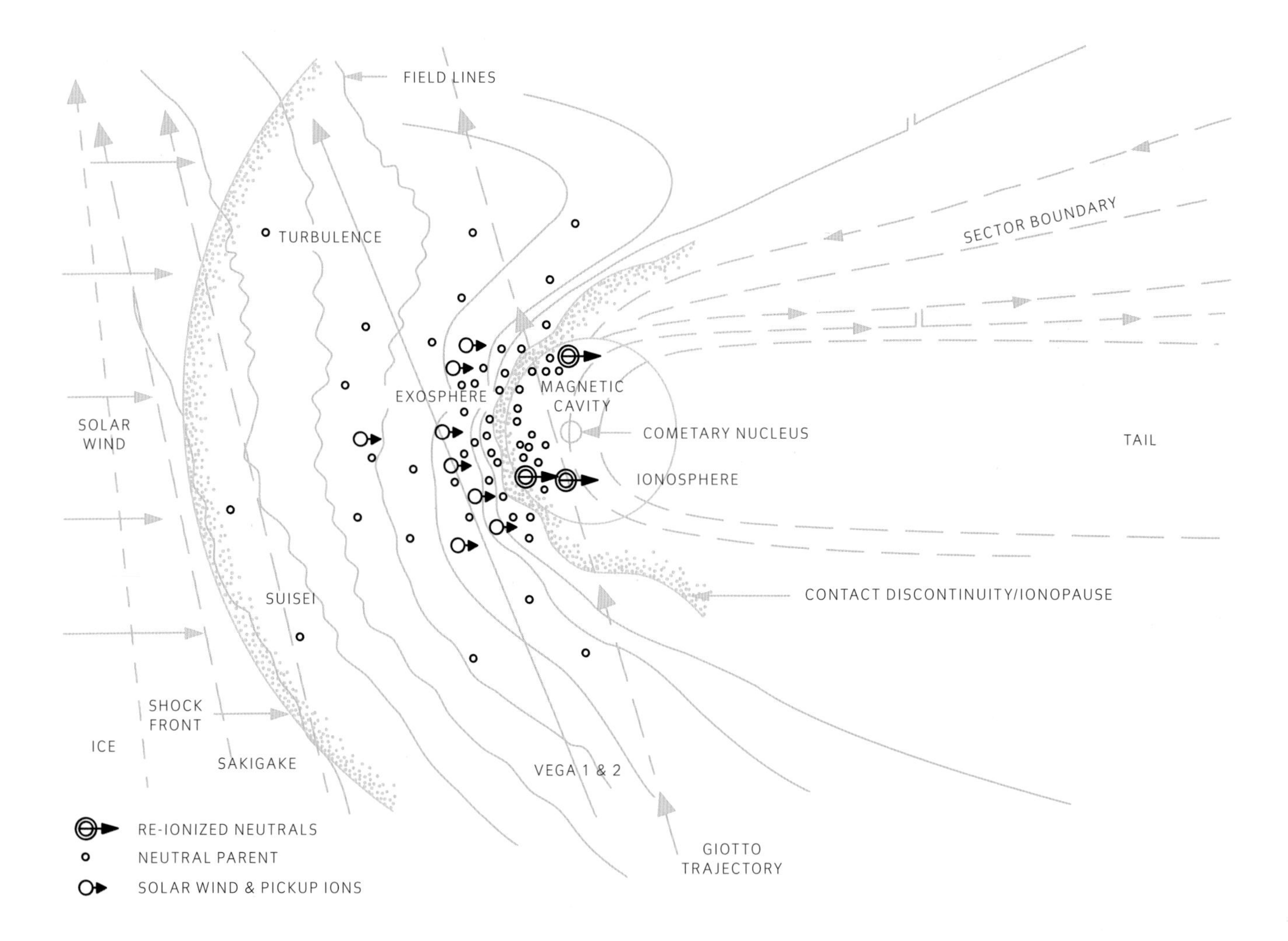

The environment around Halley's Comet in 1986, showing the approaches of the armada of five spacecraft that it encountered on its way toward the Earth.

Named after English astronomer and polymath Edmond Halley (1656–1742), who first computed its orbit, Halley's Comet is perhaps the best-known long-period comet in astronomy. Passing close to Earth once every 76 years, it often gives a spectacular show in the night sky before rounding the Sun and returning to the outer solar system. In the twentieth century, the comet twice passed close to the Earth: in 1910 before the advent of space exploration, and in 1986. Scientists around the globe began working on plans to study the comet more than a decade in advance of this second visit. NASA planned to use the existing International Cometary Explorer (ICE) probe, a joint project with the ESA, to make ultraviolet light observations of the comet from afar, but the ESA, the Soviet space program, and the Japanese space program all sent specific probes close enough to Halley's Comet to capture images of it.

The principal mission was ESA's Giotto, launched on July 2, 1985, and sent to intercept Halley's Comet. The probe made its closest approach on March 13, 1986, coming within 375 miles (596 km) of the comet. Because Giotto was scheduled to pass so close to the comet's debris trail, ESA engineers believed that the probe would not survive the encounter. They were wrong; it weathered the high-speed cometary particles just fine, and subsequently embarked on an extended mission to comet Grigg-Skjellerup on July 10, 1992. ESA finally ended Giotto's science mission on July 23, 1992, although the probe itself remains in solar orbit to this day.

Giotto benefited greatly from data obtained from ICE, which allowed ESA to target Giotto toward the nucleus more precisely. Neither the Soviet probes, Vega 1 and Vega 2, nor the two Japanese probes, Sakigake and Suisei, got anywhere near as close as Giotto, although all six probes collected valuable data on the comet and, with the exception of ICE, sampled different regions of the cometary environment.

The lack of a bespoke NASA probe to Halley's Comet was a missed opportunity for the organization, but one that served to illustrate the turbulent times the agency faced in the late 1970s. In the wake of the recent Voyager missions to the outer planets, and the ongoing development of the Space Shuttle, there was little political will or public desire for the additional expense of a comet exploration mission. NASA was also riven with bickering over what type of comet exploration mission to undertake, what instruments to send, and who would direct the effort. One American space scientist, John Simpson of the University of Chicago, even ended up designing the Dust Particle Detector for the Soviet Vega spacecraft.

Despite all these problems, NASA space scientists greatly benefited from, and had full access to, all the data collected from the various probes sent to the comet.

Viva los Vega

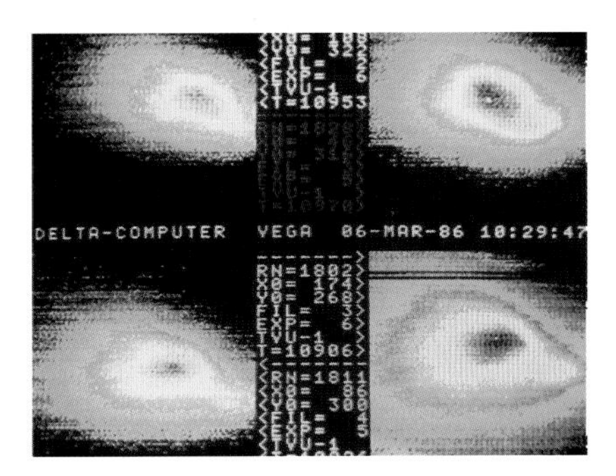

These four false-color images of Halley's Comet were the first taken of its nucleus. They were captured by the Soviet probe Vega 1 near its closest approach to the comet on March 6, 1986. Vega 1 passed less than 5,600 miles (9,000 km) from the nucleus, and was followed three days later by Vega 2. Halley's Comet was a secondary objective for the probes after they first performed a flyby of Venus, and delivered landers to the Venusian surface.

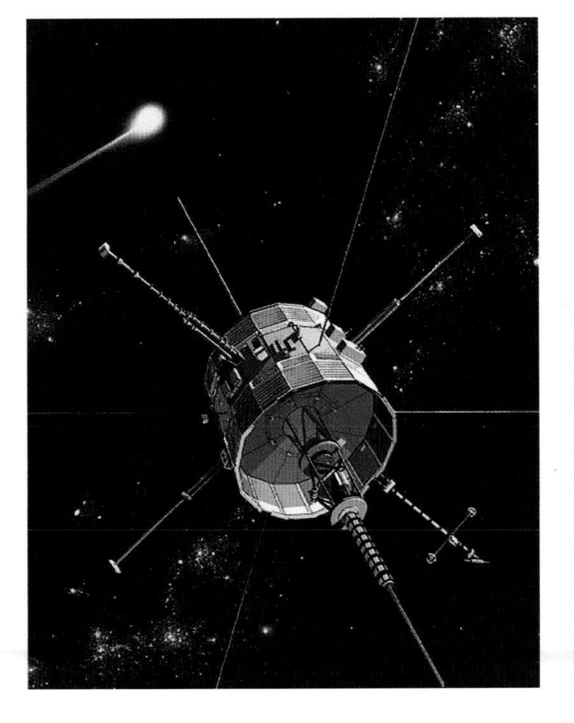

LEFT Concept art of the NASA/ESA International Cometary Explorer probe, which was redirected to study Halley's Comet in 1986.

BELOW This enhanced image from ESA's Giotto probe shows the potato-shaped nucleus of Halley's Comet, the first photographic image of a cometary nucleus ever captured.

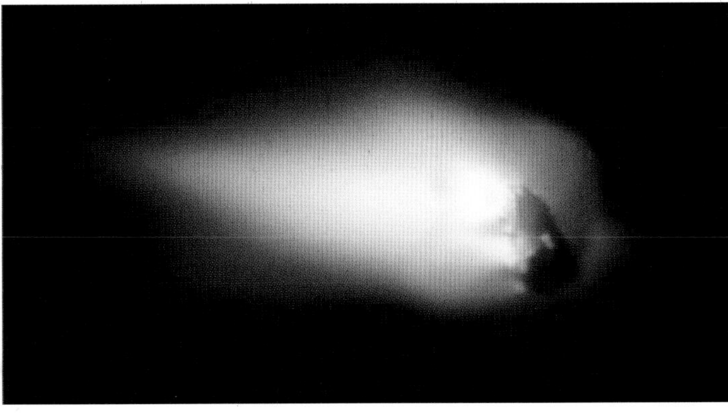

In Search of Jupiter's Secrets

"Every 12 years Jupiter returns to the same position in the sky; every 370 days it disappears in the fire of the Sun in the evening to the west, 30 days later it reappears in the morning to the east."

GAN DE (*CA*. 400–*CA*. 340 BCE), CHINESE ASTRONOMER CREDITED WITH MAKING JUPITER OBSERVATIONS RECORDED WITH DATE AND COORDINATE INFORMATION

The gas giant Jupiter is the largest planet in the solar system. Its diameter of 88,846 miles (142,984 km) is more than 11 times that of Earth. Located 483,780,000 miles (778,570,000 km) from the Sun, it is more than five times farther than Earth from our solar system's center. But despite being visited by six spacecraft since the 1970s, including an extended mission from the NASA probe Galileo between 1995 and 2002, Jupiter remains a planet of mysteries.

The exploration of Jupiter began with Pioneer 10 and Pioneer 11, which performed flybys of the planet, as did Voyager 1 and Voyager 2. These four spacecraft returned information to Earth that has revolutionized our understanding of the science of the solar system, helping resolve some key questions, such as the mass of Jupiter and how its system operates, while raising intriguing new ones about the origin and evolution of the planet and its moons. Together, the quartet of spacecraft took well over 100,000 images of various aspects of the Jovian system, including its rings, tracked its satellites, and took millions of magnetic, chemical spectra, and radiation measurements. The two most recent probes to visit the planet, Galileo (p. 314) and Juno, have explored the Jovian system in even greater detail. To date, 67 moons have been identified around Jupiter. The four largest, Callisto, Ganymede, Europa, and Io, are known as the Galilean moons after the astronomer Galileo Galilei, who first discovered them. The smallest of Jupiter's moons appear to be little more than asteroids captured in Jupiter's gravity.

Jupiter is the largest, most massive, and fastest-rotating planet in the solar system. Its great size and massive gravity suggest that Jupiter was the first planet to form around the Sun, while the other planets emerged out of the debris left over. Jupiter's mass and gravity are also thought to have enabled it to retain a significant share of the gas from the interstellar cloud that gave rise to the Sun. Jupiter is therefore viewed as holding all of the ingredients that went into the formation of our solar system, and may yet provide clues not available elsewhere about how our solar system evolved.

This diagram shows only a small selection of the 67 moons circling Jupiter, maintained in orbit by the massive influence of the planet's gravity.

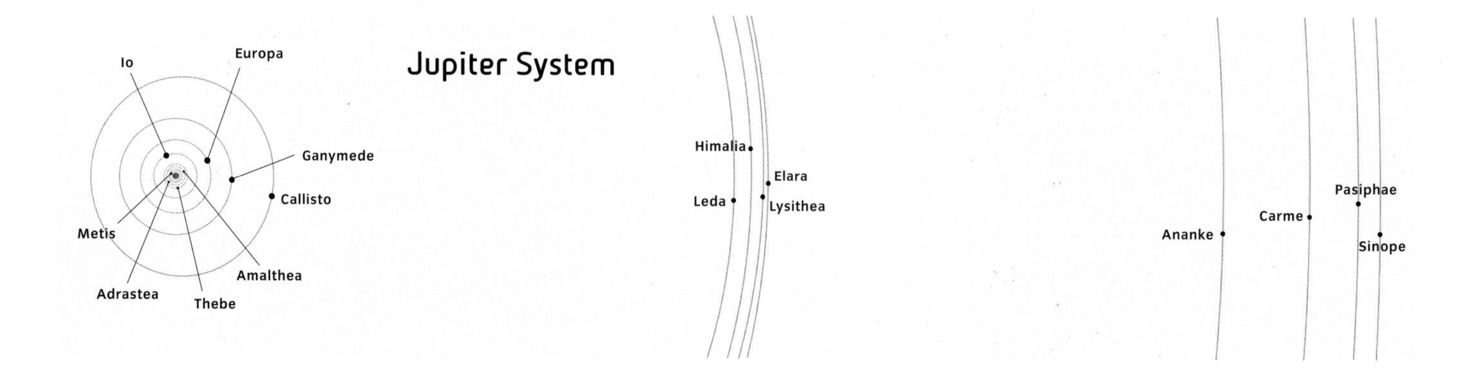

ABOVE A composite image of the comet Shoemaker-Levy 9, broken into 21 fragments and hurtling toward an impact with Jupiter.

LEFT This composite image shows the relative sizes of the four largest moons of Jupiter. From top to bottom: Europa, Io, Callisto, and Ganymede.

Galileo to Jupiter

> *"We learned mind-boggling things. This mission was worth its weight in gold."*
>
> CLAUDIA ALEXANDER, GALILEO PROJECT MANAGER

NASA undertook its first sustained exploration of Jupiter on October 18, 1989, when its Galileo spacecraft was deployed during the Space Shuttle mission STS-34, and set on its gravity-assisted journey that took it past Venus and Earth, on its way to Jupiter. On arrival in December 1995, Galileo became the first spacecraft to orbit one of the gas giants of the outer solar system. The spacecraft circled the planet, and returned invaluable scientific data on the density and chemical makeup of Jupiter's cloud cover.

Prior to reaching its destination in 1995, Galileo had become a source of concern for its mission controllers, because its large high-gain telecommunications antenna failed to unfurl fully as intended, forcing NASA's scientists and engineers to use the spacecraft's low-gain antenna instead. As this was far less powerful, it required NASA to reconfigure its Deep Space Network of Earth-based receivers to compensate for the weaker signal, and the spacecraft to use its onboard tape recorder to help compress data for transmission. Over the course of its operational life, the spacecraft returned enormously significant scientific data, including evidence of subcrustal oceans on Europa, Jupiter's large ice-rock moon. Galileo also conducted close-up inspections of the planet's moons Ganymede, Callisto, and Io. While passing by the latter, the spacecraft observed magnificent eruptions from Io's volcano Loki, the largest such active volcano in the solar system.

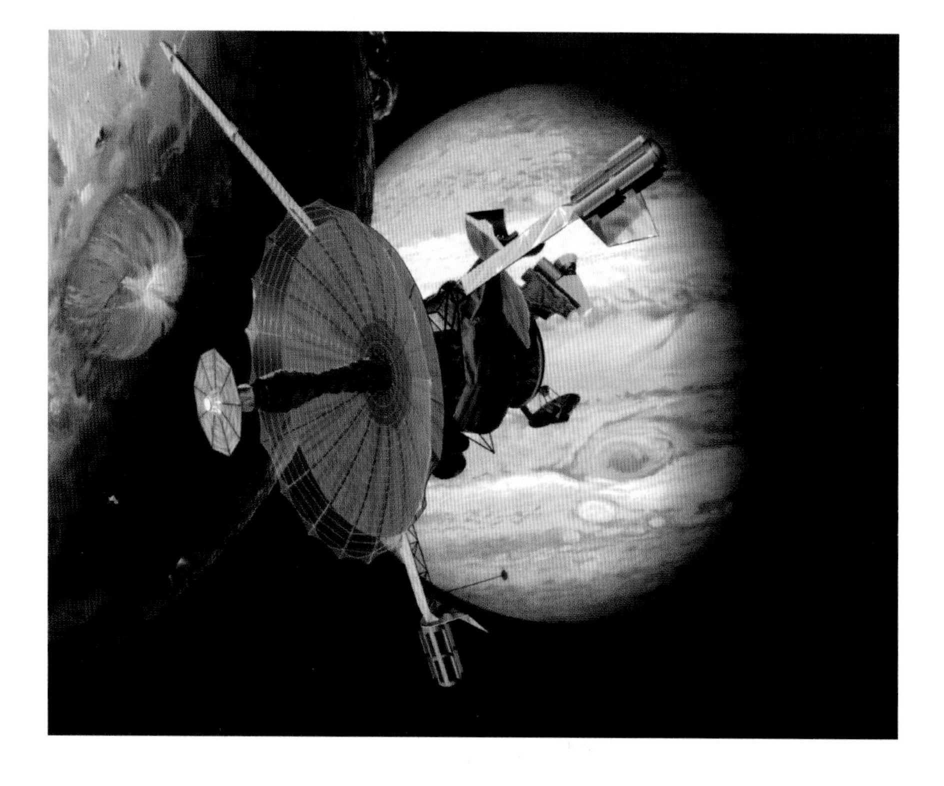

This concept art shows the Galileo spacecraft flying past a volcanic eruption on the Jovian moon Io, while Jupiter itself lurks in the background.

The Galileo spacecraft also carried a small descent probe that was dropped into Jupiter's dense atmosphere on December 7, 1995. Its instruments began relaying data to the main orbiter on the chemical composition of the atmosphere, the nature of the cloud particles, and the structure of the cloud layers. It also monitored the atmosphere's radiative heat balance, pressure, and dynamics, as well as the composition of its ionosphere. The descender lasted only about 45 minutes before it was destroyed in the planet's atmosphere.

Galileo's mission has led to a reinterpretation of scientists' understanding of Jupiter, its moons, and the outer solar system more generally. A list of Galileo's most important discoveries includes the following:

- Evidence for a liquid water ocean under the surface of Jupiter's moon Europa.
- The discovery of a satellite circling the asteroid Ida 243.
- The discovery of an intense interplanetary dust storm.
- The discovery of an intense new radiation belt approximately 31,000 miles (50,000 km) above Jupiter's cloud tops.
- Detection of Jovian wind speeds in excess of 400 mph (643 km/h).
- Helium abundance in Jupiter is very nearly the same as its abundance in the Sun.
- Io's surface has changed substantially, due to continuing volcanic activity, since the Voyager 2 flyby.
- Preliminary data support the tentative identification of intrinsic magnetic fields for Jupiter's moons Io and Ganymede.

Galileo's main mission came to an end on February 28, 2003, after a final playback of scientific data from its tape recorder was received by NASA scientists on Earth. The mission team then programmed the spacecraft's onboard computer to send it plunging into Jupiter's atmosphere, bringing its remarkably successful mission to a dramatic conclusion.

Galileo Specifications

Weight
Total: 5,648 pounds (2,562 kg)
Orbiter: 4,901 pounds (2,223 kg)
Descent probe: 747 pounds (339 kg)

Power Source Radioisotope thermoelectric generator (RTG)

Launch Vehicle
Space Shuttle Atlantis, STS-34

Launch October 18, 1989

Entered Service December 8, 1995

End of Mission
September 21, 2003, deorbited

Exploring Europa

1997–2001

This Jovian moon could harbor both subterranean seas and, possibly, some form of simple life.

One of the Galileo's most exciting pieces of data concerned Europa, one of the largest of Jupiter's moons, and its capacity to harbor life. Somewhat smaller than the Earth's Moon, Europa is covered with a crust of water ice, frozen by a bone-chilling surface temperature of -260°F (-162°C). The tidal tug-of-war, created by the gravitational pull between Jupiter and its various moons, generates an internal source of heat. That heat may have created a subterranean ocean on Europa, and a favorable environment for the development of life.

Many scientists and science fiction writers have speculated that Europa, along with Mars and Saturn's moon Titan, are the three bodies in the solar system most likely to possess environments in which life could exist. The possibility of liquid water on Europa has prompted many space scientists to advocate for a future lander mission to Europa that would be capable of exploring under the moon's icy surface.

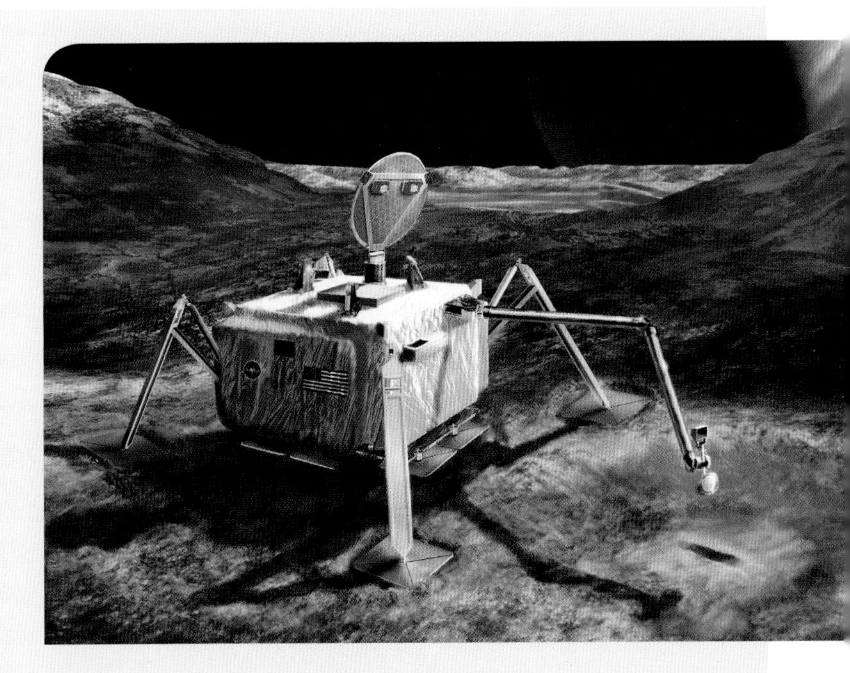

This NASA concept art shows a possible design for a potential future mission to land a robotic probe on the surface of Europa. The lander is shown with a sampling arm extended in preparation to probe beneath the moon's surface crust.

Saturn and Its Moons

"I still remember... the first time I pointed the telescope at the sky and I saw Saturn with the rings. It was a beautiful image."

UMBERTO GUIDONI, ITALIAN ASTROPHYSICIST AND THE FIRST EUROPEAN SPACE AGENCY ASTRONAUT TO VISIT THE ISS

ABOVE RIGHT Taken by NASA's Cassini spacecraft as it approached Saturn on October 26, 2007, this image shows the planet and its largest moon, Titan. Saturn's moon Mimas is also visible as a bright dot close to the planet and beneath its rings.

OPPOSITE This diagram shows some of the rings of Saturn, together with some of the planet's smaller moons.

Saturn, like Jupiter, has long been a target of exploration. The second largest planet in the solar system, Saturn boasts spectacular rings that have fascinated humans for centuries. They are lettered roughly in the order in which they were discovered, with the Italian astronomer Giovanni Cassini (1625–1712) discovering the first gap in Saturn's ring system in 1675. This split the rings into an A division and a B division. A further division in the A-ring was discovered by German astronomer Johann Encke (1791–1865), and announced in 1837. Other astronomers have identified further divisions over time.

Pioneer 11 was the first spacecraft to visit Saturn. It discovered further, fainter rings around the planet, beyond those discernible from Earth, and also established a safe flightpath through debris orbiting the outermost regions of Saturn's gravitational influence for the Voyager 1 and Voyager 2 probes that would follow it. Those missions revealed that Saturn has the lowest density of all the planets, and that the yellow and gold bands in its atmosphere are the result of super-fast winds. Those winds can reach speeds of up to 1,100 mph (1,800 kph) around the planet's equator, due in part to the rising heat from the planet's interior. Additional discoveries of the Pioneer and Voyager missions include the following:

- Saturn is the flattest planet. Its polar diameter is 90 percent of its equatorial diameter, largely due to its low density and fast rotation. Saturn turns on its axis once every 10 hours and 34 minutes, giving it the second-shortest day in the solar system behind Jupiter.
- Saturn's upper atmosphere is divided into bands of clouds; the top is formed mostly of ammonia ice, but below that there is water ice, and, farther down, layers of cold hydrogen and sulfur ice.
- Saturn has oval-shaped storms similar to Jupiter's.
- Deep inside Saturn is a hot core surrounded by a layer of metallic hydrogen.
- Saturn has the most extensive ring system in the solar system. The rings are made mostly of ice and dust. While they extend as far as 75,000 miles (120,000 km) from the planet, they are only about 66 feet (20 m) thick.

Saturn also has many moons, with an official tally of 60 with confirmed orbits. Of these, only 52 have been named thus far. The vast majority of Saturn's moons are small, rocky bodies without atmospheres or regular orbits. One moon in particular, Titan, has long fascinated scientists because of its atmosphere and landscape containing hydrocarbon lakes and river networks, and some even speculate that the moon could be capable of supporting life. Interest in Titan prompted NASA to plot a course for Voyager 1 that included a close encounter with Titan, and inspired the ESA to develop the Huygens probe (p. 318). In addition to Titan, scientists are interested in further exploring Saturn's moon Enceladus, which is thought to possess liquid water beneath its frozen surface, a key indicator that it, too, may be capable of supporting life.

Put a Ringlet on It

This composite image, captured by Voyager 2, focuses on Saturn's C Ring, with the B Ring visible at top and left. The spacecraft's photopolarimeter was able to observe the planet's rings at a much higher resolution than previous probes, and to discern many more ringlets within the larger structure of Saturn's ring system.

Rings of Saturn

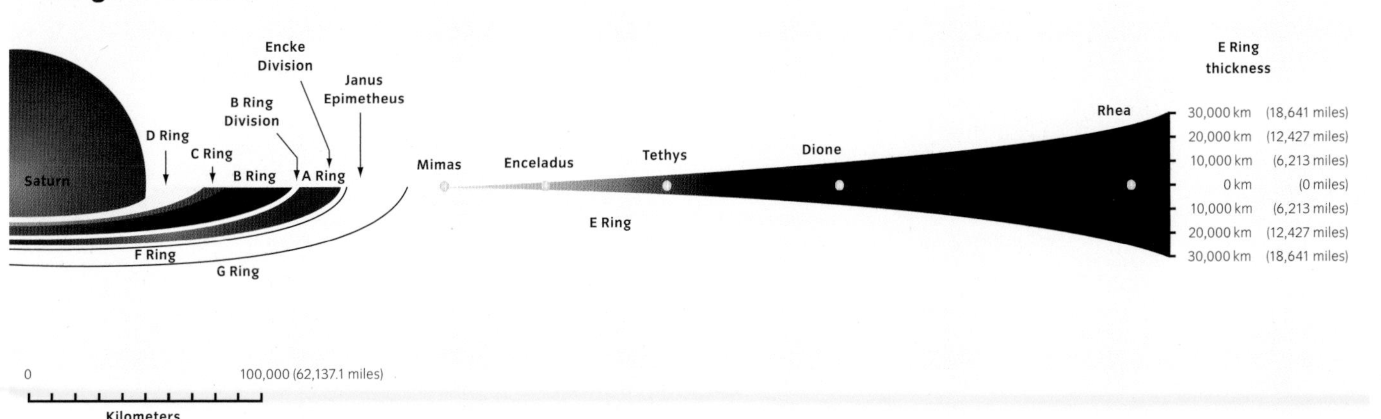

Cassini to Saturn

The most significant mission undertaken thus far to study Saturn and its moons has been Cassini-Huygens, a joint project between NASA, the ESA, and the Italian Space Agency (ASI) that proved remarkably successful over its 20-year operational lifespan. Launched in 1997, NASA's Cassini probe became the first spacecraft to orbit Saturn upon its arrival on July 1, 2004. It carried with it ESA's Huygens lander, which it dropped onto the surface of Saturn's moon Titan on January 15, 2005.

Cassini is the third-largest interplanetary probe to have been built to date; only the failed Soviet Phobos 1 and Phobos 2 spacecraft were larger. It relied on three plutonium-238 radioisotope thermoelectric generators (RTGs) for power and carried 12 science instruments to collect a wide range of information about the environment around Saturn and its moons. These instruments took images across the infrared, visible, and ultraviolet light portions of the electromagnetic spectrum, measured dust particles, and measured Saturn's magnetosphere. Cassini collected data on the planet and its environs between 2004 and 2017, orbiting Saturn at a distance of between 8.2 and 10.2 Astronomical Units (AU) from Earth.

This concept art shows what the Cassini spacecraft may have looked like as it orbited Saturn.

Cassini Specifications

Length
22 feet (6.8 m)

Width
13 feet (4 m)

Weight
Orbiter: 4,740 pounds (2,150 kg)
Huygens Probe: 770 pounds (350 kg)

Power Source
Radioisotope thermoelectric generator (RTG)

Launch Vehicle
Titan IV (401) B B-33

Launch
October 15, 1997

Jupiter Closest Approach
December 30, 2000

Distance
6,122,323 miles (9,852,924 km)

Jupiter Orbital Insertion
July 1, 2004

Huygens Titan Landing
January 14, 2005

End of Mission
September 15, 2017,
deorbited into Saturn

A huge, ferocious storm churning through the atmosphere in Saturn's northern hemisphere overtakes its own wake as it encircles the planet in this true-color view from NASA's Cassini spacecraft taken on February 25, 2011.

The top ten list of scientific discoveries from the Cassini-Huygens mission are:

1. The identification of 44 new moons orbiting Saturn. This brings Saturn's total moon tally to 62, of which only 53 have been officially named.
2. Saturn's moon Enceladus may have microbial life. Cassini found evidence there of liquid water and complex organic chemicals.
3. Titan, Saturn's largest moon, has an internal, liquid water-ammonia ocean. With its open lakes and active weather system, including rainfall, Titan is the most Earth-like body in our solar system.
4. Saturn's F-ring contains more than 500 "mini-jets," strange objects that are over 2,600 feet (800 m) long and leave glittering emission trails through the ring.
5. Titan's surface is covered by what may be pebbles of water ice. These objects were captured in the first pictures the Huygens probe sent back from Titan's surface.
6. Saturn's poles are home to giant, swirling storm systems.
7. Cassini detected small amounts of propylene, a type of plastic made from hydrocarbons, in Titan's lower atmosphere.

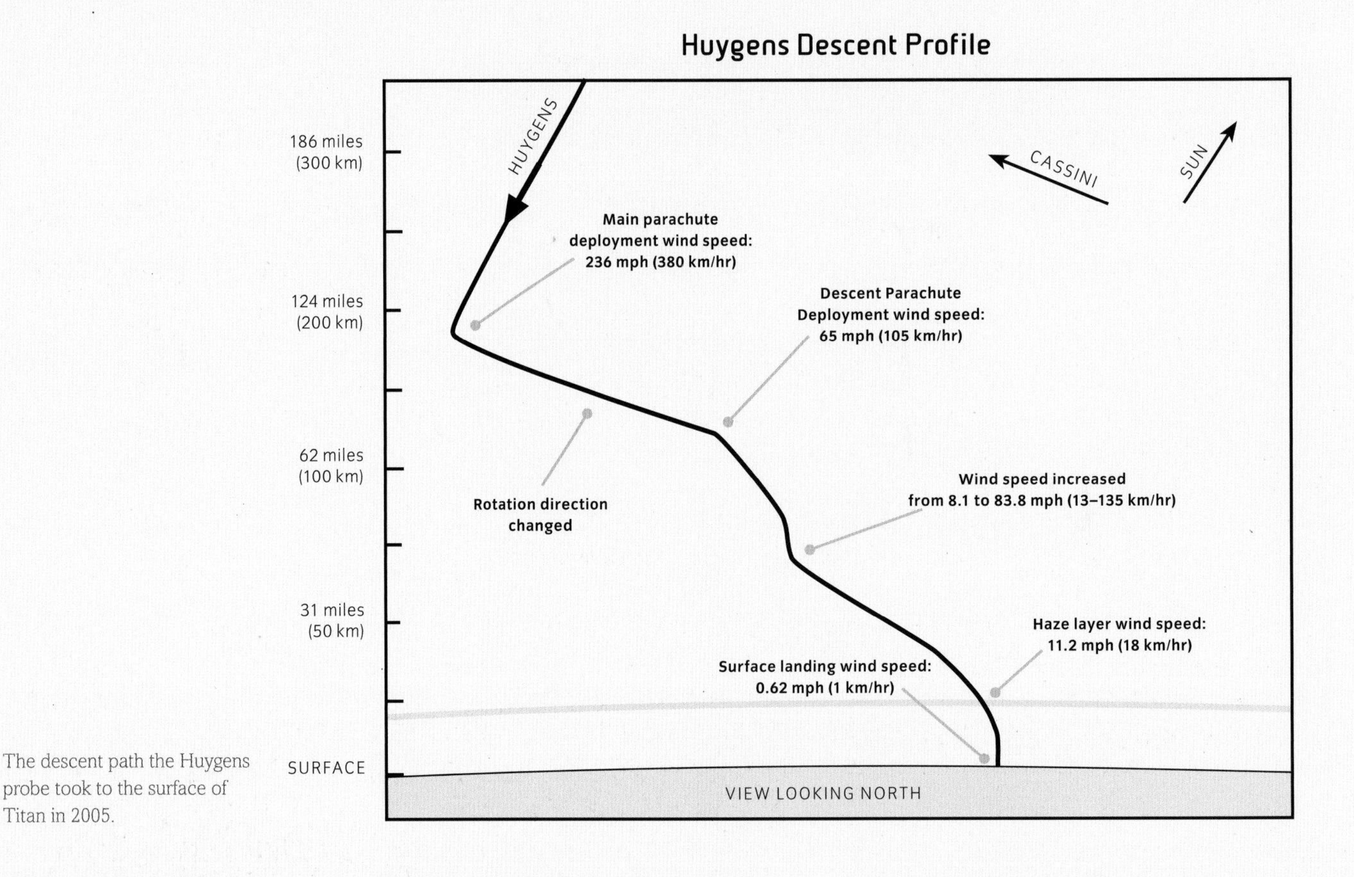

The descent path the Huygens probe took to the surface of Titan in 2005.

Landing on Titan

2005

The mission to land a probe on the most distant body from the Earth attempted to date.

Huygens was ESA's first outer planetary mission. Named after Dutch astronomer Christiaan Huygens (1629–95), who discovered Saturn's moon Titan in 1655, the probe's purpose was to study the geophysical properties of the satellite. But while it would take the spacecraft a long time to reach its destination, its mission was not designed to be a long one: after descending through Titan's atmosphere on a parachute, Huygens was expected to operate for only 153 minutes before its batteries failed. In reality, the probe exceeded expectations, and continued working for more than 210 minutes, including 90 minutes on Titan's surface.

Huygens was equipped with a total of nine instruments, including the Surface Science Package (SSP), which helped to ascertain the composition and physical properties of Titan's surface at the probe's landing site. It also included an accelerometer, tilt sensors, a thermal properties assembly, acoustic properties sensors, and instrumentation to measure fluid on the surface. Key science objectives for the mission were the study of Titan's organic chemistry, and the origin, nature, and distribution of organic compounds on the moon. From imagery and data analysis, mission scientists were able to determine that globules of water ice were present on the moon's orange surface.

Huygens also demonstrated that hydrocarbons exist in abundance on Titan, particularly methane. The moon appears to have oceans of liquid methane, and some images captured by the probe suggest islands and a mist-shrouded coastline. Imagery from the spacecraft before landing confirmed the existence of permanent liquid hydrocarbon lakes at Titan's poles, which are among the chemicals necessary for the evolution of life. Establishing whether or not life currently exists on Titan will make further exploration of the moon a priority for future missions.

8. Cassini detected an exosphere, the outer element of an atmosphere, around Saturn's moons Dione and Rhea.
9. Enormous and long-lasting planet-engulfing thunderstorms erupt from Saturn's atmosphere, affecting the climate of the gas giant for many years. One such eruption occurred in 2010, spawning billowing white ammonia and water ice clouds that were blown around by Saturn's powerful winds. Eventually this storm engulfed the whole northern hemisphere of the planet.
10. Like Earth, Saturn experiences seasons due to the 27° tilt of its axis of rotation. These seasons affect the planet's climate and cloud systems. When Cassini arrived, Saturn's wintry northern hemisphere appeared blue due to a lack of clouds and hazes.

"A mission of this ambitious scale represents a triumph in international collaboration."

EARL MAIZE,
CASSINI PROJECT MANAGER

Scientists received their last contact with Cassini on September 15, 2017, when, at the culmination of its mission, it plummeted into the giant planet. It was a spectacular ending to a spectacular mission. Using its attitude control thrusters, the spacecraft kept its antenna pointed at Earth as long as possible. This way the probe could continue transmitting data that detailed the composition of Saturn's upper atmosphere during its suicidal plunge. Once the atmospheric torque grew too strong, Cassini was no longer able to maintain its orientation and the spacecraft began to tumble, losing radio contact with Earth. After that the probe just burned up in Saturn's upper atmosphere.

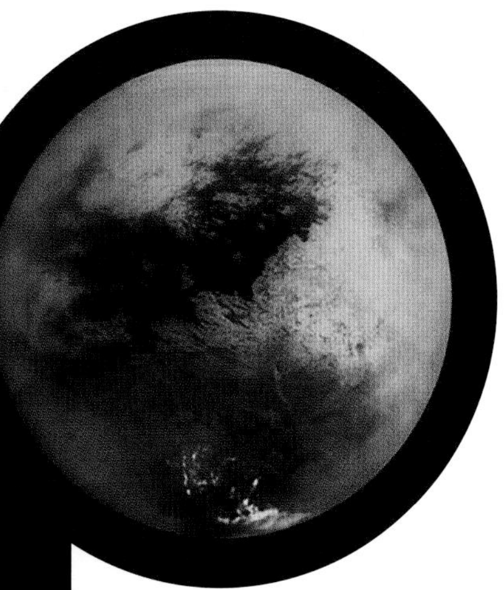

ABOVE A collage portrait of Titan formed of nine images captured by Cassini during its first close flyby of the moon on October 26, 2004.

LEFT Saturn's moons Janus and Prometheus captured by Cassini from just beneath the planet's ring plane.

Potential Life in the Outer Solar System

"*On Titan the molecules that have been raining down like manna from heaven for the last four billion years might still be there, largely unaltered, deep-frozen, awaiting the chemists from Earth.*"

CARL SAGAN

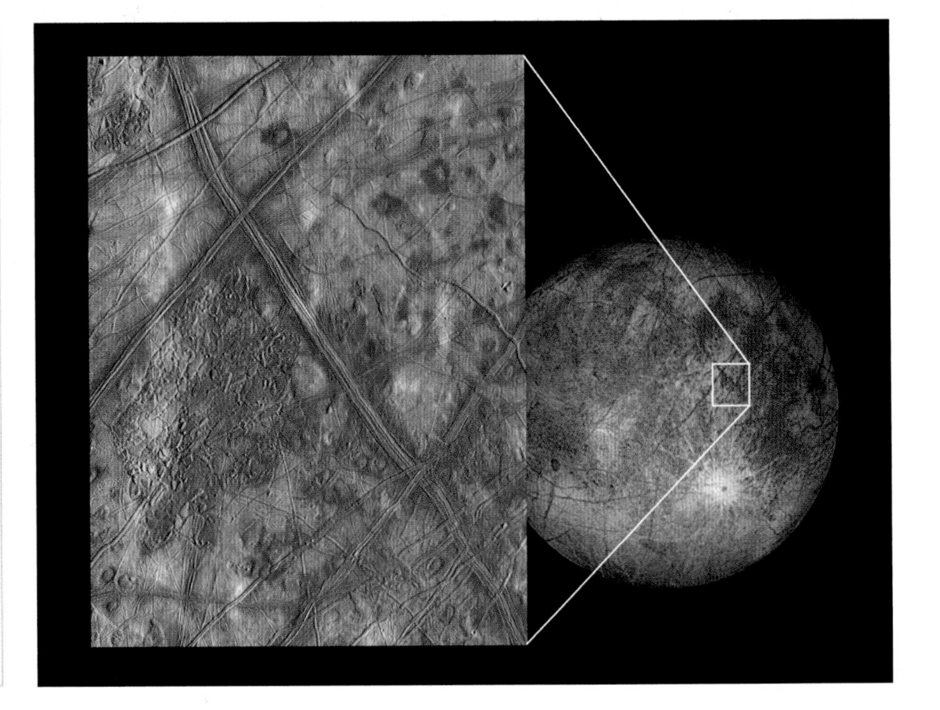

ABOVE The cracks in Europa's crust in these images from NASA's Galileo spacecraft suggest that the moon might possess a subterranean ocean.

OPPOSITE LEFT Ligeia Sea is the second largest known body of liquid on Saturn's moon Titan. It is filled with liquid methane and ethane.

OPPOSITE FAR LEFT Some early concept art for NASA's proposed Europa spacecraft, planned for a launch sometime in the 2020s.

Since the beginning of the space age, humanity has been searching for evidence of life beyond our own world. As of 2018, scientists have not yet found any firm evidence for life elsewhere in our solar system or beyond, but they do have some promising leads.

Building from the data returned by Pioneer 10, Pioneer 11 (p. 294), Voyager 1, and Voyager 2 on their flying visits to Jupiter and Saturn, as well as the more extended missions of the Galileo probe to Jupiter, and Cassini/Huygens to Saturn, scientists are now speculating about the possibility of simple life existing under the methane pools of Titan and the ice of Europa. Both of these are striking targets for future exploration. Europa, a satellite of Jupiter somewhat smaller than the Earth's moon, is covered with a crust of water ice, frozen by a bone-chilling surface temperature of -260°F (-162°C). The tidal tug-of-war created by the gravitational push/pull between Jupiter and its various moons has probably given Europa an internal source of heat through friction. This may have, in turn, created a subterranean ocean on the moon that could be a favorable environment for the development of some kind of simple life.

In the absence of sunlight, which would likely be blocked by Europa's surface ice, any life on the moon would need an alternative source of energy. In the 1970s, marine scientists on Earth discovered dense populations of sea creatures at the bottom of the ocean. The creatures clustered around hydrothermal vents, cracks in the ocean floor that provided a source of heat and chemical nutrients, including sulfur and carbon dioxide. The existence of these creatures thriving under the most hostile conditions startled scientists. It caused them to widen their assumptions about the range of environments under which life can exist. Scientists now believe that isolated areas, such as the seas of Europa, could potentially support life, despite their location far outside of the habitable zone of the inner solar system (p. 292). The next spacecraft to visit Europa should first look for further evidence of liquid water oozing up between cracks in its crust, while a lander mission may eventually seek to melt through the ice, and explore whatever it discovers beneath.

One of the most exciting observations made on Titan during the Cassini-Huygens mission was the discovery of liquid hydrocarbons on its surface. The moon has an extremely dense atmosphere, with frigid temperatures of -290°F (-179°C), but, like Earth, Titan's atmosphere is composed primarily of nitrogen, with significant additional concentrations of methane and argon. Methane can form a variety of heavier organic compounds, and some scientists believe that there is sufficient organic material on the moon to produce chemical evolution similar to that which occurred on Earth when life first evolved there. It will require more probes and further study before scientists can determine with any degree of certainty if there is potential for life on Titan.

The consequence of knowing that life begins, or possibly even began, on other bodies in the solar system would be profound. If life begins easily, and thrives in even hostile environments, the universe could be teeming with biological activity.

Looking down on Titan for Traces of Life

This mosaic of images is comprised of data from the descent imager/spectral radiometer on the Huygens probe overlaid on a radar image of Titan's surface captured by the Cassini orbiter. The red "X" marks Huygens's eventual landing site, which it investigated for evidence that the moon could support life.

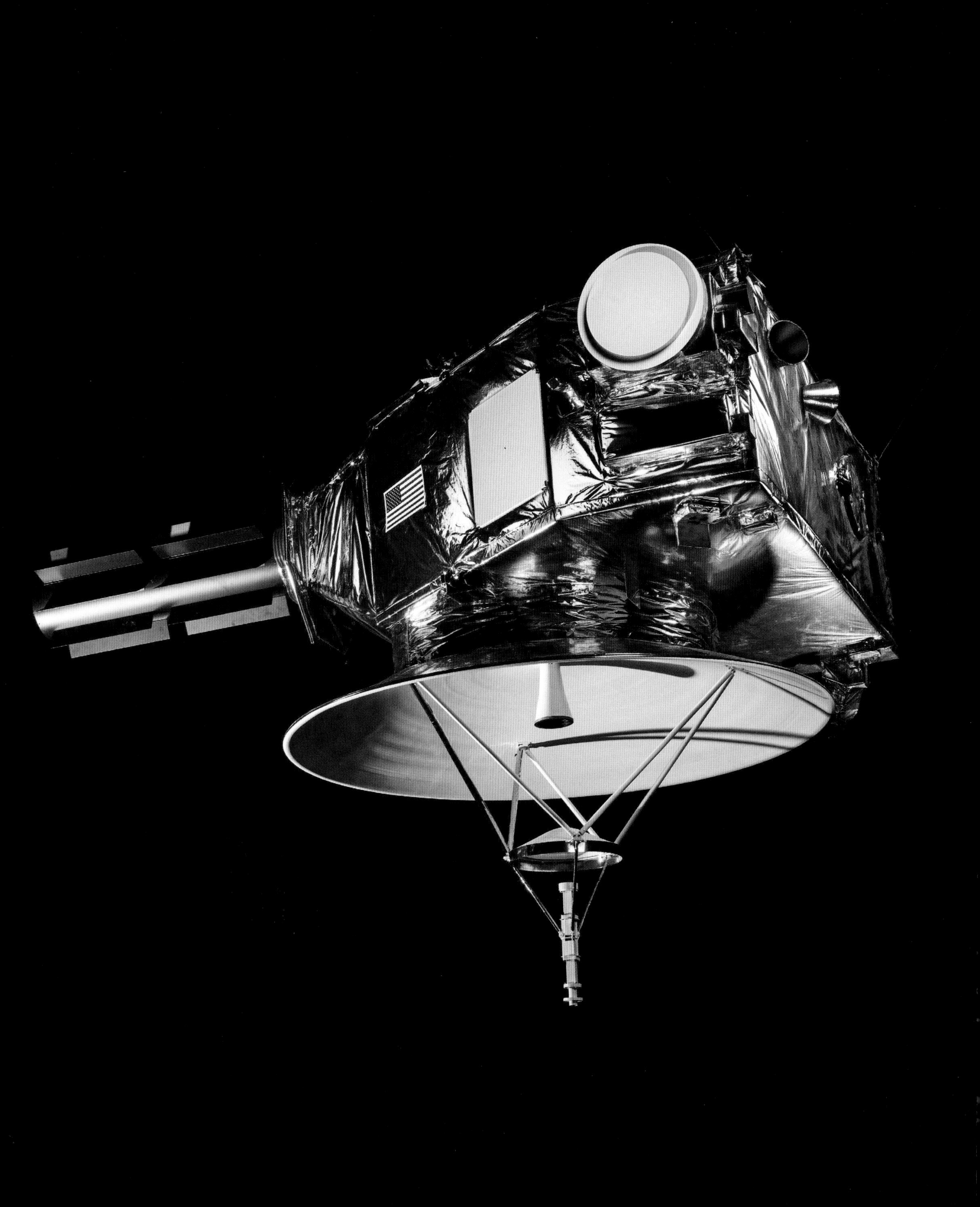

New Horizons to Pluto and the Kuiper Belt

The New Horizons mission, launched on January 19, 2006, marked NASA's first expedition to Pluto, its moon Charon, and several other objects in the Kuiper Belt. It was also a significant long-distance undertaking. The New Horizons spacecraft took almost 15 years to reach its target. During the first 13 months of its voyage, mission scientists performed numerous diagnostic checks, instrument calibrations and tests, and trajectory corrections on the New Horizons probe, as well as rehearsals for an encounter with Jupiter on February 28, 2007.

Similar drills and tests were conducted regularly throughout the probe's post-Jupiter cruise to Pluto, which lasted from March 2007 to June 2015, increasing in frequency as New Horizons approached its destination. On July 14, New Horizons passed Pluto at a distance of only 6,200 miles (10,000 km). Soon after it passed within 16,000 miles (27,000 km) of Charon. The probe continued to study the two bodies through to mid-August 2015 but, because of the long time lag between collection and sending data back to Earth, it took until mid-April 2016 for NASA mission controllers to receive the last of the data from these observations.

Thereafter, New Horizons began an extensive exploration of the Kuiper Belt that will last until at least 2020, involving two encounters of objects ranging from about 25–55 miles (40–90 km) in diameter. The mission plan calls for the spacecraft to acquire the same sort of data it collected on Pluto and Charon. Again, owing to the vast distances involved, data from these encounters would not reach Earth until about two months after each one had concluded.

OPPOSITE This model of New Horizons was created by the Johns Hopkins University Applied Physics Laboratory and donated to the Smithsonian Institution's National Air and Space Museum in 2008.

BELOW This chart shows all the known Trans-Neptunian objects (bodies outside the orbit of Neptune). Kuiper Belt objects are shown in blue tints, and objects in the more distant Scattered Belt region are shown in red tints. The vertical axis indicates the inclination of the object's orbit, with the colored horizontal lines indicating the minimum and maximum distance from the Sun in Astronomical Units (AU).

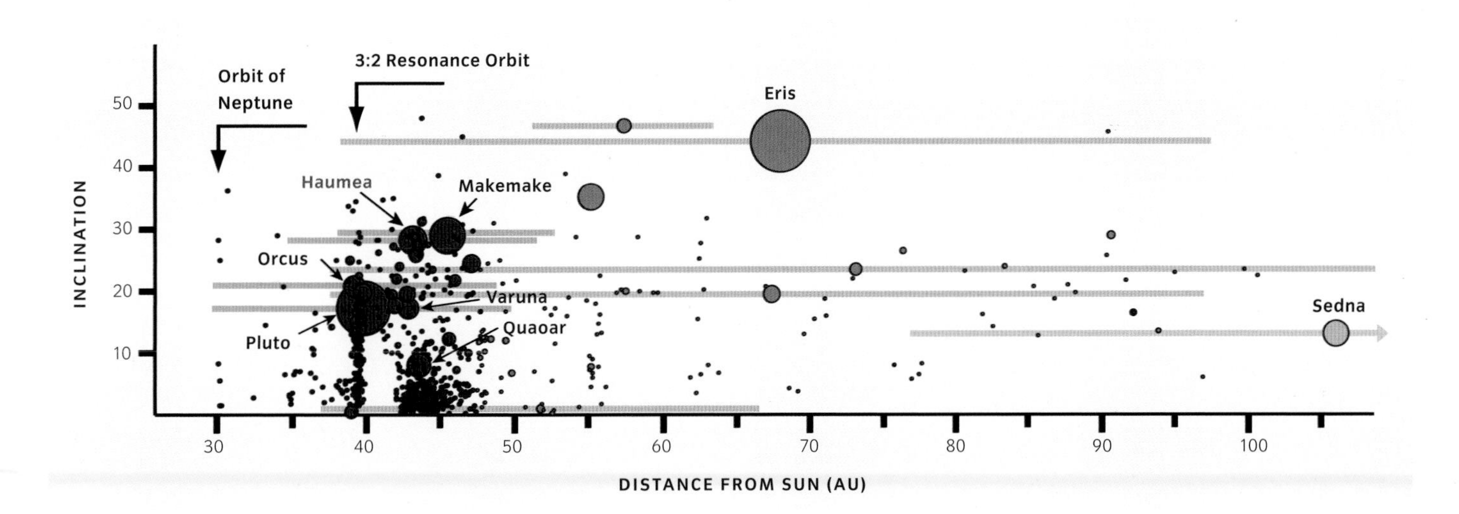

New Horizons Specifications

Length
9 feet (2.7 m)

Width
7 feet, 3 inches (2.2 m)

Height
6 feet, 11 inches (2.1 m)

Weight
1,054 pounds (478 kg)

Power Source
Plutonium-238 radioisotope thermoelectric generator (RTG)

Launch Vehicle
Atlas V 551

Launch
January 19, 2006

Pluto Closest Approach
June 14, 2015,
distance 7,800 miles (12,500 km)

End of Mission
2020

New Horizons also carried nine souvenirs of Earth, including some of the remains of Clyde Tombaugh, the astronomer who discovered Pluto, and a piece of SpaceShipOne.

The 1,054-pound (478-kg) New Horizons spacecraft contained an array of seven instruments designed to answer a number of questions about Pluto. These included: what elements exist in its atmosphere? How do they behave? What does the surface of Pluto look like? Does it possess any big geological structures? How do particles ejected from the Sun, also known as the solar wind, interact with Pluto's atmosphere? The spacecraft's instruments directly measured Pluto's geophysical properties with a degree of overlap, so that the results of each individual instrument could be verified.

The key science achievement of New Horizons' Pluto encounter was the rough age-dating of Pluto's surface through counting the craters visible on its surface. The lower-than-expected count indicated that Pluto has been geologically active throughout much of the past four billion years, with recurring eruptions that covered over old craters. Had this not been the case, the expectation would have been that the surface would have appeared much more pockmarked by impacts like Pluto's moon Charon, which sported many more craters. The visible scars of billions of years of impacts suggest that Charon has not been geologically active for a long time, if it ever was at all. New Horizons also found that Pluto has water ice on its surface, and possesses a more complex chemical composition than had previously been suspected.

Finally, New Horizons' charged-particle instruments revealed that the interaction region between the solar wind and Pluto's vestigial atmosphere is confined to the side of Pluto facing the Sun. This suggests that the Sun's gravitational pull is still significant, even in the Kuiper Belt region.

New Horizons being prepared for launch at the Kennedy Space Center in 2006.

Should Pluto Be a Planet?

1930–2006

The passionate science debate that saw the number of recognized planets in the solar system change.

ABOVE This computer-generated image shows some of the best-known objects found outside Neptune's orbit. Top row (left to right): Xena and its moon, Gabrielle; Pluto and Charon; and Makemake. Bottom row: 2003 EL61 Haumea and its moons; Sedna; and Quaoar.

ABOVE RIGHT Astronomer Clyde Tombaugh using the blink comparator he employed in his discovery of Pluto in 1930.

A debate has raged for many years among astronomers about the classification of Pluto. When American astronomer Clyde Tombaugh first identified Pluto in 1930, he was in no doubt that what he had found was a planet. Indeed, he had been expecting to find such a heavenly body. Astronomers had already detected the gravitational influence of some kind of mass beyond Neptune's orbit, and had even gone so far as to name its cause Planet X. Unfortunately, Pluto's small size could not account for the expected gravitational pull on its own: there had to be other bodies in the same vicinity. This area became known as the Kuiper Belt, after Dutch-American astronomer Gerard Kuiper (1905–1973).

Over time, scientists studying the Kuiper Belt began to find other objects, many of a similar size to Pluto, raising the question of its proper classification. Pluto's status continued to be a point of debate among astronomers. Astrophysicist Neil deGrasse Tyson (1958–) famously excluded Pluto from the solar system display at the Hayden Planetarium in New York City in the early 1990s. Other scientists, such as New Horizons principal investigator Alan Stern (1957–), objected to the move, largely based on the historical precedent that Pluto had already been recognized as a planet.

The discovery of the dwarf planet Eris in 2005, a body even more massive than Pluto, along with two other large Kuiper Belt objects, Makemake and Hauemea, and a similar body in the asteroid belt between Mars and Jupiter named Ceres, further challenged the notion of a nine-planet solar system.

In response, the International Astronomical Union (IAU) called a meeting in 2006 to discuss whether to declare all these new bodies planets, or to simply demote Pluto. After much passionate argument and debate, the IAU announced that Pluto would be given the newly created classification of dwarf planet, alongside Eris and the other similar worlds.

Voyager at the Heliopause

After both Voyager 1 and Voyager 2 had completed their primary missions in 1989, the two spacecraft were handed a new task: exploring the heliopause, the boundary between the limit of the Sun's influence and interstellar space. As part of this extended assignment, dubbed the Voyager Interstellar Mission (VIM) by the scientists at NASA, the probes were to establish the outer limits of the Sun's gravitational and magnetic fields, as well as the parameters of the solar wind, the flow of charged particles thrown out in all directions by the Sun.

This model, transferred to the Smithsonian Institution's National Air and Space Museum in 1978, shows an early design concept for the Voyager spacecraft.

At the beginning, no one knew how successful this extended mission would be. The two spacecraft involved had a finite power source, and had already been operating for 12 years. Also, their instruments had been designed to explore Saturn and Jupiter, and not specifically to monitor the Sun's influence, but they were both moving farther and farther away from the Sun, traveling in different directions.

Voyager 1 is currently more than 141 Astronomical Units (AU) away from the Sun, escaping the solar system at a speed of about 3.6 AU per year, roughly 35° above the ecliptic plane, in which all the planets in our solar system orbit. Voyager 2 is a mere 114 AU away from the Sun, escaping the solar system at a speed of about 3.3 AU per year, 48° below the ecliptic plane.

On their journeys out of the solar system, both spacecraft are measuring the diminishing influence of the Sun. In the region in which the Sun's influence gives way to interstellar space, termed the termination shock, the speed and direction of travel of the charged particles that comprise the solar wind begin to change. Voyager 1 crossed the termination shock at 94 AU in December 2004,

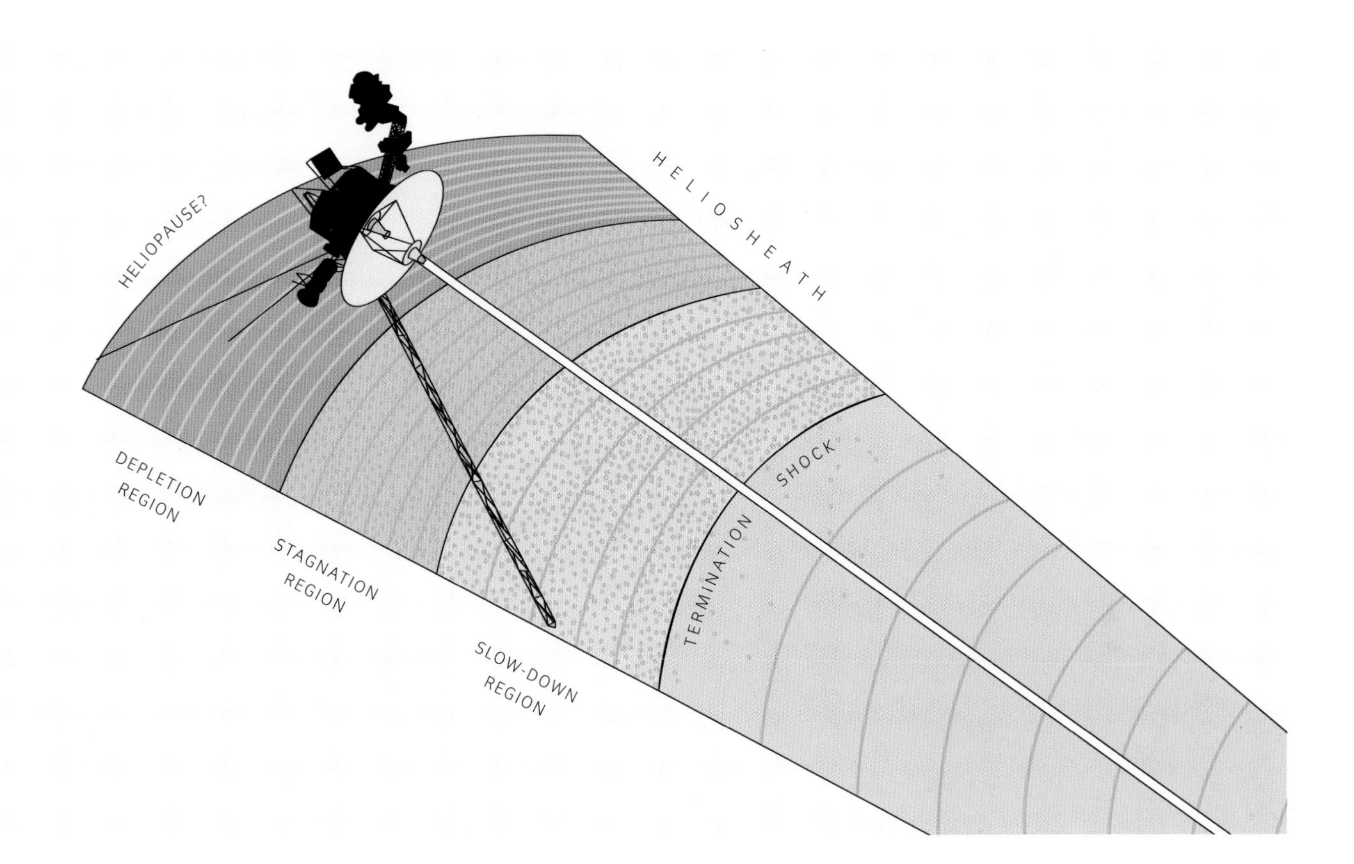

This diagram shows NASA's Voyager 1 spacecraft exploring the "depletion region" at the outer limits of our heliosphere, the bubble of material the Sun blows around itself and the whole solar system. In this region, the magnetic field lines generated by the Sun (yellow arcs) are piling up and intensifying while the low-energy charged particles of the solar wind (green dots) have disappeared.

with Voyager 2 following it at 84 AU in August 2007. This extremely small dataset, consisting of just two measurements, seems to indicate that there is a considerable variation in where the termination shock begins. Beyond the termination shock, the Voyager spacecraft entered new transitional regions between our solar system and interstellar space. In the first of these, the slow-down region, the speed of the charged particles streaming out from the Sun slowed down, while the number of particles detected increased. Then, in what scientists refer to as the stagnation region, around 113 AU from the Sun, the charged particles cease their outward movement, and sporadically reverse direction. In the depletion region, there are fewer detectable charged particles, and more cosmic rays, although the Sun's magnetic influence still appears to be restricting their direction.

Finally exiting the heliopause may take years, since no one knows exactly how far it extends. The Voyager spacecraft may or may not remain active through this process. To prolong operational life, NASA engineers have cut power to the probes' cameras and many of their instruments. The current estimate is that the Voyager probes will have enough electrical power to remain operational, and enough thruster fuel to keep their antennas oriented toward Earth, until at least 2020. At some point beyond that, when the pair eventually run out of fuel, or power, or both, they will no longer be able to supply scientists with data, bringing to a close humanity's longest-range space exploration mission to date.

> *"I do not know where I'm going, but I'm on my way."*
>
> CARL SANDBERG, AMERICAN POET

NASA
esa

The Hubble Space Telescope and the Universe beyond the Solar System

As early as the latter 1960s, NASA started to research a plan to build a large space-based telescope as a means of seeing distant stars and galaxies with greater clarity than is possible when viewing them through the various gases in the Earth's atmosphere. The project underwent a number of changes during its development, but it eventually became the Hubble Space Telescope (HST).

Initially designed for launch on a Saturn rocket, it was eventually redesigned to be deployed and serviced by the smaller Space Shuttle. To compensate for the change in dimensions, NASA made use of the latest military-grade optical techniques. The result was a space telescope with a 94-inch (238-cm) mirror that cost $2 billion to develop and build.

The HST was placed in orbit in April 1990, an event greeted with much excitement and anticipation by astronomers across the globe. Hubble was expected to deliver a quantum leap forward in astronomical capability, allowing astronomers to view galaxies as far as 15 billion light years away with greater resolution than ever before. Unfortunately, there was a problem with a key component. The mirror had what was termed a "spherical aberration," a defect only one-twenty-fifth the width of a human hair, which prevented the HST from focusing incoming light into a single point. Consequently, the images it produced were blurry.

At first, many believed that the defect would cripple the telescope, and NASA received considerable negative publicity. But soon scientists found a way to use computer enhancement to compensate for the defect, and engineers planned a shuttle mission to install additional equipment to correct the issue.

Since it first became operational, images from the HST have led to a large number of scientific discoveries. For example, the images of galaxy formation captured by the HST suggest that elliptical galaxies developed into their present shapes remarkably rapidly, while spiral galaxies evolved over a much longer period of development. The HST also discovered a new dark spot on Neptune, captured

Hubble Space Telescope Specifications

Length 43 feet, 3 inches (13.2 m)

Width 13 feet, 9 inches (4.2 m)

Weight 24,490 pounds (11,110 kg)

Power Source Solar (2,800 watts)

Launch Vehicle
Space Shuttle Discovery, STS-31

Launch April 24, 1990

End of Mission
Ongoing until at least 2020

OPPOSITE The Hubble Space Telescope at the Lockheed assembly plant being readied for transportation to the Kennedy Space Center for launch.

"We often frame our understanding of what the [Hubble] space telescope will do in terms of what we expect to find, and actually it would be terribly anticlimactic if in fact we find what we expect to find.... The most important discoveries will provide answers to questions that we do not yet know how to ask and will concern objects we have not yet imagined."

JOHN BAHCALL, AMERICAN ASTRONOMER

images of stars beginning to form in the stellar nursery of the Eagle Nebula, and documented in colorful detail the births and deaths of various celestial objects.

The HST provided visual proof that pancake-shaped dust disks around young stars are common, and showed for the first time that jets of material rising from embryonic stars emanate from the centers of disks of dust and gas, turning what was previously just a theory of star and planet formation into an observed reality.

The telescope further identified Supernova 1987A, the closest exploding star to the Earth observed in the past four centuries. The HST has captured pictures of a collision between a wave of material ejected from the doomed star and a ring of matter surrounding it. In the next decade, astronomers expect even more material to hit the ring, illuminating the surrounding region, literally throwing light on the exploding star's history.

Hubble has even undertaken a census of 27 nearby galaxies, and found black holes in three of them. Accordingly, the conditions that allow black holes to form are now viewed as much more common than previously thought. Although several astronomers had previously found massive black holes dwelling in galaxies the size of the Milky Way or larger, these new results suggest that virtually all galaxies of any size or type possess at least one black hole within them. This discovery goes a long way toward expanding scientific knowledge about the conditions and processes that led to the formation of our galaxy and solar system.

However, perhaps the telescope's greatest achievement to date comes from images taken in 1998 of light emitted by supernovae, such as the exploding star Sanduleak -69° 202a. By measuring the speed of the light from these novae, scientists calculated that the universe's expansion is speeding up.

These two images of the M100 Galactic Nucleus were both taken by the Hubble Space Telescope. The image on the left was taken with the Wide Field Planetary Camera-1 in 1990 while the image on the right was taken with the Wide Field Planetary Camera-2 in 1994 after the telescope received its first servicing.

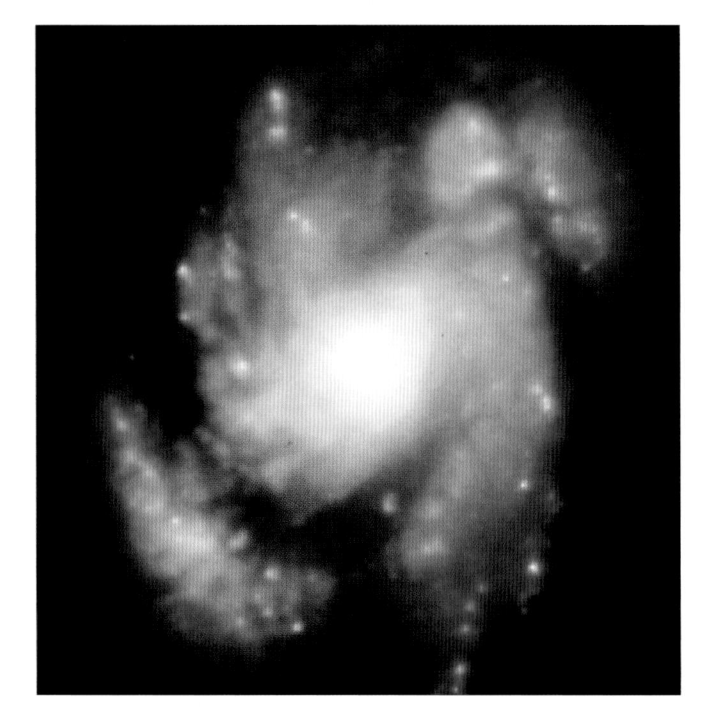

Servicing Hubble

1993–2009

Like an automobile, space telescopes require regular maintenance to maximize their working life span.

In December 1993 NASA sent the Space Shuttle Endeavour to the HST on a vital repair mission to insert corrective equipment into the telescope and to service other instruments. The mission, which required five separate space walks, was a resounding success. The first reports from the newly repaired HST indicated that the clarity of images being obtained had improved significantly. This and a further four servicing missions conducted between 1993 and 2009 further enhanced the telescope's performance.

Arguably the most significant astronomical instrument in use, the HST has facilitated the observation of the effects of black holes and the identification of galaxies that are farther and further away, in distance and in time, than was previously possible. It has also gathered data that has helped scientists calculate a rough estimate for the age of the universe.

Anchored on the end of the Canadarm Remote Manipulator System, astronauts Jeff Hoffman (foreground) and Story Musgrave service Hubble during the STS-61 mission in 1993.

Hubble Servicing Missions

Mission Number	Shuttle Flight	Dates	Comments
SM1	Endeavour STS-61	December 2–13, 1993	First maintenance on HST; astronauts installed equipment to correct the flaw in Hubble's primary mirror and other instruments.
SM2	Discovery STS-82	February 11–21, 1997	Astronauts installed new instruments to extend HST's wavelength range to near infrared for imaging and spectroscopy and replaced various failed or degraded spacecraft components to increase efficiency and performance.
SM3A	Discovery STS-103	December 19–27, 1999	An urgent mission undertaken when the fourth of six gyros failed on November 13, 1999. HST required three to maintain stability. Hubble entered safe mode until the gyros were replaced. This required NASA to split the third servicing mission into two parts to more quickly bring HST back into operation
SM3B	Columbia STS-109	March 1–12, 2002	Astronauts replaced HST's solar panels and installed the Advanced Camera for Surveys (ACS), which took the place of its Faint Object Camera, the telescope's last original instrument.
SM4	Atlantis STS-125	May 11–24, 2009	In final servicing mission astronauts installed two new scientific instruments: Cosmic Origins Spectrograph (COS) and Wide Field Camera 3 (WFC3). Two failed instruments, the Space Telescope Imaging Spectrograph (STIS) and the Advanced Camera for Surveys (ACS), were brought back to life by the first ever on-orbit repairs. To prolong its life, new batteries, new gyroscopes, and other modifications were also completed.

The Great Observatories

Around the same time that Voyager 1 and Voyager 2 were reshaping our knowledge of the solar system, space scientists were investigating humanity's understanding of the wider universe beyond its boundaries. The traditional field of astronomy was reinvigorated by the technology of the space age, particularly with the advent of orbiting resources, such as the space telescopes. In addition to greatly enhanced capabilities for observation in the visible light spectrum, space agencies across the world supported the development of a wide range of x-ray, gamma ray, ultraviolet, infrared, microwave, cosmic ray, radar, and radio astronomical projects. These efforts have collectively facilitated the most systematic efforts in history to explain the origins and development of the universe.

"In a new adventure of discovery no one can foretell what will be found, and it is probably safe to predict that the most important new discovery that will be made with flying telescopes will be quite unexpected and unforeseen."

LYMAN SPITZER JR., AMERICAN THEORETICAL PHYSICIST AND ASTRONOMER

By the early 1980s, satellite-based astronomical observatories had helped to generate a major change in the larger field of astronomy, and had reordered thinking on the subject. Many ground-based astronomers were at first reluctant to use space-based telescopes, but after witnessing the clearer results that such telescopes produced, those same astronomers soon came to embrace the technological advance. But the addition of more space-based astronomy equipment was just the beginning. NASA's Great Observatories program has produced four major space-based projects between 1990 and the present. Each one was designed to conduct astronomical studies over different wavelengths of the electromagnetic spectrum: visible light, gamma rays, x-rays, and infrared, and each could function in isolation, but when used in conjunction with each other, the four observatories

An artist's illustration of the Chandra X-ray Observatory in orbit.

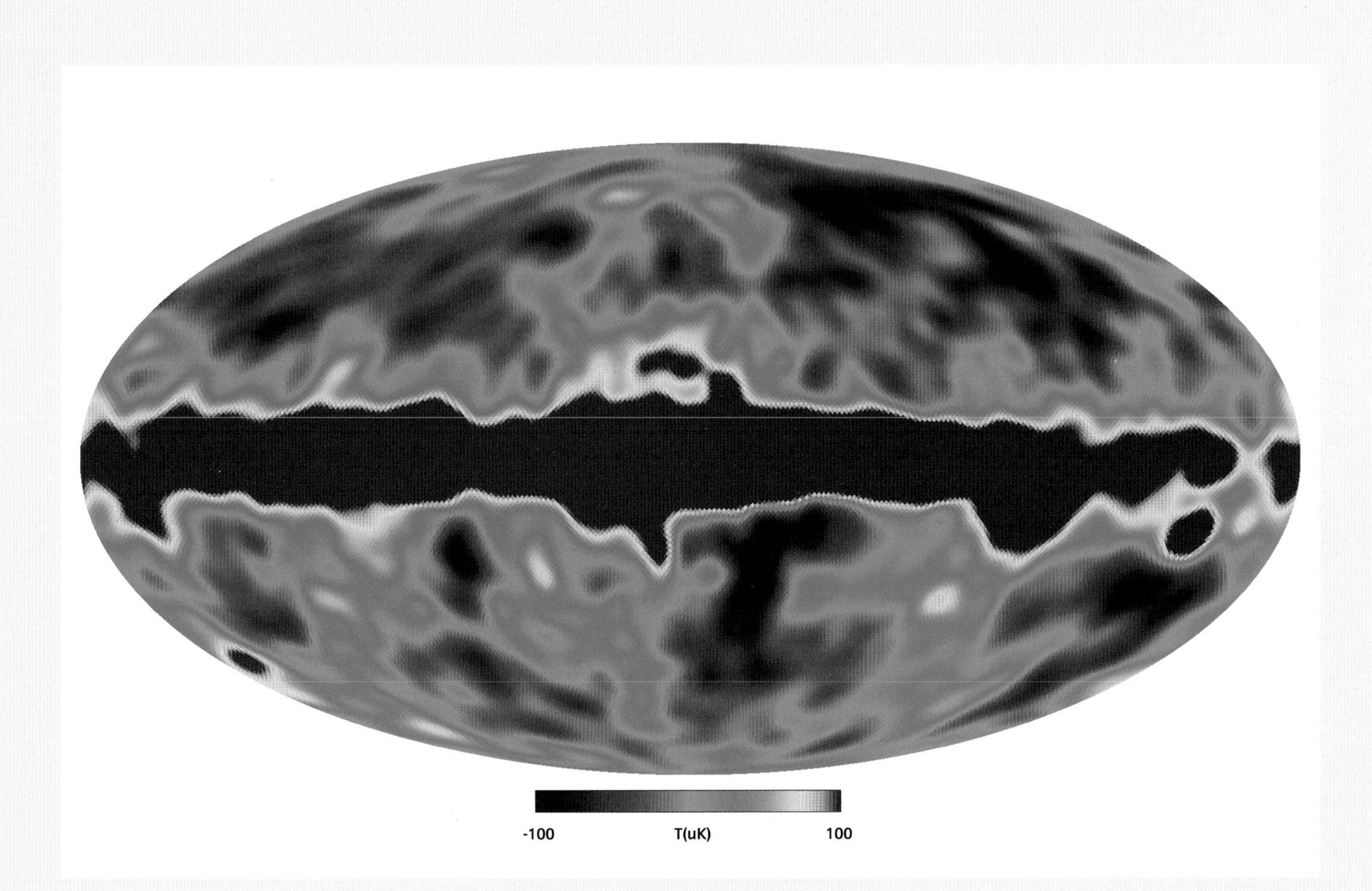

The Cosmic Background Explorer

1989–93

The probe that explored the origins of our universe by examining the cosmic background radiation.

Shortly before its Great Observatories program, NASA launched another spacecraft with a mission focused on exploring the wider universe in 1989. The Cosmic Background Explorer (COBE) was designed to search for traces of microwave radiation left over from the Big Bang, the event that is thought to have been the origin of the universe. Over its short operational life, the spacecraft found two key pieces of evidence that supported the Big Bang theory. The first of these was measuring the temperature of the microwave background radiation, and the second was establishing the relationship between the tiny variations in the remaining heat in the microwave background and the possible location for the Big Bang event that caused it.

COBE's principal investigators, George Smoot III (1945–) and John Mather (1947–), received the Nobel Prize in Physics in 2006 for their work on the project. According to the Nobel Prize committee, "the COBE project can also be regarded as the starting point for cosmology as a precision science." Mather, who had primary responsibility for COBE's measurements of cosmic background radiation, was the first NASA scientist to win a Nobel Prize, while Smoot was responsible for measuring the variations in its temperature.

This image shows the minute variations, within ± 100 microKelvin, of cold and hot regions in the microwave background detected by COBE.

allowed astronomers to study the same object in the cosmos at divergent spectral wavelengths in greater depth than ever before.

The first of the Great Observatories, and certainly the most famous, was the Hubble Space Telescope (HST) (p. 331). The second was the Compton Gamma Ray Observatory (CGRO). Named after gamma ray physicist Arthur Compton (1892–1962), CGRO was deployed on April 5, 1991, from the Space Shuttle Atlantis on mission STS-37. It operated until 2000, and collected scientific data on some of the most violent physical processes in the universe, such as supernovae and black holes, and extremely high-energy emissions from stellar bodies beyond this solar system. Its four instruments—Burst and Transient Source Experiment (BATSE), the Oriented Scintillation Spectrometer Experiment (OSSE), the Imaging Compton Telescope (COMPTEL), and the Energetic Gamma Ray Experiment Telescope (EGRET)—explored the high-spectral region of gamma rays that are commonly emitted by stars. Among other findings, CGRO conducted the first all-sky survey of all objects visible to its orbital position, the most comprehensive survey yet of the galactic center of the Milky Way, discovered a possible antimatter source, and definitively showed that the majority of gamma-ray bursts must originate in distant galaxies, and therefore must be enormously energetic. After ten years of operation, CGRO deorbited on June 4, 2000, and burned up in Earth's atmosphere.

The Chandra X-ray Observatory (CXO), named after the Indian-American astrophysicist Subrahmanyan Chandrasekhar (1910–95), was the third of the Great Observatories. It was deployed from the Space Shuttle Columbia on mission STS-93 on July 23, 1999. Focused on observing black holes, quasars, supernovae, dark matter, and high-temperature gases throughout the x-ray portion of the electromagnetic spectrum.

The Spitzer Space Telescope, named for theoretical physicist Lyman Spitzer (1914–97), was the last of the Great Observatories. It monitored celestial bodies at the infrared area of the spectrum, something that cannot easily be done from Earth-based observatories because the planet's atmosphere blocks most interstellar infrared radiation. Launched on August 25, 2003, Spitzer detects infrared energy radiated by objects in space between wavelengths of 3 and 180 microns.

OPPOSITE An artist's impression of the Spitzer Space Telescope.

BELOW In September 2017, NASA's Spitzer Space Telescope provisionally detected the faint afterglow of the explosive merger of two neutron stars in the galaxy NGC 4993. The left panel is a color composite image from the Spitzer, rendered in cyan and red. The center panel is a median-filtered color composite showing a faint red dot at the known location of the merger. The right panel shows the residual data after subtracting out the light of the galaxy from an older image of the same section of space that predates the merger.

Exoplanets and the Search for Another Earth

Holiday in the Suns

NASA's JPL created this poster to advertise its discovery of the exoplanet Kepler-16b. While the poster portrays the exoplanet as a rocky world, and a potential holiday destination, NASA admits that it could equally be a gas giant. However, like Tatooine in the 1977 film *Star Wars*, Kepler-16b does orbit binary stars, and would witness similar double sunsets.

> *"By monitoring auroral activity on exoplanets, we may be able to infer the presence of water on or within an exoplanet; now, it's not going to be easy—it's not as easy as Ganymede and Jupiter, and that wasn't easy. It may require a much larger telescope than Hubble, it may require some future space telescope, but nevertheless, it's a tool now that we didn't have prior."*
>
> HEIDI HAMMEL, AMERICAN ASTRONOMER

Throughout human history, the belief has persisted that humans are not alone in the universe, but until relatively recently there was little hard evidence on which to base this notion. Then two key findings came together that provided some scientific support for the idea that life might exist beyond Earth. Firstly, researchers unearthed evidence that life was able to persist in the extreme environments around geothermal vents at the bottom of the oceans and even within rocks. These examples presented, according to the popular science communicator Bill Nye, "compelling evidence…that the environmental limits for living things are set pretty far apart." Secondly, beginning in 1995, the first planets outside our solar system—or exoplanets, as they are often called—were discovered, fueling limitless speculation about the number of planets that might exist in the universe.

Using new instruments, technologies, and techniques, a loose confederation of scientists around the world is engaged in detecting and cataloguing a rising number of exoplanets around other stars. The first such planet to be discovered, 51 Pegasi B, later named Dimidium, was detected as a result of observations undertaken by Michel Mayor (1942–) and Didier Queloz (1966–), of the Observatoir de Genève, Switzerland, as reported in the science journal *Nature* on October 6, 1995. Dimidium is 50 light-years away from Earth in the constellation Pegasus. Far from Earth's twin, this world is more akin to Jupiter, and orbits closer to its parent star than Mercury is to the Sun.

The Kepler Observatory

2009–

The space observatory that has confirmed the existence of over 1,000 exoplanets so far and is still going strong.

Since 1995 several thousand planets have been discovered within a few hundred light years of Earth, and the Kepler spacecraft has identified thousands of other candidates out to 1,600 light years away from our home planet, using a photometric technique of transit, which involves capturing images of stars over long periods, and studying them for the change in emitted light caused by a planet passing in front of them.

Named for German astronomer Johannes Kepler, the Kepler mission was the brainchild of NASA scientist William Borucki (1939–), but outside of the delivery of the probe itself, the key to the success of the wider Kepler mission has been the organization of a worldwide network of hundreds of scientists it fostered. With access to many of the largest Earth-based telescopes in the world, these scientists surveyed and classified the stars by their potential to host planetary systems as candidates for further study by Kepler. They also conducted crucial follow-up observations to confirm Kepler-detected planetary candidates.

Launched on March 7, 2009, the Kepler spacecraft has discovered 1,013 confirmed exoplanets to date. Extrapolating from this, scientists are now confident that our galaxy hosts tens of billions of Earth-sized planets that could reside in the habitable zones of their parent stars. This important result lays the foundation for future studies of exoplanet atmospheres that would take us closer to identifying planets that could support complex life elsewhere in the universe.

Kepler Specifications

Width 8 feet, 10 inches (2.7 m)

Length 15 feet, 5 inches (4.7 m)

Weight, Empty
2,320 pounds (1,052.4 kg)

Power Source
Solar, 1000 watts

Launch Vehicle Delta II

Launch March 7, 2009

End of Mission Ongoing

William Borucki (front left) and the Kepler mission team receive the Smithsonian Institution's National Air and Space Museum 2015 Trophy for Current Achievement at a ceremony in Washington on March 25, 2015, from the museum's Director General John Dailey.

Transitory Success

Pictured here, the New Mexico Exoplanet Spectroscopic Survey Instrument (NESSI) has its own way to identify and analyze exoplanets. Termed transit spectroscopy, the observatory uses its near-infrared spectrographic system to search for atmospheric traces of unseen exoplanets orbiting distant stars. NESSI is trained on a target star, which it observes for a period of time. When an exoplanet is spotted crossing in front of and behind the star, the observed light is passed through the spectrometer, which breaks it apart to reveal the chemical composition of the exoplanet's atmosphere.

As of November 30, 2017, 3,558 confirmed exoplanets have been discovered. Of these, there are 1,414 that are Neptune-like, another 1,186 that are smaller gas planets, 881 terrestrial-like planets, 52 super-Earths (Earth-like worlds that are up to ten times the Earth's mass), and 25 that have yet to be characterized. Scientists are finding more exoplanets all the time. Indeed, there are currently 4,496 possible exoplanets that require additional observation to confirm their status. While none have yet been discovered that are exactly like Earth, scientists believe that it is only a matter of time before one is found, especially as eager planet-hunters continue to develop new techniques for identifying and verifying exoplanets.

One of the most distant exoplanets discovered so far, the gas giant OGLE-2014-BLG-0124L, located about 13,000 light-years from Earth, was spotted in 2014 by the Spitzer Space Telescope and the Mount John University Observatory using a technique called gravitational microlensing. This approach reveals the presence of unseen planets through observing the effects their gravity has on other far-off objects. In essence, gravitational microlensing allows scientists to detect objects that range from the mass of a large planet to the mass of a star, regardless of the light they emit, by measuring how their gravity bends light around them.

Although Earth-sized planets are still too small to be detected directly, even through gravitational microlensing, it appears that "super-Earths" are common. The first super-Earth was discovered in 2005, and others, such as Gliese 581d and Gliese 581c, have masses of eight and five Earths respectively. In 2009, Bulgarian astronomer Dimitar Sasselov (1961–) proposed a three-band classification of super-Earths: those composed mostly of rocky materials, though subject to considerably higher pressures than on Earth depending on mass; more exotic water worlds with deep oceans, and carbon-rich planets in which there is more carbon than oxygen. Only time will tell how widely applicable this classification will prove. What is certain at this point is that, just as super-Earths exist, an exoplanet that is even closer to Earth in size and scale will almost certainly be discovered eventually.

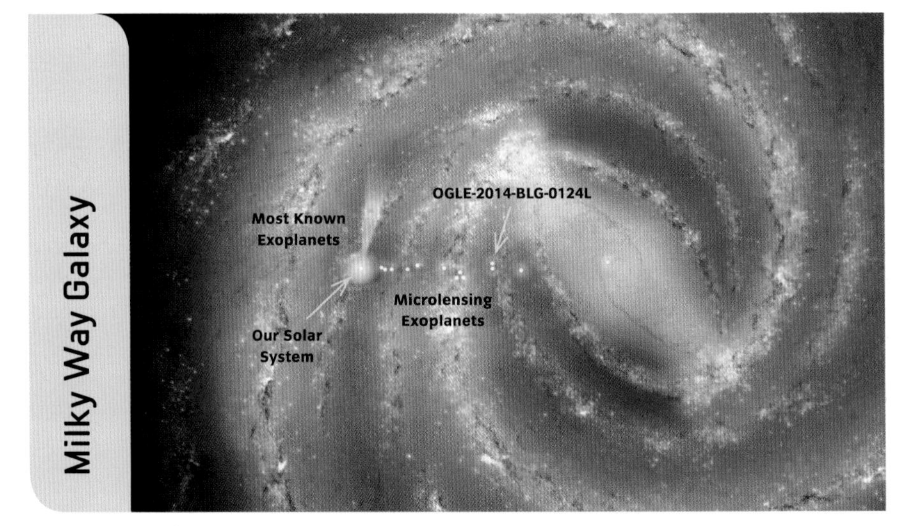

RIGHT In this diagram, planets discovered with microlensing are shown in yellow. The farthest lies in the center of our galaxy, 25,000 light years away.

OPPOSITE A newly discovered exoplanet, Kepler-452b, is the closest match so far for an Earth-like planet. This artist's impression of a planetary lineup shows five habitable-zone planets with similarities to our own world. Following on from Earth at the top these are Kepler-186f, Kepler-62f, Kepler-452b, Kepler-69c, and finally Kepler-22b.

The SETI Program

The Robert Byrd Green Bank Telescope, the world's largest fully steerable radio telescope, in Green Bank, West Virginia, sometimes used for the SETI program.

In 1960, American astronomer Frank Drake (1930–) went to West Virginia and tuned a dish at the National Radio Astronomy Observatory to 1,420 MHz and pointed it at Epsilon Eridani, a Sun-like star some ten light years away. Drake theorized that if intelligent life existed elsewhere in the universe, it would probably communicate using electromagnetic signals, just has humans do. This led him to conclude that detecting regular signals of extraterrestrial origin would provide the strongest evidence yet that intelligent life exists elsewhere in the universe. To his astonishment, Drake soon identified a strong signal of unknown origin, but upon further investigation he was disappointed to learn that he had not in fact stumbled upon proof of extraterrestrial intelligence, but a secret Earth-based military broadcast.

Drake was not alone in either his desire to find proof of extraterrestrial intelligence, nor in his conviction that searching the heavens for signals was an effective means for identifying possible evidence of an alien civilization. During the 1970s, a group of American space scientists successfully lobbied their nation's government for funds to help them pursue such a search. They called their efforts "SETI," the Search for Extraterrestrial Intelligence. In the 1990s, these space scientists and their allies at NASA petitioned the American government to accelerate the search by funding Project Cyclops, a $20 billion network of 1,500 ground-based antennas scanning the skies for potential alien signals. Their proposal was not well received. Not only did the SETI scientists fail to secure financial backing for their new project, they also lost their existing government funding. This forced them to pursue private financing through a new non-governmental body called the SETI Institute.

Scientists engaged in the search through the institute would purchase time on ground-based radio telescopes of all sizes to collect transmissions from nearby stars and analyze the data for patterns that could only be made by intelligent beings. For each observation of a single star, signal detection computers would examine tens of millions of channels within a 10 MHz bandwidth. Computers would filter out signals from ground-based sources on Earth and also eliminated signals from the growing number of telecommunications satellites. The remaining results were then compared to the universe's natural background noise. Any strange signals would trigger a special procedure, called FUDD, or follow-up detection device. Two separate radio telescopes a significant distance apart would track the signal and try to verify what had been heard. So far, no signal has ever been verified through the FUDD procedure. A signal from an extraterrestrial civilization could come tomorrow; one may never come at all.

> *"A single message from space will show that it is possible to live through technological adolescence.... It is possible that the future of human civilization depends on the receipt of interstellar messages."*
>
> CARL SAGAN

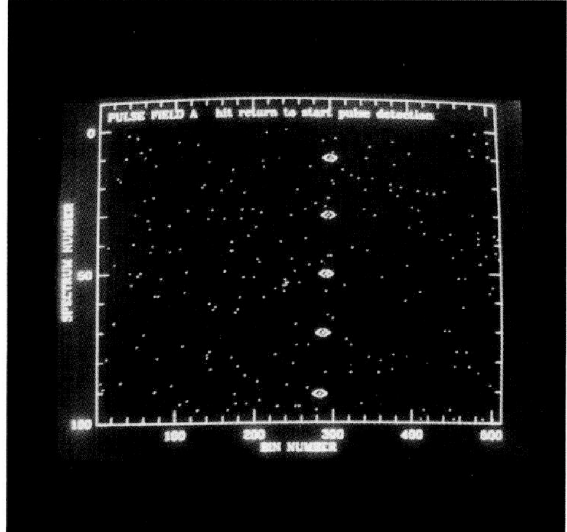

LEFT The Very Large Array Radio Telescope in Socorro, New Mexico, has provided vast quantities of data for SETI observations.

ABOVE The SETI Multichannel Spectrum Analyzer instrument from Stanford University was used in the first SETI sky survey.

> *"The existence of life beyond Earth is an ancient human concern. Over the years, however, attempts to understand humanity's place in the cosmos through science often got hijacked by wishful thinking or fabricated tales."*
>
> JILL TARTER, AMERICAN SCIENTIST AND FORMER DIRECTOR OF THE SETI INSTITUTE

Despite the lack of any positive results, since the search began in 1960, the SETI project has taken a small but significant place in American culture, spawning a variety of popular accounts of their scientific pursuit. Motion pictures, such as *Species* (1995), *The Arrival* (1996), and *Contact* (1997), have romanticized the search for extraterrestrials. Meanwhile, the SETI Institute's radio program, "Are We Alone?," hosted by astronomer Seth Shostak, translates the science behind the search into engaging nuggets understandable to all.

SETI's groundbreaking citizen science program, SETI@home, a volunteer computing project established in 1999, presented anyone with the opportunity to run as a computer background process analysis of individual SETI information packets to find patterns in the data. With over 200,000 active participants as of 2017, SETI@home volunteers have reviewed scans of over 67 percent of the sky observable from the Arecibo radio telescope.

In 2015, SETI began a $100 million campaign to raise funding for a ten-year initiative called Breakthrough Listen, which seeks to harness new instruments and techniques for the ongoing SETI search. Whether or not the organization reaches its funding goal, there is virtually no one associated with space activities around the globe who does not believe civilizations exist beyond Earth. The same is true of many space scientists. They fervently believe intelligent extraterrestrial life exists, but that it has just not been discovered yet. It remains widely hoped among the space exploration community that scientists will eventually be able to determine the frequency with which life evolves across the universe. Such a revelation would likely serve to both focus SETI's ongoing search efforts and also further increase the public's interest in the search.

RIGHT SETI scientist Seth Shostak explains to an audience at NASA that SETI is not looking for gray aliens.

FAR RIGHT Former SETI director Jill Tartare, the inspiration for the main character in *Contact* (1997), at work in her office.

The Drake Equation

1961

The method space scientists use to quantify what they don't know about intelligent extraterrestrial life.

Defending the search for extraterrestrial life in a 1975 issue of *Scientific American*, Carl Sagan and Frank Drake (1930–) announced "our best guess is that there are a million civilizations in our galaxy at or beyond the Earth's present level of technological development." This estimate is predicated on a famous equation that Frank Drake concocted in 1961. It is expressed as follows:

$$N = R_* f_p n_e f_l f_i f_c L$$

In the equation N is equal to the number of technological civilizations where the other variables are:

- R_* The average rate of star formations in our galaxy.
- f_p The fraction of formed stars that have planets.
- n_e For stars that have planets, the average number of planets that can support life.
- f_l The fraction of those planets on which life actually develops.
- f_i The fraction of planets bearing life on which intelligent, civilized life has developed.
- f_c The fraction of these civilizations that have developed communications technologies that release detectable signals into space.
- L The length of time over which such civilizations release detectable signals.

At this time, each and every one of these variables remains unknown, but Drake didn't design his equation as a math problem to simply be solved; instead it serves as a framework to organize our thoughts on the scale of the task involved in finding active, technologically advanced extraterrestrial civilizations. Its use in museums and science centers has helped increase public engagement with both science in general and the ongoing search for extraterrestrial intelligence in particular.

ABOVE Frank Drake demonstrating his equation, which sets out the framework for establishing how many developed civilizations there might be in the universe.

NASA

10

Transterrestrial Expectations

From the start of the space age in the twentieth century, space scientists and engineers developed the means to reach beyond our planet's atmosphere, first sending artificial satellites (p. 90), and then humans (p. 114) into Earth orbit. National programs of space exploration followed, with Moon landings (p. 148), space stations, and probes sent to every planet of the solar system and on toward interstellar space. But the greatest challenges lie ahead.

One of the first will be to find ever cheaper and more effective means to get into orbit. This could be through improved rocket technology (p. 341), some kind of new reusable winged space plane (p. 346), or by even more exotic feats of engineering (p. 353). This will probably be developed by a private company, rather than by a national space agency.

A return to the Moon (p. 361), and a first human voyage to Mars (p. 364) are already established as clear, near-term goals, as is the continued quest to find evidence for even the simplest form of extraterrestrial life. Such efforts will require resources, and a further challenge will be to secure more of these from outside the Earth, with solar power generation and asteroid mining (p. 366) key areas of investigation. Over the longer term, the goals get even more ambitious. Humanity may journey the massive distances between stars (p. 372) in the hope that our descendants find and colonize Earth-like planets (p. 376). Such long-distance space exploration is unlikely to occur during our lifetimes, but the advances made in the next few decades will take us closer, as scientists continue to increase our knowledge of our universe and our place within it. We may even learn some valuable lessons about how to live better on our small and precious world through our explorations beyond it.

OPPOSITE An artist's rendering of the Mars Ice Home concept developed by researchers at NASA's Langley Research Center in Hampton, Virginia. It is a large inflatable torus, a shape similar to an inner tube, which is surrounded by a shell of water ice.

SPACEX

Next-Generation Space Access

> *"Rockets are cool.*
> *There's no getting around that."*
>
> ELON MUSK,
> FOUNDER AND CEO OF SPACEX INC

Among the biggest problems space exploration faces are the limitations imposed by the launch vehicles used to transport crews and payloads into orbit. Even the most efficient launch vehicles used today are expensive, unreliable, and difficult to maintain. Fortunately, rocket engineers have been working on these difficulties, and they confidently predict that the next generation of cheaper, more reliable, and easier-to-maintain rockets will begin routine operations by 2020, with new launch vehicles for placing humans in orbit entering service soon afterward. New launch vehicles under development include updated versions of existing rockets, such as Lockheed Martin's Atlas series, the Orbital ATK Antares rockets, the ESA's Ariane 6, the Boeing Company's Delta IV, and SpaceX's Falcon 9. The corporate rocket manufacturers behind these launch vehicles want to increase their share of the space access market, but will have to compete, not only with each other, but also with launcher vehicles under development by the Indian, Chinese, and Russian space agencies, particularly in the lucrative satellite launch market.

Since the start of the space age, rocket engineers have worked to minimize the chance of losing a payload and launch vehicle to around two in 100 flights. However, the goal is to reduce the risk even further, to just one loss per 10,000 missions. A NASA risk analysis put the odds of a launch catastrophe involving the Space Shuttle at one in 450. As a point of contrast, military pilots face the probability of a catastrophic loss at one in 22,000 flights, and commercial airline passengers take a risk of one in 10 million.

SpaceX's Falcon 9 rocket and the Dragon spacecraft at Launch Complex 39A at NASA's Kennedy Space Center in Florida in preparation for a resupply mission to the ISS scheduled for February 18, 2017. This was the first launch from the Kennedy Space Center since the Space Shuttle fleet retired in 2011.

Commercial Launch

On May 31, 2012, the aerospace manufacturer and space transport company SpaceX became the first private enterprise to carry supplies to the ISS. Pictured here in the wake of this historic achievement while standing in front of the Dragon Capsule is the company's founder and CEO Elon Musk (right), accepting congratulations from NASA administrator Charlie Bolden.

Can costs be cut while increasing rocket reliability sufficiently to make the space launch market significantly more affordable? The answer depends on the ability of engineers to squeeze improved performance out of liquid-fueled rockets through innovative design modifications. New engines under development are smaller and more efficient, increasing the thrust available for the same fuel consumption. New alloys enable engines to withstand greater heat, and to be less susceptible to erosion; they thus achieve longer life cycles, and require less maintenance.

Next-generation launchers would also benefit from improvements to their thermal protection systems, employing lighter, more heat-resistant materials. If rocket engineers can achieve some or all of these improvements to increase the efficiency of commercial launch vehicles by a factor of ten, then entry to space becomes much cheaper, and all kinds of new exploration possibilities emerge.

An Orbital ATK Antares rocket lifts off from NASA's Wallops Flight Facility in Virginia on a cargo resupply mission to the ISS on November 12, 2017.

Reducing the Cost of Space Access

A variety of private space companies have recently demonstrated their capacity to launch payloads into space more cheaply than older launch vehicles, but will they be able to reduce the cost of sending humans into orbit?

LEFT The Blue Origin New Shepard reusable rocket lifts off from the proving grounds in Van Horn, Texas on January 22, 2016.

ABOVE Blue Origin's founder and owner, Jeff Bezos, plans to make the crew capsule that his company is developing, pictured behind him, both reusable and compatible with several different rockets. This may facilitate major savings in launch costs.

With their different designs and payload capacities, it can be difficult to compare the efficiency of various rockets on a strict like-for-like basis. However, the most accessible metric through which to measure the capacity of space launch vehicles is simply to calculate how much each costs to launch per pound of payload it lifts into orbit. For the last 30 years, the typical launch cost ratio has remained roughly $10,000 per pound to orbit, with some margin up or down, depending on the launcher, and the exact deal negotiated to deliver a payload. The glittering prize for providers of space access around the world is to lower the costs of reaching orbit to around $1,000 per pound.

Exact prices are often considered commercially sensitive information, and not typically released to the public, yet the launch costs for some rockets have been announced over the years. For example, the Delta 7920/5 costs roughly $5,461 per pound of payload sent into space; the Atlas IIAS is slightly more expensive, at $5,768 per pound, while the average Soyuz launch costs around $2,731 per pound, and the Long March costs $2,771 per pound.

In the last decade, the rocket development activities of entrepreneurs such as Elon Musk (1971–) and Jeff Bezos (1964–) have had a significant impact on launch costs through the development of new, largely reusable,launch systems. Musk's SpaceX quotes a price of $62 million per launch for its Falcon 9 rocket. Each rocket can carry 50,265 pounds (22,800 kg) of payload, equating to a launch cost of roughly $1,233 per pound of payload to be sent into orbit.

However, these launch costs only really apply for non-human flights. Humans lower the available payload capacity of any space launch vehicles, due both to their own weight and the extra systems and safety features required to sustain them. These systems and features require extra safety checks, which add further expense. Flight controllers are also less willing to take risks with the lives of a human crew.

Both SpaceX and rival company Orbital ATK have launched, landed, and later reused the first stages of their respective rockets. While these results are impressive, it remains uncertain whether these companies will be able to reduce the cost of human spaceflight in the near future.

Space Launchers beyond the Next Generation

A Magnetic Launch Assist System track and test vehicle at NASA's Marshall Space Flight Center.

Could even more advanced launch vehicles appear before 2040? Absolutely! Beyond the current and next-generation launchers that should come into service before 2020 (p. 341), engineers are already working on new and improved methods of reaching orbit, projected to materialize before 2040. Such vehicles could potentially make spaceflight as routine as air travel, and open the space frontier more widely than ever before.

Among the key goals for such space transport would be to improve safety to the point where the probability of losing a space vehicle is no worse than one in a million, a ratio that approaches the safety level of contemporary commercial jetliners. Mission planners would want further to reduce the cost of spaceflight to about $100 per pound of payload sent into orbit. They would also like to see the development of reusable launch vehicles that could be prepared for launch by a handful of people in a day or two, allowing thousands of flights per year. Such specifications are easy to state, but hard to achieve.

Launch vehicles of the 2040s will probably still utilize liquid-fueled rocket engines, but the efficiency of the technology should have improved, perhaps to the point where, instead of carrying an internal oxidizer, the engines would be designed to burn oxygen gathered from the atmosphere. This would significantly reduce the mass of propellant required by the rocket to fly. The mass could then potentially be reallocated to allow the rocket to carry more mass in its payload.

One possible new launch vehicle design would be single-stage-to-orbit space planes, such as the British Skylon, already in development. Essentially, these would use a combined-cycle engine—a hybrid of air-breathing jet and rocket engine that injects fuel into liquid oxygen harvested from the highest regions of the Earth's atmosphere. Atmospheric oxygen would be compressed by the forward momentum of the vehicle. NASA has tested a variety of its own space plane designs over the years, including the X-43 scramjet-powered research vehicle, which achieved a speed of Mach 9.6 on November 16, 2004. Data from that research have been factored into other space access projects.

Away from combustion engines, magnetic levitation, or MagLev, is an alternative launch method with significant potential. A MagLev system would use linear electric motors to accelerate a space capsule to more than 1,000 mph (1,600 km/h) along a track potentially as long as 60 miles (110 km). The track would then turn upward, perhaps following the contour of a mountain, allowing the spacecraft to be flung into orbit. Small-scale tests of the technology carried out to date suggest that a MagLev launch system should definitely be explored further.

Other ideas involve the development of a Beamed Energy Launch Vehicle (BELV). Such a "lightcraft" would be propelled into space by ground-based lasers and satellite-based microwave emitters. The laser operators would shoot a concentrated beam of light at the bottom of a spacecraft disk. Intense heat would cause movement in the air adjoining the disk, propelling the spacecraft upward. An orbiting satellite could then beam microwaves toward the ascending vehicle. The BELV would focus this energy, creating an "atmospheric spike" that deflects oncoming air. The resulting vacuum would further help the vehicle to rise.

Space exploration is certainly not lacking in engineering ingenuity, but whether or not any of these ideas reach fruition, only time will tell.

ABOVE The X-43 during a test flight over the Pacific Ocean in 2001.

BELOW LEFT This concept art shows a flying-saucer-like lightcraft delivering payload to orbit.

BELOW RIGHT A diagram showing how a laser-driven lightcraft could work.

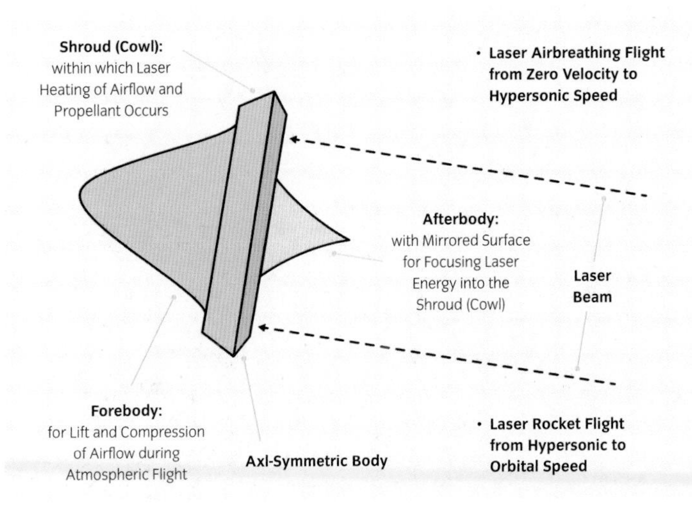

Boarding the Next Space Plane

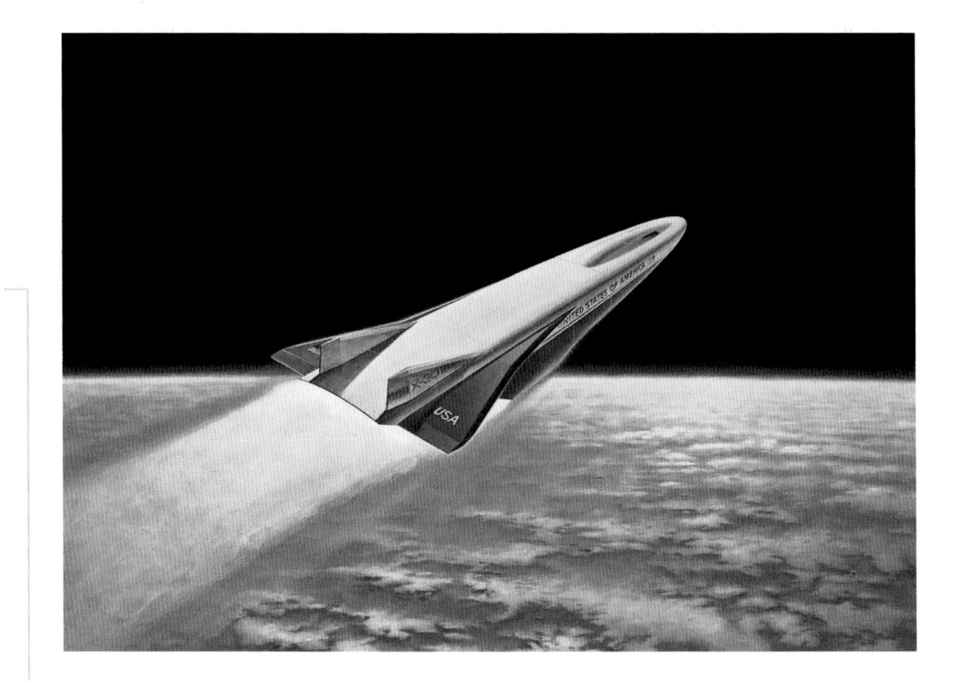

> *"Since the Wright brothers, we have examined how to make aircraft better and faster. Hypersonic flight is one of those areas that is a potential frontier for aeronautics. I believe we're standing in the door waiting to go into that arena."*
>
> ROBERT MERCIER, UNITED STATES AIR FORCE RESEARCH LABORATORY DEPUTY FOR TECHNOLOGY, HIGH SPEED SYSTEMS DIVISION

ABOVE This concept artwork show the X-30 experimental space plane flying through Earth's atmosphere on its way into orbit.

In the early 1980s, officials at NASA and the United States Department of Defense explored the possibility of developing a hybrid air and space plane that could carry ordinary people between New York City and Tokyo in about an hour. The concept was simple: an aerospace vehicle that could take off like a conventional airplane from an ordinary runway. Using jet engines, it would ascend to an altitude of 45,000 feet (13,700 m), where it would attain supersonic speeds. Then, the pilots would start the craft's scramjet engines. The scramjet would obtain and compress oxygen from the atmosphere to serve as an oxidizing agent to combust the plane's liquid hydrogen fuel, pushing the vehicle to hypersonic speeds. The space plane would then rise to the edge of space, although not quite reaching orbit, before beginning its descent to land like a conventional plane. The whole design and approach of the vehicle drew heavily on the ideas and concepts of Eugen Sänger (p. 43) in the mid-twentieth century.

The project eventually became the National Aero-Space Plane (NASP), designated the X-30. The NASP program initially proposed building two research craft, at least one of which was to be capable of also flying single-stage-to-orbit. The NASP was to use a multicycle engine that shifted from a jet configuration to ramjet, and then to scramjet as the vehicle attained higher altitudes and velocities.

Neither of the NASP research vehicles achieved anything approaching flight status, but the project did contribute significantly to the development of various advanced materials capable of repeatedly withstanding extreme temperatures, both high and low. By 1990, NASP researchers had made significant progress in the development of a variety of new, high-endurance materials, including titanium aluminides, titanium aluminide metal matrix composites, and coated carbon-carbon composites, all of which were important for operating in, and returning from, the extreme environment at the edge of the Earth's atmosphere. Government and contractor laboratories fabricated and tested large titanium aluminide panels under approximate vehicle operating conditions, and NASP contractors fabricated and tested titanium aluminide composite pieces. While the NASP never flew, follow-on research offers promise for future suborbital space planes operating globally by the 2040s.

One design that may benefit from NASP research is currently being developed by the United States Air Force. The X-51, or Waverider, is a scramjet-powered potential space plane that can reach Mach 5 (3,300 mph; 5,300 km/h) and an altitude of 70,000 feet (21,000 m). On May 1, 2013, a small-scale test version of the Waverider flew at hypersonic speeds for 140 seconds, the longest-duration powered hypersonic flight achieved to date.

Another space plane currently under development is the SpaceLiner from the German Aerospace Center. The SpaceLiner is a two-stage, reusable, sub-orbital, point-to-point vehicle with a remotely-piloted first stage, and a piloted second stage that could carry cargo and up to 50 passengers around the globe. Since development started on the project in 2005, seven different designs have been produced, but significant design and performance improvements still need to be made before the vehicle has a hope of entering into commercial service.

ABOVE Concept art of the X-51A Waverider hypersonic research vehicle.

LEFT Scientist Martin Sippel presents a scale model of the rocket-propelled SpaceLiner at the German Aerospace Center in Bremen, Germany.

The Future of Orbital Space Planes

This concept art shows the X-33 making rendezvous with the ISS, and demonstrates the expectations NASA and Lockheed Martin had for their space plane design.

With the demise of the American NASP project in 1992 (p. 346), the quest to get a space plane into orbit entered an exciting new phase of research and testing. Beginning development in 1995, the X-33 space plane was a joint project between NASA and the Lockheed Martin Corporation. The collaborators set themselves an ambitious timetable to produce a flying prototype by 2001. Once the design had been proven, Lockheed vowed to scale up the X-33 into a launch vehicle capable of transporting humans into space. Dubbed the VentureStar, Lockheed viewed the project as a potential replacement for the Space Shuttle. Unfortunately, the X-33 never flew, and was eventually canceled because of technological and budget problems.

Yet, despite the difficulties space plane developments continue to face, a significant number of spaceflight advocates still view this as the optimal technology for getting humans into orbit. Science journalist Clay Dillow, writing in *Popular Science* in 2010, voiced the position of many human spaceflight enthusiasts when he stated: "A reusable space plane design [is] the cheaper and safer way to move crews to and from the ISS; its 'blended lifting body' allows it to move from its orbital trajectory as it reenters and place its point of landing where the pilot wishes."

NASA has remained open-minded in its approach to a replacement system for the Space Shuttle. NASA's Commercial Crew Development (CCDev) program, a project designed to stimulate the establishment of a range of privately owned and operated spacecraft that could carry human crews into orbit, accepted both rocket-based and space plane-based proposals. The first four firms to be awarded contracts under the program are Blue Origin, Sierra Nevada Corporation, SpaceX, and Boeing. As of 2018, SpaceX is the nearest to achieving success in launching crews with its Falcon family of rockets and Dragon capsule. However, perhaps the most exciting prospect for space plane designs among these companies lies with Sierra Nevada's rocket-powered Dream Chaser, which is currently expected to make its first flights in the latter part of the 2020s.

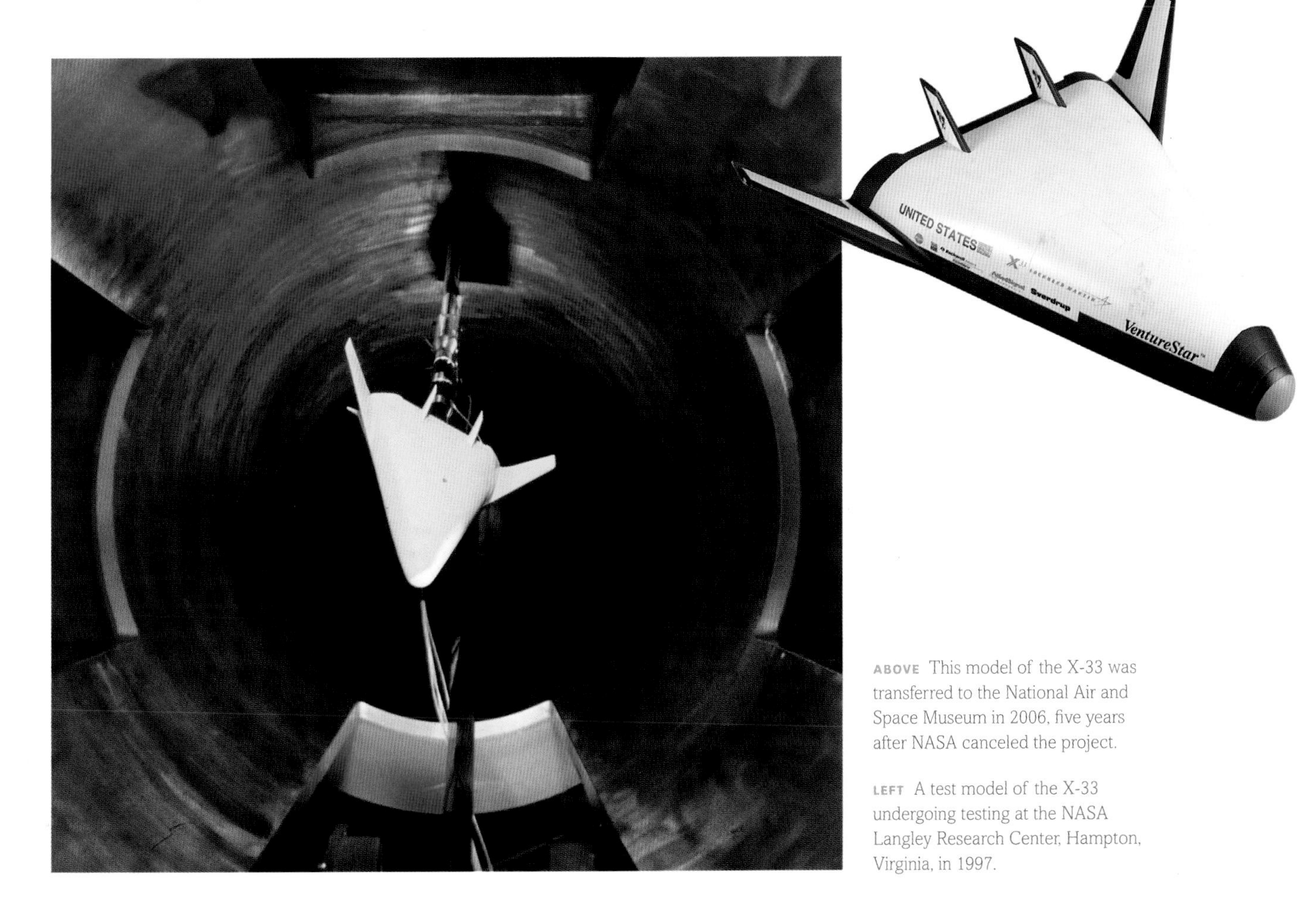

ABOVE This model of the X-33 was transferred to the National Air and Space Museum in 2006, five years after NASA canceled the project.

LEFT A test model of the X-33 undergoing testing at the NASA Langley Research Center, Hampton, Virginia, in 1997.

The X-37 Orbital Test Vehicle

1998–

A contender to be the next space plane to carry humans into orbit.

ABOVE LEFT An Orbital Test Vehicle taxis on the runway in June 2009 at Vandenberg Air Force Base, California.

ABOVE RIGHT Concept artwork of the Orbital Test Vehicle in space.

Another key candidate for the first crewed space plane is the X-37, also known as the Orbital Test Vehicle (OTV). This program originated at NASA in August 1998 as a response to a call to develop the "Future-X," a space plane designed to validate emerging hypersonic technologies with an end goal of reducing the cost of space access. The announcement specifically called for the development of propulsion, propellants and tanks, thermal protection systems, avionics, and structures—all especially thorny technology issues that would need to be solved by any vehicle delivering significant efficiencies in space access. NASA signed a contract with Boeing in July 1999 for a four-year cooperative agreement to develop what became known as the OTV advanced flight demonstrator.

The program was transferred to the United States Air Force in 2003, and flew an orbital test mission beginning on April 22, 2010, which circled the Earth until December 3, 2010.

A second OTV launch took place on March 5, 2011. The spacecraft's successful operation in orbit suggested that the OTV program held great promise to, in the words of the Secretary of the Air Force Michael Donley (1952–), complete "risk reduction, experimentation, and operational concept development for reusable space vehicle technologies, in support of long-term developmental space objectives." It has flown several times since 2011.

A consensus seems to be emerging that the OTV holds promise as the future space plane that might carry robotic and, perhaps after modification, human missions. Arthur Grantz (1961–), chief engineer of the Experimental Systems Group at Boeing Space and Intelligence Systems, builders of the OTV, bragged in late 2011 that a version of this space plane could carry five to six astronauts or cargo into low-Earth orbit with significant cost savings and enhanced safety. There may yet be a space plane taking astronauts into orbit.

The Dream Chaser is not the only space plane currently under development. The British company Reaction Engines is continuing its long-term effort to build the Skylon space plane, a single-stage-to-orbit vehicle, by using the Synergistic Air-Breathing Rocket Engine (SABRE). Although fundamentally a private venture, the British government pledged an additional £60 million to the Skylon project on July 16, 2013, and the ESA has started backing the development work with the aspiration that it could take some of its astronauts into orbit in the future. According to the current schedule, Skylon is due to start undertaking test flights no later than 2025.

In Japan, JAXA also has an orbital space plane under development, the HOPE-X, which is still in the design phase. The same is true for Russia, whose Cosmoplane would be capable of transporting passengers into orbit. Indeed, most of the space agencies of the world have some work underway on an orbital space plane. Currently, few of these have plans that have developed beyond the engineering concept stage, but it is likely that at least one of them will become reality, and take humans into orbit at some point in the next 20 years.

BELOW LEFT Sierra Nevada's Dream Chaser space plane seen here under construction on February 5, 2011.

BELOW RIGHT TOP A Dream Chaser space plane being readied for an approach and landing test on November 11, 2017, at NASA's Armstrong Flight Research Center, Mojave, California.

BELOW RIGHT BOTTOM The SABRE in place on a Skylon space plane. Designed by British company Reaction Engines Ltd., this unique engine will use atmospheric air in the early part of the flight before switching to rocket mode for the final ascent to orbit.

LunOx

The Space Elevator

Beyond the current rockets and even future space planes, by the start of the twenty-second century humans may be using an entirely different technology to get into orbit: the space elevator. Space elevators involve extending a strong but ultra-lightweight cable from the ground to a counterweight orbiting in Geostationary Earth Orbit (GEO), some 22,200 miles (35,500 km) above the Earth's equator. The whole edifice would orbit the Earth in sync with the planet's rotation, so the top of the elevator would maintain a stationary position over its base. Electromagnetic vehicles could then travel the length of the cable up into orbit. It is even plausible that large reels with high-strength cables could be employed to winch capsules in exactly the same way regular elevators function in high-rise buildings. To better understand the concept of a space elevator, think of the game of tetherball. In this game, a ball is attached to a pole with a rope. Think of the rope as the cable, the pole as Earth and the ball as the weight. Now, imagine that the ball is put into perpetual spin around the pole, so fast that it keeps the rope taut. This is generally how a space elevator would work. The weight at the end of the cable would spin around the Earth, keeping the cable taut. Spacecraft would simply ride up the cable.

As early as 1895, Russian space theoretician Konstantin Tsiolkovsky (1857–1935) suggested humans could reach orbit with what he called a "Celestial Castle" in Geosynchronous Earth Orbit attached to a tower on the ground. Another Russian engineer, Yuri Artsutanov (1929–), wrote some of the first modern explorations of

"First we'll develop the technology. In 50 years or so, we'll be there. Then, if the need is there, we'll be able to do this."

DAVID SMITHERMAN, TECHNICAL MANAGER, FUTURE SPACE PROJECTS, MARSHALL SPACE FLIGHT CENTER

LEFT This illustration of a working space elevator is based on a concept described by Arthur C. Clarke and is now known as the Clarke Clipper.

OPPOSITE NASA concept artwork for another possible space elevator design.

Developing a vehicle for traversing a model space elevator was the goal for engineering students participating in the European Space Elevator Challenge held in Garching, Germany, on September 13, 2016.

space elevators in 1960. In the West, the space elevator concept was first raised to prominence among the spaceflight engineering community through a technical paper written in 1975 by Jerome Pearson (1938–) of the United States Air Force Research Laboratory. This paper inspired British science fiction author Arthur C. Clarke (1917–2008) to introduce the concept to a broader audience. In his 1978 novel, *Fountains of Paradise*, engineers construct a space elevator on a mountain.

An operational space elevator would spark a rapid decline in the cost of sending anything to orbit. There would no longer be any need for a rocket to escape the Earth's gravity. Vehicles moving up the elevator would probably need to draw their initial launch energy from an electrical-power station on the ground, but, as they moved farther up the cable, they would require less electrical energy, relying instead on the centrifugal force produced by the spinning counterweight to pull them into orbit. By the time the vehicles reached the end of the cable, they could be moving as fast as 6.79 miles per second (10.93 km per second). Vehicles traveling at these speeds would be moving fast enough to reach Mars in days or weeks.

Building a space elevator would be no small task. It would require a cable that extends more than 89,000 miles (144,000 km) from the equator into space, and an equally formidable counterweight, possibly a repositioned asteroid. Nevertheless, the potential benefits for space exploration would be immense.

Space Elevator Cable System

This diagram of a lunar space elevator shows the elaborate cable system required. One end of the cable is anchored in the surface at the equator while the opposite end, far out in space, has a counterweight reaching up into the vicinity of the L-1 and L-2 Lagrangian points in the Earth-Moon orbital system.

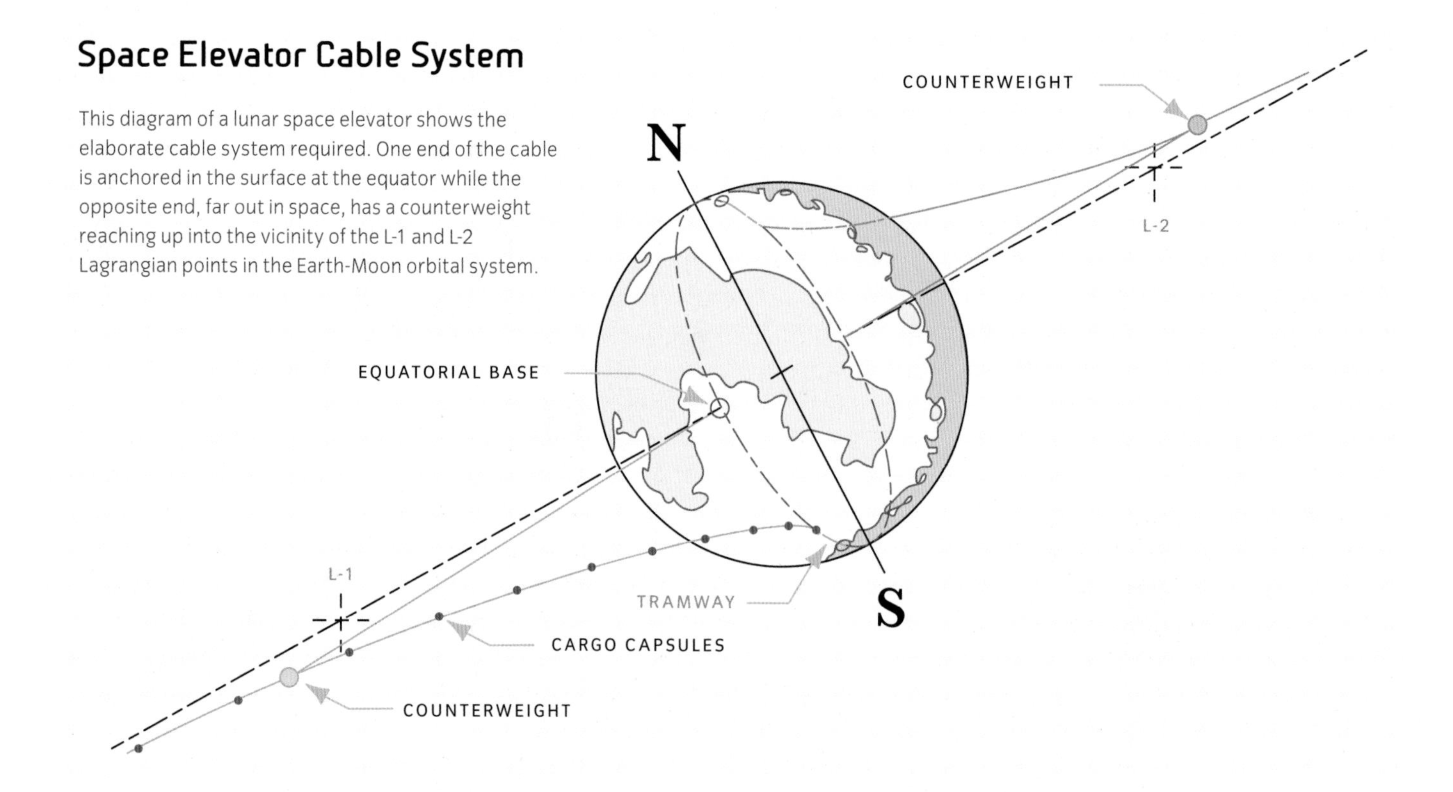

Technologies Enabling a Space Elevator

This small loop of carbon nanotubes created by Odysseus Technologies was submitted to a NASA innovation competition in 2010. Such super-strong building material may one day be employed in the building of a space elevator.

There are many challenges to be overcome before a working space elevator could be constructed, but there are some who believe such a mammoth engineering feat could be accomplished in less than 100 years. In theory, such a space transportation system could also allow trips into space for a few thousand dollars, roughly the same as a long-distance plane journey today.

Space elevators represent an intriguing alternative means of space transport, but much of the technology required to build one has not yet been fully developed. Feasibility research into the concept indicates that there are five key areas in which the requisite technologies will need to mature before anyone can seriously contemplate building such a system. They are as follows:

1. Research into carbon nanotubes has revealed a material that is 100 times stronger than steel. Until the discovery of carbon nanotubes, scientists lacked a strong material that was light enough to build a cable that could span the necessary distances space required by a space elevator. Carbon nanotubes have a tensile strength of 200 GPa. For comparison, graphite, quartz, and aluminum each have a tensile strength of just over 20 GPa.
2. Tether technologies developed for space transportation systems will need to advance new construction materials and methods in the deployment and control of long structures. Essentially, scientists will need to work out how to get their cable into orbit and deploy it downward so it can be attached to the ground.
3. Tall tower technology must advance to foster the development of towers in excess of a mile (1.6 km) for commercial applications. The tallest buildings on Earth today are barely half this height.
4. Electromagnetic, or MagLev, propulsion (p. 345) is needed for high-speed ground transportation and launch assist along the space elevator cable.
5. Space infrastructure for transportation, utilities, and facilities out to GEO will be needed to support construction for the space elevator's orbital end.

Making Earth Orbit a Normal Realm of Human Activity

ABOVE LEFT Astronauts Akihiko Hoshide and Karen Nyberg on STS-124 work the controls to maneuver the Kibo Japanese Pressurized Module to dock with the ISS to facilitate further space science.

ABOVE RIGHT Cosmonauts Aleksandr Kaleri and Valeri Korzun pose with a Protein Crystal Growth experiment on Mir in 1997.

Since the race to the Moon of the 1960s, the area from 62 to 310 miles (100 to 500 km) above the Earth has ceased to be a frontier. It is just another environment of human operations, albeit one that is both harsh and hostile, and which requires plenty of specialist equipment and resources for humans to traverse safely. Humans have learned how to operate in Earth orbit, and to exploit the region for purposes as diverse as espionage, weather monitoring, telecommunications, and astronomy. They have made the feats of space exploration undertaken in the 1960s, which were once viewed as full of risk and heroism, much more routine and mundane. In essence, the greatest success of modern astronauts and cosmonauts is that they have made a job that countless children and adults around the world would still love seem, in some ways, a bit boring.

The initial wave of space exploration was funded by national and transnational governments, just as various monarchs and rulers of earlier ages funded expeditions to explore new territories around the globe. In the early years of the twenty-first century, just as in earlier times, this national patronage is increasingly giving way to private commercial investment. As the range of investors, and the variety of space applications they pursue, continue to increase, humanity will continue to benefit from routine commercial activity in space.

The 1968 Stanley Kubrick film, *2001: A Space Odyssey*, has mesmerized millions over the years with its complex, sometimes surreal, story. But perhaps the film's most accessible aspect for viewers is the vision of routine, accessible, private sector space operations it presents. The now-defunct American airline Pan Am operates a commercial service into orbit and other private firms provide telecommunications and accommodation services.

Space travel is now tentatively edging toward a similar era of commercialization. The ISS has already energized the development of private orbital science experimentation. But there is also a real possibility that human-tended laboratories and private habitats could be established in orbit near the ISS. Commercial organizations could become tenants of an orbital research park, taking advantage of the microgravity to pursue research in materials science, fluid physics, combustion science, and biotechnology. And if scientists and technicians can travel to habitat modules in the heart of this orbital research park, why not those traveling for pleasure? Space tourists have already visited the ISS (p. 232). A dedicated space tourism facility seems almost inevitable.

Virgin Galactic has already begun soliciting for clients for suborbital space tourism trips, but the partnership of Bigelow Aerospace, which builds orbital habitats for science experiments, and the emerging private space launch company SpaceX, has fantastic potential for orbital tourism. As private entrepreneurs continue to develop the market for near-Earth space applications, the role of government will probably become less dominant over the next 100 years. Space activities that have few immediate commercial applications, such as deep space exploration, will likely remain the province of state-sponsored research activities, but widespread human spaceflight near Earth could become the province of the commercial sector as the twenty-first century progresses.

The Bernal Sphere

First proposed in 1929 by scientist John Desmond Bernal (1901–70), Bernal spheres are artificial space habitats located inside hollow spherical shells about 9.9 miles (16 km) in diameter. A sphere would create its own artificial gravity through spinning on a central axis, and receive sunlight to be refracted around the installation through windows at one of its poles. Its projected 20,000–30,000 inhabitants would live in dwellings on the interior walls of the sphere.

Through his company Bigelow Aerospace, entrepreneur Robert Bigelow has built and flown his inflatable self-contained Genesis modules on the ISS. These modules could also be placed in orbit for commercial use.

Solar-Powered Satellites and the Human Future

Concept artwork showing the as-yet untested SolarDisk platform designed to direct the Sun's rays down to power stations on Earth, where they would be used to generate electricity.

Someday beyond 2040, people in space may build, service, and operate solar power satellites providing energy to Earth. In the last decade, scientists across the globe have reached consensus on the reality of global warming. The search is now on for alternative forms of sustainable energy. One of the most compelling forms of renewable energy is solar radiation, which is plentiful in the region of the solar system in which the Earth orbits. Solar power is clean, ecologically sustainable, and more plentiful than any other energy source on Earth. With this in mind, advocates of solar power have recommended the construction of solar power stations orbiting the Earth and the Moon. Such stations could use large mirrors to concentrate and reflect solar energy toward Earth, where the normal photovoltaic conversion process could take place at ground stations. This is an appealing idea, especially for use in remote locations that do not have access to electricity through power lines.

At present, the cost of generating electric power from space is not competitive with Earth-based sources. On an annual basis, people on Earth generate nearly 20 billion kilowatt-hours of electrical energy from all sources, and this grows at a rate of about 2 percent per year. Conversion of even a small percentage to space-based production would be a boon, provided that it can be harvested at a cost competitive with other forms of energy. This is not currently feasible, but if costs to get into space can be reduced significantly, as is currently the aim (p. 343), there is every reason to believe that solar power satellites might eventually be more prevalent.

Since the 1970s, scientists have investigated two concepts that might provide moderate amounts of low-cost solar power. One of these, called SunTower, consists of a modular, self-assembling satellite that would operate in an orbit relatively close to the Earth. The SunTower arranges a series of solar collectors along a very tall pole, and reflects the sunlight to a ground station, where it can be used to power a turbine making electricity. A second concept, SolarDisk, would place a large reflective disk in geostationary orbit to reflect sunlight down to ground stations, where again they could be used to power turbines to make electricity. Solar energy advocates would like to test both concepts by 2035.

A further solar power satellite concept was proposed by former NASA engineer John Mankins (1956–) in 2008. Called SPS-ALPHA, Mankins's design used wireless power transmission, not unlike that found in modern telephone and other electronics recharging systems that operate without wires, constructed from tens of thousands of small elements that could deliver thousands of megawatts remotely and affordably to discrete markets on Earth. As Mankins himself states: "You don't need hundreds of billions of dollars to see if these new systems are economically viable." He seeks to demonstrate his idea on a small scale within the next decade.

All of these concepts are in the very early stages of development, and their long-term feasibility has not yet been proven. The technologies needed for space-based solar power generation are neither readily available at present nor easy to develop. Engineers may require as much as half a century to deliver a workable space-based system of solar-based electricity generation. However, as the population on Earth continues to grow, power generation may yet become a pivotal commercial space activity in the future.

> *"We could make a very small version of SPS-ALPHA and slightly modify the transmitter before launch to send radio signals rather than a microwave beam. Then the power station becomes a high-power communications satellite."*
>
> JOHN MANKINS, AMERICAN PHYSICIST

Concept designs for the SPS-ALPHA, a further possible design for an economical solar power satellite.

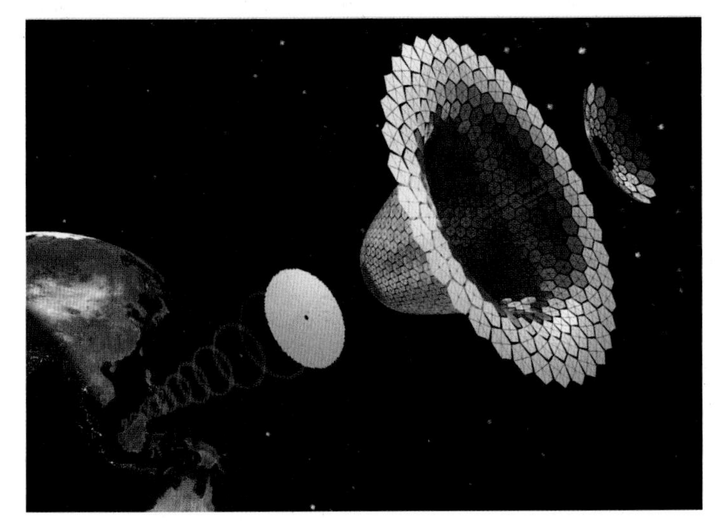

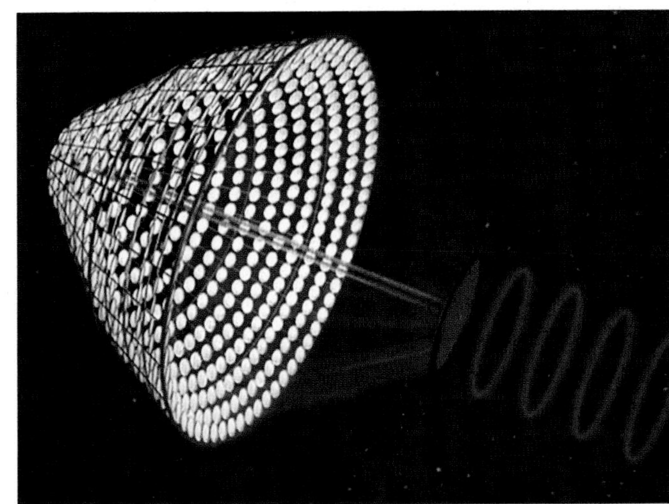

Pat Rawlings '89

Lunar Research Station

Sometime beyond 2050, a permanent human presence, likely part of some international collaborative effort, will probably be established on the Moon. Indeed, it would be surprising if this did not take place. All the technology needed to reach the Moon, land, and return to Earth is already known and was proven in 1969 (p. 148). But future lunar missions may not be direct flights like those of the Apollo program. Instead, they are more likely to take advantage of the lower launch costs associated with reaching and transferring out of low Earth orbit (LEO). The ISS, or more likely its successor, would be configured to berth a lunar transit vehicle, which would make regular trips to and from the lunar surface. Alternatively, a transit vehicle derived from space station technology could be placed in a permanent orbit cycling between the Earth and the Moon. Such an endeavor would probably cost around $100 billion over the construction and first ten years of operation, but the investment would almost certainly reward any organization or company that made it.

With regular and routine access to the Moon, humans could swiftly construct an outpost on its illuminated side. The Moon has a wealth of resources to help supply a self-sufficient base. Ice has already been detected at the Moon's polar regions by the Clementine spacecraft of 1994, and confirmed more recently by the Lunar Prospector and Lunar Reconnaissance Orbiter. From ice, humans could create water, oxygen, and hydrogen. The last named could be used to produce rocket fuel, and to generate electricity. Even at the poles, solar arrays could be positioned at the top of craters to provide an additional source of energy for the half-month that the Sun faces that section of the Moon.

A lunar research station could potentially expand the tradition of international cooperation established on the ISS. Various nations could be part of the endeavor. However, this would also require a level of coordination between various governments, and perhaps even private firms. As territorial claims on the Moon are prohibited by the 1967 Outer Space Treaty, diplomats and other officials representing the spacefaring governments would have to draft agreements regulating the uses of the Moon, and establishing property rights for entrepreneurs.

OPPOSITE This concept art shows a lunar surface crane removing a newly arrived habitation module from an expendable lunar lander. The crane operator will place the module on the flat trailer for hauling back to the main base.

BELOW This poster is thought to have been created on behalf of Lockheed Martin to promote a proposal for a lunar base.

Why Return to the Moon?

Establishing a Moon base would likely be a costly enterprise for whoever undertakes it, but could the benefits humanity might reap from such a mission outweigh the expense?

There are essentially six compelling reasons for humanity to return to the Moon ahead of any potential mission to Mars:

1. The Moon is only three days' travel from Earth, as opposed to Mars, which takes nearly a year to reach. The relatively short journey goes some way to mitigating the consequences of unforeseen accidents, and other difficulties that may be encountered on a longer voyage.
2. The Moon offers an ideal test bed for technologies and systems required for more extensive space exploration.
3. The Moon provides an excellent base for astronomy, geology, and other sciences, enabling the creation of critical building blocks of knowledge that may prove useful for traveling into the farther reaches of the solar system.
4. A cooperative international return to the Moon will stimulate the technological and engineering capabilities of all nations involved.
5. It increases the imperative to develop low-cost energy generation and other technologies that will have applications both on the Moon and on Earth.
6. The Moon could provide a base for systems that could be used to destroy near-Earth asteroids and comets as well as any other threats to the Earth.

In addition to a colony on the Earth-facing side of the Moon, the far side would be an excellent location for a Moon-based observatory. A lunar astronomy platform would offer many of the advantages of Earth-based observatories, particularly in terms of stability, but with the added advantages of space-based telescopes, particularly avoiding the Earth's distorting atmosphere. Such an observatory would be shielded from the Sun's light, as well as from radio noise, thereby significantly reducing two areas of potential interference.

The Moon would also be an excellent place to study the high-energy particles of the solar wind, as well as cosmic rays from outside the solar system. Earth's magnetic field and atmosphere deflect many of these particles and rays. The Moon has virtually no atmosphere, and it spends most of its time outside of Earth's magnetosphere. Detectors placed on the Moon could get a complete profile of solar particles, perhaps shedding further light on the processes taking place inside the Sun, as well as intergalactic sources of cosmic rays.

Even something as simple as establishing the dates when various craters on the Moon were formed could provide a unique picture of how the flux of meteoroids in the vicinity of Earth has changed over time. This impact history may be critical to understanding the extent of meteor and asteroid impacts, and may even help humans better to predict the odds of a lethal meteor strike on Earth (p. 309).

Finally, the Moon could be mined for valuable mineral and chemical resources. The Moon's ice may even be converted into rocket propellant that could be used to take space travelers to Mars and beyond.

A future Moon base may serve as the perfect location for a large lunar-based radio telescope, as depicted in this concept art.

This cutaway image of a lunar base of the future suggests what might be accomplished on the Moon. Depicted here is a large inflatable habitat that contains an operations center, a pressurized lunar rover, a fully equipped life sciences lab, a geology lab, hydroponic gardens, a wardroom, gym, private crew quarters, and an airlock.

The rights to mine the Moon will need to be resolved before we seek to establish operations there. However, such negotiations are unlikely to occur until the establishment of a Moon base is an imminent reality. The first lunar base will probably look a lot like the kind of research bases found across Antarctica today, and would likely be staffed by an international crew of scientists, engineers, and technicians. Such an international endeavor would likely necessitate the following:

- An international agreement on the use of the Moon.
- Stimulation of high-technology capabilities in all nations involved in the program.
- Advancement of virtually every science and engineering discipline in all of the spacefaring nations.
- Development of low-cost energy and other technologies.
- Conceptualization of a new vision of the future in which a closed Earth is replaced with a boundless universe, open to all.

The Moon, like the continent of Antarctica before it, may yet become an international area, with scientists from many nations in permanent residence.

Potential Benefits of International Space Exploration

ABOVE Artwork showing a possible international Mars mission led by China.

OPPOSITE LEFT Illustration of a possible international mining mission on Saturn's moon Titan.

OPPOSITE RIGHT Concept art showing the landing of an international team of astronauts on Mars. The larger American flag signifies possible American leadership of the project, with the flags of other participating nations underlining the project's international character.

Thanks to the various crews of the ISS, there has been an unbroken human presence in space since 2000. This international project has fostered an unprecedented level of cooperation between the space agencies of various participant countries. The goal for future missions will not merely be to maintain this level of presence and interaction, but also to develop stronger international bonds, as humanity moves on farther to the planets together.

Using the space station as a base camp, humans may well be able to return to the Moon (p. 361), and establish a permanent human presence there. From the Moon, the next most likely goal would be for an international coalition of nations, corporations, and other entities to pursue human exploration of other places in the solar system. Possible candidates for such an expedition include a number of bodies in the asteroid belt, as well as some of the moons of Jupiter and those of Saturn, but the most obvious and closest target would be Mars. While human expeditions to Mars have been mooted since before humans had even set foot on the Moon, such a venture would not be without controversy. Opinion polls conducted in the United States have long indicated that there is little support for public spending on such a mission.

Because of the cost, technological sophistication, and broad risk, it seems unlikely that any one nation would undertake an expedition to Mars, except as part of an international consortium. A number of spacefaring nations, including the United States, China, and Russia, could all conceivably take the lead in such a project. However, a transnational approach, such as that pioneered by the ESA, would help to spread the risk, cost, and difficulty of the endeavor.

All the promise held out for the ISS in gaining scientific knowledge and technological development may well pale in comparison to the very real possibility of enhanced international relations through a continued unified mission of exploration of the wider solar system. The same may be true of the costs involved: the expense, though enormous, is nevertheless a small price to pay for better international relations. With the ISS, spending a significant share of the public treasuries of each participant nation was eminently better, and more scientifically fruitful, than spending the equivalent sum on weapons of destruction. For all the difficulties and complexities involved in working with a large group of international partners, the knowledge gained in such cooperative programs has served all participant nations well.

These benefits would only be further increased in a grand international mission of explorations beyond Earth's orbit. It is quite possible that, 100 years in the future, the most significant aspect of any of this human exploration mission may be that it has brought various nations together in the peaceful pursuit of ongoing, large-scale, cooperative, international space exploration.

"Throughout the history of spaceflight and the study of effects of the space environment to the human body, we have accumulated enough knowledge to be able to move over to the next step: getting ready for interplanetary missions, for interplanetary exploration."

ROMAN ROMANENKO,
RUSSIAN COSMONAUT

Mining the Solar System

One of the greatest opportunities to fuel future space exploration is through mining various bodies in the solar system. Many space exploration experts recognize that, armed with some sophisticated mining equipment, humans could extract a variety of precious metals and other useful commodities from the Moon and other bodies of the solar system. The most valuable resources would be those raw materials needed for the construction of space stations, extraterrestrial bases, and for the benefit of the human population back on Earth.

The first mineral sources to be exploited in space may well be found on asteroids. While most asteroids are located in a belt between Mars and Jupiter, a few pass closer to the Earth than the Moon, making near-Earth asteroid mining both feasible and attractive. Mining asteroids is also easier than mining the Moon or planets, because asteroids tend to contain pure metal that is easier to extract, and requires little processing or purification. Two asteroids, known as 1985 EB and 1986 DA, have already been identified as containing nickel and iron, and both approach Earth at a distance roughly equivalent to that between the Earth and the Moon. On a future approach, a team of space miners could undertake a limited-duration resource extraction project on the asteroids. Another possible candidate is the near-Earth asteroid 1993 BX3. This asteroid weighs several million tons, and may well contain significant resources that humans might want to exploit. Scientists estimate that 2,000–3,000 mineral-rich asteroids intersect the Earth's orbital path on a regular basis.

This concept art shows a potential mining settlement on asteroid 90 Antiope.

Many commercial space exploration advocates view resource extraction as a key area of future operations, but despite countless feasibility studies that suggest asteroids could be rich in resources, the mineral exploitation of near-Earth space is probably still decades away. It will remain that way until the high costs of reaching target asteroids, extracting their resources, and transporting them back to Earth are reduced sufficiently to make asteroid-mined resources competitive with their Earth-mined equivalents.

This 1977 concept art shows a mining mission on an Earth-approaching asteroid.

Property Rights in Space

1967

The International Outer Space Treaty was drawn up before humanity set foot on the Moon, and to date it remains the closest thing to a space law humanity has yet developed.

Since the 1967 International Outer Space Treaty prohibits property ownership in space, any potential asteroid mining activities would face significant legal hurdles. Furthermore, it is by no means clear what jurisdiction, if any, could be prevailed upon to settle any questions of asteroid mineral rights. The legal framework set up in the Outer Space Treaty has thus far discouraged the assertion of property rights outside the Earth, and the exploitation of space resources more generally. However, as private companies develop their own reach, and exploit resources on the Moon, this position will need to be revisited.

Some legal experts propose that spacefaring nations could grant the open use of resources on the Moon or planets. This would provide a legal route for private companies to pursue asteroid mining operations within the jurisdictions of these governments, with the national courts of those countries available to defend the rights of the mining companies without anyone needing explicitly to exert property rights. With the establishment of such a practice, the precedent allowing the exploitation of space resources would then be set in international space law.

This might work, but until such a practice is tested in the courts, no one will know for certain. What is clear is that, when it becomes profitable to extract resources from the Moon, space entrepreneurs will find a way to do so, whether there is an existing legal framework or not.

Realizing Lunar Tourism

ABOVE This computer-generated image depicts a possible future hotel on the Moon.

OPPOSITE RIGHT Dutch architect Hans-Jurgen Rombaut created this design for a proposed lunar hotel.

OPPOSITE FAR RIGHT The Apollo 11 landing site will be a compelling tourist attraction once tourists start regularly visiting the Moon.

The first space tourists were multimillionaire adventurers who paid the Russian officials for access to both space and the ISS. While those with the means will undoubtedly continue to pay for similar space access in the future, the next generation of space tourists will probably be given more modest space excursions for a much smaller price tag; mostly, they will engage in commercial suborbital tourism flying on spacecraft such as SpaceShipTwo (p. 242). However, eventually more aggressive forms of space tourism will follow, and once any kind of Moon base is established it may well become a hub for wealthy and adventurous tourists.

The development of lunar tourism will involve dealing with a number of issues. The first of these is cost. Will there be enough multimillionaires interested in visiting the Moon to build a sustainable lunar tourism market, or will it require the development of lower-cost options to drive enough Moon boots through the doors to make lunar hotels feasible? The second issue will be the safety and reliability of the technologies employed in any Moon-based tourism facility. Finally, there is the question of the level of comfort that Moon tourism will be able to provide. Certainly, the uniqueness of the experience will be enough to inspire an initial wave of visitors, but it will most probably be the quality of the amenities that will be critical to making the Moon attractive for repeat visitors.

In its initial years, lunar tourism would probably look much like tourism to Antarctica does today. It would probably start with one or two berths at the first research stations on the Moon; those facilities would have been supported by governmental entities, or through contracts with private companies for habitation and support services. Having that critical infrastructure in place would make it easier for privately-owned facilities to be built for tourism at a reasonable cost.

"What I've been experiencing here has been an amazing journey so far, from takeoff to arrival to adaptation. There's so much to learn, there's so much to discover, there's so much to look at."

GUY LALIBERTÉ,
CANADIAN SPACE TOURIST

Moving on from accommodation, Tranquility Base, the landing site of Apollo 11, will undoubtedly become a major tourist attraction for Moon visitors of the future. At present, the whole historic site remains exactly as it was when Neil Armstrong and Buzz Aldrin blasted off to rendezvous with the Apollo 11 Command Module for the trip back to Earth in 1969. The presence of visitors, even in small and tightly controlled numbers, would subvert the historic purity of the site. But demand is likely to prevail in the end, and some kind of access to that site, and those of every other spacecraft to have made lunar landfall during the pioneering years, is probably inevitable. At present, few take planning for the preservation of the Apollo landing sites on the Moon seriously, but that will undoubtedly change when the first missions to return to the Moon get underway.

Lunar tourism is probably not on the immediate horizon in the near term, but we are rapidly reaching the point, perhaps as soon as 2050, when getting to the Moon will be the ultimate destination of choice for those with the means and the desire to get away from it all in the biggest possible way.

Colonies in Space

BELOW Physicist Gerard O'Neill and his wife, Tasha.

OPPOSITE LEFT In 1976 and 1977 NASA sponsored summer studies on the feasibility of space colonies. As part of these studies, it commissioned artist Rick Guidice to prepare depictions of these facilities, including the example here.

OPPOSITE RIGHT This concept art of a colony in space shows how the rotating structure might simulate a sky to allow access for sunlight.

In addition to the possibilities of migrating to Mars or living in a mining colony on an asteroid, there has long been a dream of building larger, free-floating, non-orbital colonies in space. The most compelling advocate for these colonies was Princeton University physics professor Gerard O'Neill (1927–92), who urged migration from Earth to help reduce pollution, preserve natural resources, and relieve the problems of overpopulation.

During the early 1970s, O'Neill, infected with the enthusiasm of the Apollo program, undertook a set of studies aimed at answering the question "Is a planetary surface the right place for an expanding technological civilization?" His research suggested that the possibilities for human colonies in free space—not on any planet or moon in the solar system—seemed limitless. He calculated the technical issues of energy requirements, area, size and shape, atmosphere, gravitation, and sunlight necessary to sustain a colony. O'Neill planned for colonists to live on the inside of gigantic cylinders or Bernal spheres (p. 357). These installations would hold a breathable atmosphere, and all the ingredients necessary for sustaining crops, livestock, and human life. They would even rotate to provide artificial gravity. While the human race might eventually build millions of these space colonies, each settlement would, by necessity, be an independent biosphere where water, waste, and other materials could be recycled endlessly. Animals and plants endangered on Earth could thrive on these cosmic arks. Solar power, directed into each colony by huge mirrors, would provide a constant source of non-polluting energy.

O'Neill optimistically argued that the first colony could be completed by 2005. He also estimated that emigration to the newly-constructed off-world colonies would reverse the population rise on Earth by 2050. O'Neill's supporters greeted his findings enthusiastically, and founded the L-5 Society, named for the fifth gravitationally stable Lagrangian point in a planetary system (p. 174), which might serve as an ideal location for such a colony to be established. The United States Congress conducted hearings, and NASA officials commissioned feasibility studies.

Alas, as inviting and idyllic as these ideas might seem, they are unlikely to come to pass any time soon, and such schemes are not a workable solution to the planet's current problems of climate change and an ever-growing population. To solve these, humans will have to learn to take care of their own planet, finding ways to inhabit its polar regions and perhaps even areas under its seas. Thoughts of abandoning humanity's home are best left for when it is technologically viable to do so.

Colonies in Space in Recent Movies

2010–18

Space colonies might not yet be within humanity's grasp, but they are a regular feature on cinema screens.

Space colonies may be a long way from science fact, but they have featured prominently in two recent science fiction films. In *Elysium* (2013), the Earth is a wreck populated by humans living in squalor, while the rich people have all migrated to orbital colonies. The plot revolves around the need for medical attention by those on Earth, and the medical facilities that exist at the orbital colony.

In *Interstellar* (2014), the Earth of the future is dying, and humanity must find another home. NASA, now an underground organization, undertakes a series of desperate interstellar missions to find another Earth-like planet that humanity could potentially colonize. The planetary quest fails, but humanity eventually develops space colonies that provide the opportunity for our species to migrate elsewhere in the solar system. Gerard O'Neill would have been proud.

In the 2014 film *Interstellar*, the solution to the problem of a worn-out, poisoned Earth is portrayed as being human migration to colonies in space.

Pursuing Interstellar Space Exploration

Interstellar Flight

There are currently four main methods that humans might use to travel beyond the solar system:

- Using either wormholes, tears in the space-time continuum that link two geographically distant points in space-time, or some kind of exotic propulsion system that enables travel faster than the speed of light (983,571,056 feet per second or 299,792,458 m/s).
- Through the use of some form of suspended animation process that slows the human metabolism, enabling humans to endure through centuries-long travel between stars.
- Through the use of multigenerational spaceships, capable of sustaining humans for multiple centuries.
- Through developing much longer human life spans.

This NASA concept art depicts a hypothetical spacecraft with the ability to warp space-time to achieve hyperfast transport to reach distant galaxies.

Because the distance between most stars is so vast, the prospects for interstellar space exploration by humans seems exceedingly remote at present. Nonetheless, the inclination to contemplate travel to other stars, both distant and nearby, remains strong. In fact, idle interest is likely to grow as a consequence of current research on the nature of the galaxy and the evolving view of our cosmic neighborhood that will certainly follow. It is easy to forecast a time when humans find an inviting Earth-like extra-solar planet (p. 376), but actually getting to such a distant location will be a different challenge entirely.

While there is growing interest in interstellar flight, the practical difficulties are immense, and will require technology that is far more sophisticated than the conventional rockets that have launched all space exploration efforts to date. Science fiction writers achieve faster-than-light speeds effortlessly, but in our current understanding of physics there is nothing that can travel faster than light.

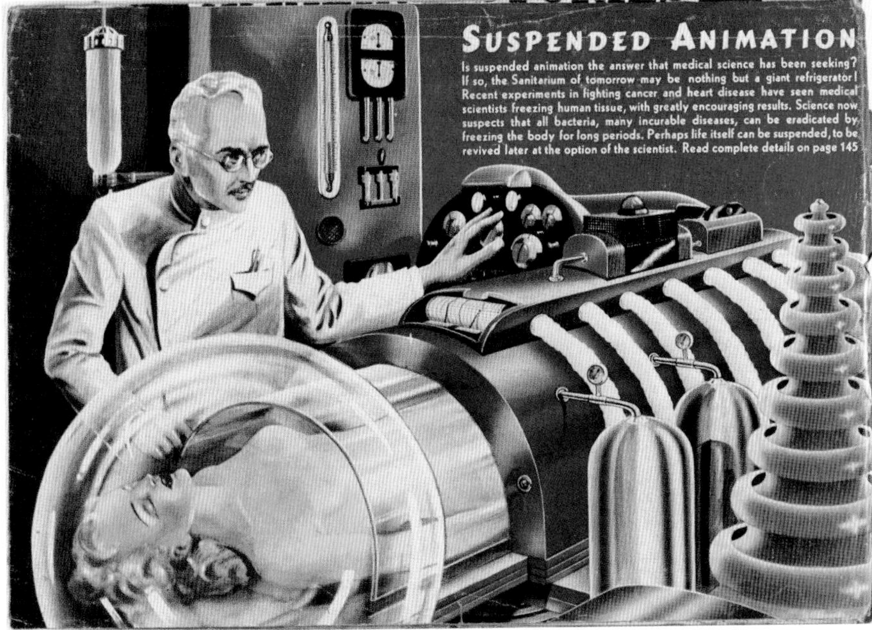

ABOVE Scientists speculate that the extreme galactic radiation sources known as quasars, like the one depicted in this artwork, might provide the means to fold space-time and permit interstellar travel.

LEFT Long-established science fiction tropes, suspended animation and advanced hibernation may be the key to long-distance space voyages of the future.

Concept art depicting a multigenerational spacecraft using a laser light sail for propulsion. A powerful laser in orbit around Earth could beam light at the sail, and the photon pressure of the laser hitting the sail would eventually accelerate the spacecraft to about half of the speed of light.

A Multigenerational Starship

The research to help build a spacecraft capable of sustaining a crew over centuries of travel has already begun.

Given the massive distances between even relatively nearby stars, it appears that any trip to such bodies would take longer than the longest human lifetime. Such a journey would require spaceships that carried somewhere between eight and ten generations of space explorers, most of whom would know nothing but life in an enclosed and entirely artificial environment. Serious proposals exist for such ships, but so far no one has demonstrated that it would be possible to create a self-contained environment that could support humans over generations on Earth, much less one that is capable of flying through the near-vacuum of space.

In the late 1980s, a group of volunteers participated in an experiment funded by Texas oil magnate Edward Bass. The experiment, which was conducted in the specially constructed enclosed environment of the Biosphere 2 dome in Arizona's Santa Catalina mountains, was designed to test technologies that might be useful for sustaining life over long periods in such environments. The experiment eventually had to be abandoned, because of a lack of food and other supplies inside the dome.

The British Interplanetary Society approached the problem from a different angle by first designing a complete interstellar spacecraft. British engineer Alan Bond (1944–) led a team that proposed building a fusion rocket using nuclear propulsion that could reach Barnard's Star, approximately 5.9 light years away from Earth, in around 50 years. Although Bond's plan envisioned a robotic vehicle, the principles would be largely the same for a multigenerational spacecraft.

Meanwhile, in 2010, the United States Defense Advanced Projects Research Agency (DARPA), in cooperation with NASA, funded the 100 Year Starship study "to explore the next-generation technologies needed for long-distance manned space travel." Intended to jump-start innovation leading to the development of technologies that would facilitate interstellar travel, the study has led to several conferences reporting on the research that has resulted. Some science teams have even worked on technology development projects associated with the study, but, at present, the idea of a multigenerational starship remains a very long way from any kind of practical application.

In his novel *Contact*, Carl Sagan resolved this difficulty by dispatching a human crew through a wormhole created on the basis of information supplied by a SETI message (p. 342).

Might breakthroughs in physics allow the creation of wormholes that evade the conventional limits of space and time? Perhaps, but at present scientists doubt that wormholes of sufficient stability could be constructed.

On the other hand, scientists have found repeatedly that, once a goal or idea is said to be impossible, someone finds a way to achieve it. At various points in the past, the best scientists of the age confidently declared that heavier-than-air flight would be impossible, that the sound barrier could not be overcome, that humans would be unable to withstand speeds in excess of 60 mph (96 km/h) on a train… the list goes on and on. If Albert Einstein (1879–1955) was correct with his theories of relativity, speeds that are faster than light are not possible, but perhaps future discoveries in physics will reverse this situation.

Suspended animation has long been a staple of science fiction, and would make it possible for astronauts to undertake a state of deep hibernation for long-duration spaceflights. In films such as *2001: A Space Odyssey*, *Alien* (1979), and *Avatar* (2009), hibernation serves as a plot device, but there is some basis for a belief that suspended animation might at least be possible in the real world. In the 1960s and 1970s, NASA scientists investigated the phenomenon they termed “depressed metabolism”—sometimes called “suspended animation”—but abandoned their research when they found that it was technically unfeasible with current or envisioned technology. Whether this technique will ever be practical for extreme-duration space voyages remains an open question.

Bimodal nuclear thermal rockets conduct nuclear fission reactions similar to those safely employed at nuclear power plants and nuclear-powered vessels on Earth. The energy might power a future multigenerational mission to a distant star.

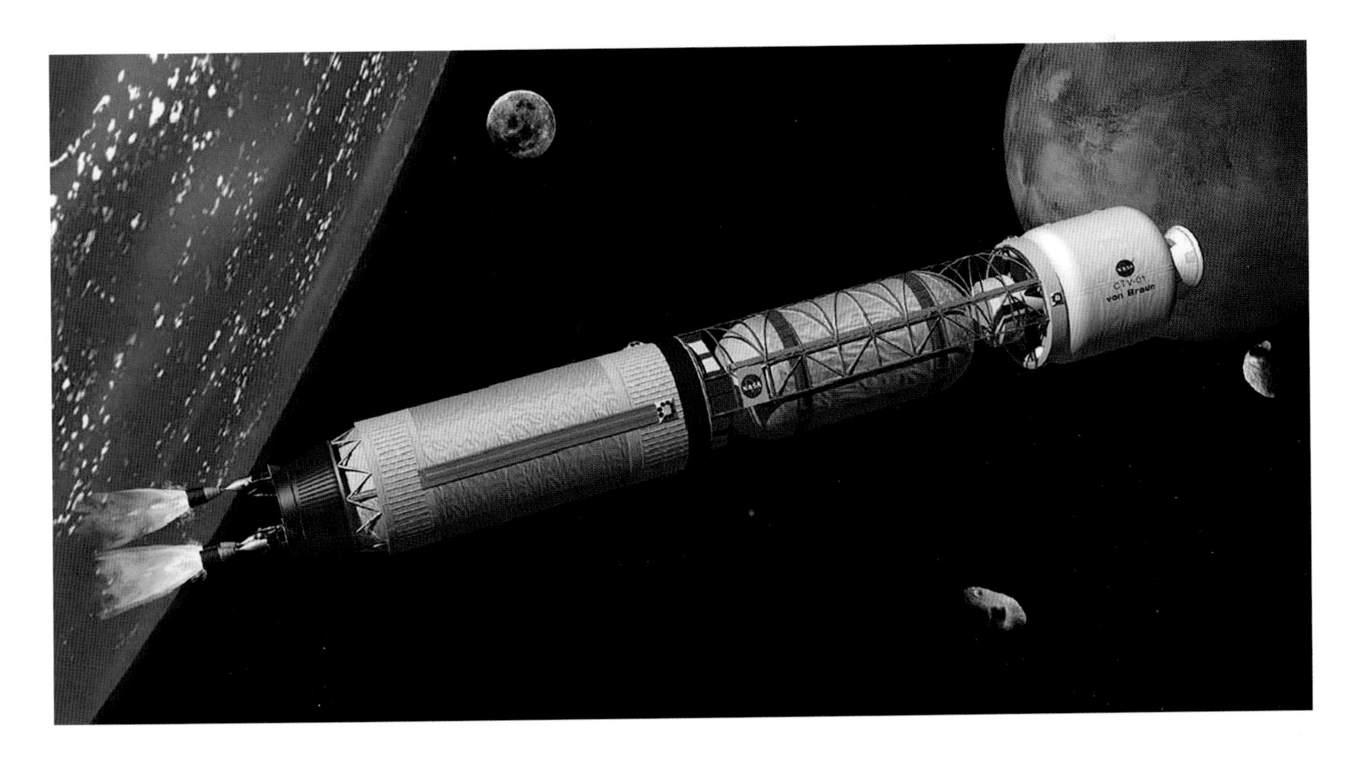

Becoming a Multiplanetary Species

"The long-term survival of the human race is at risk as long as it is confined to a single planet.... But once we spread out into space and establish independent colonies, our future should be safe."

STEPHEN HAWKING,
BRITISH THEORETICAL PHYSICIST

If humanity remains on Earth, the whole species will eventually become extinct, whether through our own mismanagement of the planet's delicately balanced ecosystem, nuclear annihilation, an expanding Sun at the end of its life, or some other catastrophe not yet predicted. To avoid this certain fate, humans must become a multi-planetary species.

Although the eventual expansion of the Sun will not happen for another five billion years, entrepreneur Elon Musk (1971–) has long warned that it is imperative that humanity escapes the confines of its home planet. There are two paths, Musk notes: "One path is we stay on Earth forever, and then there will be some eventual extinction event. I do not have an immediate doomsday prophecy, but eventually, history suggests, there will be some doomsday event. The alternative is to become a space-bearing civilization and a multi-planetary species, which, I hope you would agree, is the right way to go." Musk's answer to this problem is to colonize Mars.

There are several steps that are necessary. First, it requires the development of technologies to get humans off this planet on a long-term basis, to colonies—be they on the Moon, Mars, or farther afield—that can become independent of Earth. This won't happen overnight, but Moon bases and Mars missions will move humanity closer to becoming the multi-planetary species it needs to become to outlast the planet on which it first evolved.

Pictured here on the left is Robonaut 2, a dexterous, humanoid robot, which might eventually replace the need for humans in many areas of space exploration.

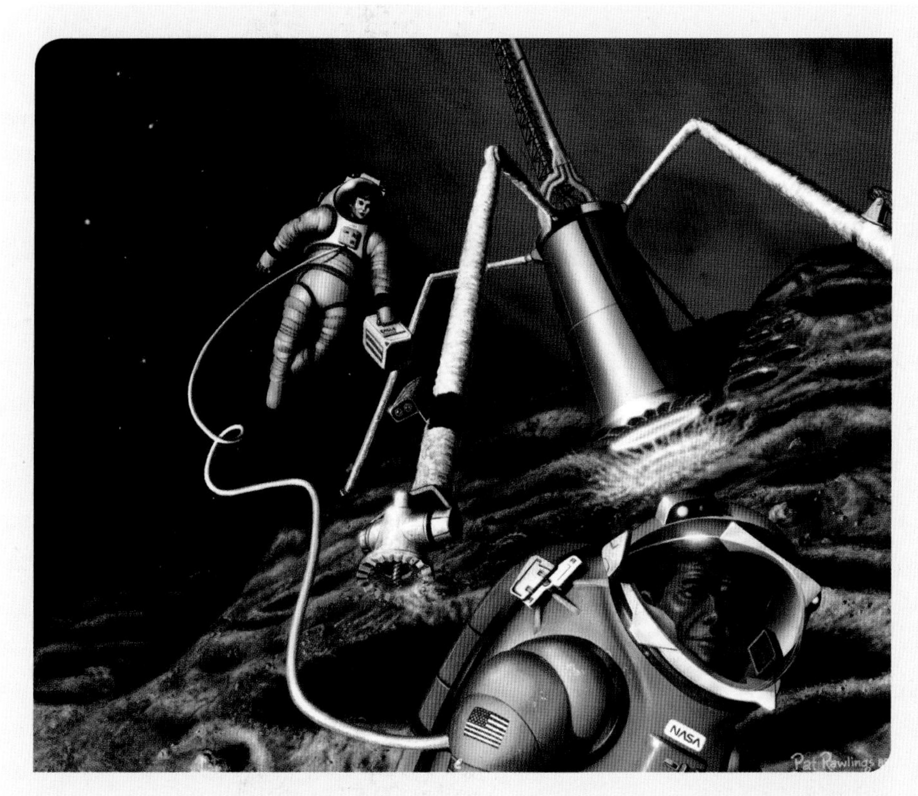

Modifying Humans for a Future in Space

Do we have the technology to make humans that are better suited to survival in space?

Spacesuits, such as those depicted in this image of men directing mining operations on the Martian moon Phobos, may eventually become obsolete.

Perhaps, instead of engineering a tiny bubble of Earth-like conditions to sustain humans on their exploration of space, might it be better to modify humans through a series of technological, genetic, and biomedical enhancements to survive in space instead?.

This idea was first broached by two NASA scientists, Manfred Clynes (1925–) and Nathan Kline (1916–83) in a 1960 study. In their work, the pair remarked: "Altering man's bodily functions to meet the requirements of extraterrestrial environments would be more logical than providing an earthly environment for him in space." To Clynes and Kline, building spacesuits and pressurized spaceships to keep human space explorers alive seemed unnecessarily complicated.

To resolve this problem, they suggested that the future astronaut attempt "partial adaptation to space conditions, instead of insisting on carrying his whole environment along." They proposed a variety of modifications, many mechanical in nature, which would allow humans to withstand radiation, the absence of atmospheric oxygen, and other hazards of space. They coined the term *cyborg* to describe this adaptation.

Such an approach may become both more socially and politically acceptable in the far future. To date, human presence in space has consisted of what might be characterized as extended camping trips, rarely exceeding a year in length. Yet space exploration advocates continue to propose the full colonization of other worlds throughout our solar system.

Humans are not well suited to very long stays beyond Earth, especially at locations with temperature conditions, radiation levels, and gravity that differ markedly from those on Earth.

Would humans born and raised in extraterrestrial locations evolve in response to different conditions? Or might humans instead embrace successive waves of advances in biotechnology to reengineer themselves?

In many ways, we are already living in a cyborg future; millions of people enjoy a better quality of life today through the incorporation of a variety of enhancements and replacements, ranging from pacemakers to prosthetic limbs. The implications of the development of further technologies aimed at enhancing humanity for a future of space exploration are fascinating. How might humanity remake itself to meet the rigors of space exploration more effectively? Should we even attempt it? These are questions that scientists may well have to answer before humanity makes its final journey to a variety of new homes among the stars.

Five Challenges for the Future of Space Exploration

RIGHT This color-enhanced image captured by NASA's Juno spacecraft shows a cloud system in Jupiter's northern hemisphere. Such probes will continue to play a significant role in space exploration for the foreseeable future.

BELOW This NASA concept art depicts a proposed Mars rover studying Martian geology.

There is no question that space exploration provides a window on the universe from which fantastic new discoveries have been made, and will continue to be made. However, there remain five core challenges that those who are engaged in space exploration will have to face in the future.

The first of these involves the political will to continue an aggressive space exploration program. At a fundamental level, this is the most critical challenge. Space exploration is a costly business, and while the whole of humanity benefits from the scientific discoveries brought about by its pursuit, securing funding, either from national governments or international corporations, remains a vital element in making future exploration projects viable.

The second challenge is the task of developing multifaceted, inexpensive, safe, reliable, and flexible access to space. Pioneers of spaceflight believed that humans could make space travel both safe and inexpensive. Despite years of effort, however, the dream of cheap and easy space access has not yet been attained. We might continue to use rocket propulsion, and with new materials and clever engineering make a launcher that is not only recoverable, but also robust enough for repeated use. We might also develop new space planes (p. 346), or other means of reaching into space (p. 353). Whichever method is used in the future, it will need to be cheaper, safer, and more reliable than the rockets in use today.

The third challenge revolves around the development of smart spacecraft and rovers to explore the solar system. Using the power of remote-sensing, data-collecting satellites to monitor and measure phenomena throughout the universe, humans could remain on Earth, but establish a virtual presence on a rich variety of planets and moons, experiencing exploration without leaving the comfort of their home world.

The fourth challenge concerns protecting the Earth and humanity itself. During the twenty-first century, humans will face three great environmental challenges: overpopulation, resource depletion (especially of fossil fuels), and environmental degradation. Without space-based resources—particularly remote-sensing satellites that monitor the Earth's biosphere—humans will not be able to control these trends. Remote-sensing satellites offer humanity a chance to monitor the health of the Earth, identify hidden resources, and spot polluters. Proposals to strip-mine the Moon and asteroids might make some people blanch, but we may have little choice if the Earth's own resources begin to run out.

The fifth and final challenge will be the sustained human exploration and development of space. The creation of a permanently occupied ISS has already been a major achievement of space exploration in the early twenty-first century. It is both a model for future international collaborations, and a bridgehead to future space exploration expeditions, including the possible colonization of the Moon and Mars (p. 370).

Since the dawn of the space age, humanity has explored a considerable area of our solar system. But what our species will discover in space during the rest of the twenty-first century will almost certainly make earlier achievements in the history of space exploration seem tiny by comparison.

> *"Limitless undying love which shines around me like a million suns, it calls me on and on across the universe."*
>
> JOHN LENNON, BRITISH SINGER/SONGWRITER

BELOW LEFT This image of Earth, showing a bright red aurora of charged particles from the Sun, was captured from the International Space Station on August 9, 2015.

BELOW RIGHT This NASA concept art shows the proposed Altair lunar lander on the Moon's surface.

OVERLEAF Mosaic assembled from dozens of images taken by the Mars Curiosity rover, looking uphill at Mount Sharp, taken on Sol 1931 in January, 2018.

Further Reading

Andrews, James T. ***Red Cosmos: Tsiolkovskii, Grandfather of Soviet Rocketry.*** College Station: Texas A&M University Press, 2009. A biography of Konstantin Tsiolkovsky.

Beattie, Donald A. ***Taking Science to the Moon: Lunar Experiments and the Apollo Program***. Baltimore, MD: Johns Hopkins University Press, 2001. A history of the lunar science undertaken during Apollo.

Bille, Matt, and Erika Lishock. ***The First Space Race: Launching the World's First Satellites***. College Station: Texas A&M University Press, 2004. A fine historical synthesis of the period between the mid-1950s and the aftermath of Sputnik, focusing on the rivalry between the United States and the Soviet Union to launch the first orbital satellite.

Boss, Alan. ***The Crowded Universe: The Search for Living Planets***. New York: Basic Books, 2009. A useful discussion of the search for planets beyond this solar system.

Burrows, William E. ***This New Ocean: The Story of the First Space Age***. New York: Random House, 1998. A solid overview of space age history during its first 50 years.

Chaikin, Andrew. ***A Man on the Moon: The Voyages of the Apollo Astronauts***. New York: Viking, 1994. One of the best books on Apollo, this work emphasizes the exploration of the Moon by the astronauts between 1968 and 1972.

Clary, David A. ***Rocket Man: Robert H. Goddard and the Birth of the Space Age***. New York: Hyperion, 2003. A superior biography of the American rocket pioneer.

Collins, Michael. ***Carrying the Fire: An Astronaut's Journeys***. New York: Farrar, Straus and Giroux, 1974. This is the best memoir of an astronaut, written by the Apollo 11 crewmember.

Conway, Erik D. ***Exploration and Engineering: The Jet Propulsion Laboratory and the Quest for Mars***. Baltimore, MD: Johns Hopkins University Press, 2015. Presents a detailed discussion of Mars exploration between the end of Viking and the Mars Phoenix mission of 2008.

David, James E. ***Spies and Shuttles: NASA's Secret Relationships with the DoD and CIA***. Gainesville: University Press of Florida, 2015. Using newly declassified documents the author reveals how NASA interacted with the national security establishment from its beginning in 1958 until the present.

Dethloff, Henry C., and Ronald A. Schorn. ***Voyager's Grand Tour: To the Outer Planets and Beyond***. Washington, DC: Smithsonian Institution Press, 2003. A fine history of the Voyager mission to the outer planets.

Dick, Steven J. ***The Biological Universe: The Twentieth Century Extraterrestrial Life Debate and the Limits of Science***. New York: Cambridge University Press, 1996. A superb history of the possibility of life elsewhere in the universe.

Gerovitch, Slava. ***Soviet Space Mythologies: Public Images, Private Memories, and the Making of a Cultural Identity***. Pittsburgh, PA: University of Pittsburgh Press, 2015. Explores the history of the Soviet human space program.

Hansen, James R. ***First Man: The Life of Neil A. Armstrong***. New York: Simon & Schuster, 2005. A superb biography of the first American to set foot on the Moon.

Hersch, Matthew M. ***Inventing the American Astronaut***. New York: Palgrave Macmillan, 2012. Demystifies the astronaut-as-hero persona developed through the 1960s and sheds light on the difficulties the earliest astronauts faced outside of their space capsules.

Hunley, J. D. ***The Development of Propulsion Technology for U.S. Space-Launch Vehicles, 1926–1991***. College Station: Texas A&M University Press, 2007. Traces the development from Robert Goddard's early rockets and the German V-2 missile through to the Space Shuttle.

Jenkins, Dennis R. ***Space Shuttle: Developing an Icon, 1972–2013***. North Branch, MN: Speciality Press, 2017, 3 vols. By far the best technical history of the Space Shuttle.

Jenks, Andrew L. ***The Cosmonaut Who Couldn't Stop Smiling: The Life and Legend of Yuri Gagarin***. Chicago: Northern Illinois University Press, 2012. A lively investigation of Gagarin's life and times.

Kalic, Sean N. ***US Presidents and the Militarization of Space, 1946–1967***. College Station: Texas A&M University Press, 2012. Discusses the national security objectives of both the United States and the Soviet Union.

Kessler, Elizabeth A. ***Picturing the Cosmos: Hubble Space Telescope Images and the Astronomical Sublime***. Minneapolis: University of Minnesota Press, 2012. An excellent book that focuses on the imagery of HST and its sublime aestheticism.

Kluger, Jeffrey. ***Apollo 8: The Thrilling Story of the First Mission to the Moon.*** New York: Henry Holt and Company, 2017. This book is a Valentine to the mission for which it is named and the astronauts who flew it in December 1968.

Koppel, Lily. ***The Astronaut Wives Club: A True Story***. New York: Grand Central Publishing, June 2013. A fine account of the lives of the wives of the first astronauts.

Krige, John. ***Fifty Years of European Cooperation in Space: Building on Its Past, ESA Shapes the Future***. Paris: Beauchesne Editeur, June 2014. Details 50 years of European collaboration in space.

Laney, Monique. ***German Rocketeers in the Heart of Dixie: Making Sense of the Nazi Past during the Civil Rights Era***. New Haven, CT: Yale University Press, 2015. Focuses on the integration of German rocket specialists and their families into a small southern community soon after World War II.

Launius, Roger D. ***Space Stations: Base Camps to the Stars***. Washington, DC: Smithsonian Books, 2003. A history of space stations—real and imagined—as cultural icons, fully illustrated with rare and evocative imagery.

Launius, Roger D and Howard E. McCurdy. ***Robots in Space: Technology, Evolution, and Interplanetary Travel.*** Baltimore, MD: Johns Hopkins University Press, 2008. A history of the debate over whether to send humans or robots to explore space.

Logsdon, John M. ***After Apollo? Richard Nixon and the American Space Program***. New York: Palgrave Macmillan, 2015. Traces in detail how Nixon charted the post-Apollo space program.

Logsdon, John M. ***John F. Kennedy and the Race to the Moon***. New York: Palgrave Macmillan, 2010. Presents the definitive examination of John F. Kennedy's role in the decision to send Americans to the Moon.

Logsdon, John M. (General Editor). ***Exploring the Unknown: Selected Documents in the History of the U.S. Civil Space Program***. 7 vols. Washington, DC: NASA Special Publication-4407, 1995–2008. An essential reference work, these volumes print more than 550 key documents in space policy.

Lord, M. G. ***Astro Turf: The Private Life of Rocket Science***. New York: Walker and Co., 2005. An inventive account of the cult of the rocket with forays into science fiction and pop culture.

MacDonald, Alexander. ***The Long Space Age: The Economic Origins of Space Exploration from Colonial America to the Cold War***. New Haven: Yale University Press, 2017. Draws fascinating parallels between large government-funded space exploration, so common since the establishment of NASA in 1958, and the *longue durée* of private-sector efforts in earlier eras.

Maher, Neil. ***Apollo in the Age of Aquarius***. New Haven, CT: Harvard University Press, 2017. A unique focus on the environmentalism emanating from the space program.

Markley, Robert. ***Dying Planet: Mars in Science and the Imagination***. Durham, NC: Duke University Press, 2005. An excellent primer on Mars and its place in popular culture.

McCurdy, Howard E. ***Space and the American Imagination***. Washington, DC: Smithsonian Institution Press, 1997. A significant analysis of the relationship between popular culture and public policy.

McDougall, Walter A. ***...The Heavens and the Earth: A Political History of the Space Age***. New York: Basic Books, 1985. Reprint edition, Baltimore, MD: Johns Hopkins University Press, 1997. This Pulitzer Prize-winning book analyzes the space race to the Moon in the 1960s. The author argues that Apollo prompted the space program to stress engineering over science, competition over cooperation, and international prestige over practical applications.

Michaud, Michael A. G. ***Contact with Alien Civilizations: Our Hopes and Fears about Encountering Extraterrestrials***. New York: Copernicus Books, 2007. Explores the political and social consequences of interstellar contact with extraterrestrials and how our species should be prepared.

Mindell, David A. ***Digital Apollo: Human, Machine, and Space***. Cambridge, MA: MIT Press, 2008. A superb account of the building of the Apollo Guidance Computer and its role in helping humanity to reach the Moon.

Moltz, James Clay. ***Asia's Space Race: National Motivations, Regional Rivalries, and International Risks***. New York: Columbia University Press, 2011. Provides the first in-depth policy analysis of Asia's 14 leading space programs, with a special focus on developments in China, Japan, India, and South Korea.

Monchaux, Nicholas de. ***Spacesuit: Fashioning Apollo***. Cambridge, MA: MIT Press, 2011. Explores layers of the spacesuit to tell the human story of its construction and use.

Morton, Oliver. ***Mapping Mars: Science, Imagination, and the Birth of a World***. New York: Picador USA, 2003. Tells the story of the landscapes of Mars.

Murray, Charles A., and Catherine Bly Cox. ***Apollo: The Race to the Moon***. New York: Simon and Schuster, 1989. Perhaps the best general account and overview of NASA's lunar program produced to date.

Neal, Valerie. ***Spaceflight in the Shuttle Era and Beyond: Redefining Humanity's Purpose in Space***. New Haven, CT: Yale University Press, 2017. Provides a basic history of the Space Shuttle.

Neufeld, Michael J. ***Von Braun: Dreamer of Space, Engineer of War***. New York: Alfred A. Knopf, 2007. The definitive biography of the German-American rocket pioneer.

Oliver, Kendrick. ***To Touch the Face of God: The Sacred, the Profane, and the American Space Program, 1957–1975***. Baltimore, MD: Johns Hopkins University Press, 2012. Emphasizes the relationship between engineering and spirituality in the space program.

Paul, Richard, and Steven Moss. ***We Could Not Fail: The First African Americans in the Space Program***. Austin: University of Texas Press, 2015. Profiles ten pioneering African Americans who worked on the American space program in the 1960s.

Pyle, Rod. ***Destination Mars: New Explorations of the Red Planet***. Amherst, NY: Prometheus Books, 2012. General account of the various robotic exploration missions to Mars to date.

Rao, P. V. Manoranjan (Chief Editor). B. N. Suresh and V. P. Balagangadharan (Associate Editors). ***From Fishing Hamlet to Red Planet: India's Space Journey***. Noida, Uttar Pradesh, India: Indian Space Research Organisation in association with Harper Collins India, 2015. A history of the Indian space program written by many of those involved.

Sagan, Carl. ***Pale Blue Dot: A Vision of the Human Future in Space***. New York: Random House, 1994. Probably the most sophisticated articulation of the exploration imperative to appear since Wernher von Braun's work of the 1950s and 1960s.

Sherr, Lynn. ***Sally Ride: America's First Woman in Space***. New York: Simon & Schuster, 2014. This biography of America's first woman in space, features insights from Ride's family and partner.

Scott, David Meerman, and Richard Jurek. ***Marketing the Moon: The Selling of the Apollo Lunar Program***. Cambridge, MA: MIT Press, 2014. An illustrated work on the efforts by NASA and its contractors to market the facts about space travel.

Shetterly, Margot Lee. ***Hidden Figures: The American Dream and the Untold Story of the Black Women Mathematicians Who Helped Win the Space Race.*** New York: William Morrow, 2016. An engaging account of the experiences of African-American women employed to calculate rocket trajectories and other statistics for NASA in the 1950s and 1960s.

Siddiqi, Asif A. ***The Red Rockets' Glare: Spaceflight and the Soviet Imagination, 1857–1957.*** New York: Cambridge University Press, February 2010. A comprehensive volume that situates the birth of cosmic enthusiasm within the social and cultural upheavals of Russian and Soviet history.

Siddiqi, Asif A. ***The Soviet Space Race with Apollo*** and ***Sputnik and the Soviet Space Challenge***. Gainesville: University Press of Florida, 2003. A superb two-volume history of the Soviet space program from its very beginnings through to the Moon race.

Singh, Simon. ***Big Bang: The Origin of the Universe***. New York: Fourth Estate, 2005. A popular account of how the Big Bang became the dominant theory explaining the origins of the universe.

Tribbe. Matthew D. ***No Requiem for the Space Age: The Apollo Moon Landings and American Culture***. New York: Oxford University Press, 2014. Offers a discussion of the Moon program as an exemplar of its time, place, and circumstance.

Tyson, Neil de Grasse. ***Astrophysics for People in a Hurry***. New York: W.W. Norton and Co., 2017. An up-to-date analysis of the basic evolution of the universe using scientific information gleaned from research from all sources, including space exploration missions.

Vertesi, Janet. ***Seeing Like a Rover: How Robots, Teams, and Images Craft Knowledge of Mars***. Chicago: University of Chicago Press, 2015. Undertakes a sociological analysis of the Mars Exploration Rover teams and how they pursued their activities over more than a decade.

Weitekamp, Margaret A. ***Right Stuff, Wrong Sex: America's First Women in Space Program***. Baltimore, MD: Johns Hopkins University Press, 2004. Tells the story of the Mercury 13, the women who underwent medical tests to demonstrate that women could be astronauts too.

Westwood, Lisa, Beth Laura O'Leary, and Milford Wayne Donaldson. ***The Final Mission: Preserving NASA's Apollo Sites***. Gainesville: University Press of Florida, 2017. Characterizes the various artifacts and detritus left on the Moon by the Apollo program. A brilliant resource explaining the laws and practices of preservation such sites, and a powerful piece of advocacy for the notion of space conservation.

Index

Page numbers in *italics* refer to illustration captions.

Picture Credits

Every effort has been made to trace all copyright owners but if any have been inadvertently overlooked, the publishers would be pleased to make the necessary arrangements at the first opportunity. (Key: **t** = top; **c** = center; **b** = bottom; **l** = left; **r** = right; **tl** = top left; **tr** = top right; **bl** = bottom left; **br** = bottom right)

2 Time Life Pictures/NASA/The LIFE Picture Collection/Getty Images **8–9** Babak Tafreshi/Getty Images **10 t** SSPL/Getty Images **10 b** Keepcases/Wikimedia Commons, CC BY-SA 3.0 **11 l** Copyright © 2001-2017 Satellite Imaging Corporation. All rights reserved. **11 r** © Digital Globe. All Rights Reserved: **12** Daniele Pugliesi/Wikimedia Commons **13** Granger/Bridgeman Images **14 l** Leemage/UIG/Getty Images **14 r** Wellcome Collection **15 t** Smithsonian Institution, National Postal Museum **15 b** Ipsumpix/Corbis/Getty Images **16** NASA/Marshall Space Flight Center **17 l** NASA/Marshall Space Flight Center **17 b** Private Collection/© Look and Learn/Bridgeman Images **18** ullstein bild/Getty Images **19 l** Private Collection/The Stapleton Collection/Bridgeman Images **19 r** Bettmann/Getty Images **20 l** NASA/MSFC **20 r** Smithsonian Institution, National Air and Space Museum, Gift of Charles D. Voy **21 l** Artokoloro Quint Lox Limited/Alamy Stock Photo **21 r** DEA PICTURE LIBRARY/DeAgostini/Getty Images **22 t** INTERFOTO/Alamy Stock Photo **22 b** Gift of Tsiolkovsky Space Museum, Russia/Smithsonian Institution, National Air and Space Museum **23 l** Smithsonian Institution, National Air and Space Museum **23 r** SPUTNIK/Alamy Stock Photo **24** NASA **25 l** US National Archives/Science Photo Library **27 l** CCI Archives/Science Photo Library **27 r** Royal Institution of Great Britain/Science Photo Library **28** Bettmann/Getty Images **29 tl** Smithsonian Institution, National Air and Space Museum, Gift of Mrs. Robert Goddard **29 tr** NASA **29 b** WorldPhotos/Alamy Stock Photo **30 t** Detlev Van Ravenswaay/Science Photo Library **30 b** Everett Collection Historical/Alamy Stock Photo **31 t** Detlev Van Ravenswaay/Science Photo Library **31 b** Library of Congress/Science Photo Library **32** Moviestore collection Ltd/Alamy Stock Photo **33 tl** Collection Christophel/Alamy Stock Photo **33 tr** Detlev Van Ravenswaay/Science Photo Library **33 b** Smithsonian Institution, National Air and Space Museum **34 r** Smithsonian Institution, National Air and Space Museum, Photo: Eric Long **35 l** Smithsonian Institution, National Air and Space Museum **35 r** Sputnik/Science Photo Library **36–7** Smithsonian Institution, National Air and Space Museum, Transferred from the U.S. Air Force **38** Sputnik/Science Photo Library **39** Fox Photos/Getty Images **40** Library of Congress, LC-USZ62-92435 **41 l** Stocktrek Images, Inc./Alamy Stock Photo **41 r** Library of Congress, LC-USZ62-135436 **42** Deutches Museum **43 t** Ralph Crane/The LIFE Picture Collection/Getty Images **43 b** Aviation History Collection/Alamy Stock Photo **44 l** Science & Society Picture Library/Getty Images **46** ClassicStock/Alamy Stock Photo **47** NASA **48** Smithsonian Institution, National Air and Space Museum **50** Smithsonian Institution, National Air and Space Museum **51 t** Transferred from the National Aeronautics and Space Administration/Smithsonian Institution **51 bl** Universal Images Group North America LLC/Alamy Stock Photo **51 br** Library of Congress, HAER ALA,45-HUVI.V,7A--32 **52** NASA **53 l** NASA **53 r** NASA **54 t** NASA **54 b** NASA **55** Smithsonian Institution, National Air and Space Museum **56 l** US government report, 1946 **56 r** © Hulton-Deutsch Collection/Corbis/Getty Images **57** US Army/The LIFE Picture Collection/Getty Images **58–9** NASA **60 l** History of Science Collections, University of Oklahoma Libraries; copyright the Board of Regents of the University of Oklahoma **60 r** Smithsonian Institution, National Air and Space Museum, Transferred from National Museum of American History **62** Mondadori Portfolio/Getty Images **63 tl** National Geographic Creative/Alamy Stock Photo **63 tr** NASA **63 b** NASA **64** © Corbis/Getty Images **65** Smithsonian Institution, National Air and Space Museum **66 t** Smithsonian Institution, National Air and Space Museum, Krause collection **66 c** Smithsonian Institution, National Air and Space Museum **66 b** Smithsonian Institution, National Air and Space Museum, Krause collection **67** Smithsonian Institution, National Air and Space Museum, Gift of Charles Johnson **68 t** NASA/Langley Research Center **68 b** NASA **69 l** NASA **69 r** Smithsonian Institution, National Air and Space Museum, Transferred from the National Advisory Committee for Aeronautics (NACA) **70** NASA **71 l** NASA **71 r** NASA **72** NASA **73** Sergei Arssenev/Wikimedia Commons, CC BY-SA 4.0 **74** European Space Agency/Cnes/Science Photo Library **75 l** Mark Williamson/Science Photo Library **75 r** Paul Fearn/Alamy Stock Photo **76 t** Smithsonian Institution, National Air and Space Museum **76 b** Smithsonian Institution, National Air and Space Museum, Photo: James Evans **77 t** Smithsonian Institution, National Air and Space Museum, Photo: NACA **77 b** Smithsonian Institution, National Air and Space Museum, Photo: US Army Air Force **78** New York Public Library/Science Photo Library **79** Sputnik/Science Photo Library **80 l** The Advertising Archives/Alamy Stock Photo **80 r** AF archive/Alamy Stock Photo **81** Science History Images/Alamy Stock Photo **82** Collier's, 22 March 1952 **83 t** Smithsonian Institution, National Air and Space Museum **83 b** Smithsonian Institution, National Air and Space Museum, Gift of William C. Estler **84** Transcendental Graphics/Getty Images **85 l** SilverScreen/Alamy Stock Photo **85 r** The Conquest of Space by Ley Willy, New York: Viking, 1949 **86–7** National Portrait Gallery, Smithsonian Institution; gift of Time magazine **88 t** Howard Sochurek/The LIFE Picture Collection/Getty Images **88 b** National Academies of Sciences **89 l** ZUMA Press, Inc./Alamy Stock Photo **89 r** NG Images/Alamy Stock Photo **90 l** Sputnik/Science Photo Library **90 r** Sputnik/Science Photo Library **91** Smithsonian Institution, National Air and Space Museum, Gift of Carl B. Cobb **92** Photo by Dmitri Kessel/Life Magazine, Copyright Time Inc./The LIFE Premium Collection/Getty Images **93 t** Sputnik/Alamy Stock Photo **93 b** © Hulton-Deutsch Collection/CORBIS/Corbis via Getty Images **94 t** Bettmann/Getty Images **94 b** Hank Walker/The LIFE Picture Collection/Getty Images **95** Bettmann/Getty Images **96** US Air Force/Science Photo Library **97 tl** National Reconnaissance Office/Science Photo Library **97 tr** Smithsonian Institution, National Air and Space Museum **97 b** Smithsonian Institution, National Air and Space Museum, Naval Research Laboratory **98** NASA/MSFC **99 l** NASA/MSFC **100** NASA/JPL/Science Photo Library **101** NASA/MSFC **102** NASA **103 l** NASA **103 r** NASA **104** NASA **105** NASA/Langley Research Center **106 l** NASA **106 r** NASA **107 l** Smithsonian Institution, National Air and Space Museum, Transferred from the National Aeronautics and Space Administration **107 r** NASA/Langley Research Center **108 t** Fine Art Images/Heritage Images/Getty Images **108 b** ITAR-TASS News Agency/Alamy Stock Photo **109 l** Sputnik/Science Photo Library **109 r** Sputnik/Science Photo Library **110** NASA **111 l** Smithsonian Institution, National Air and Space Museum, Photo: USAF **111 r** Smithsonian Institution, National Air and Space Museum **112** NASA **113** Sputnik/Science Photo Library **114 t** Sputnik/Science Photo Library **114 b** Sputnik/Science Photo Library **115** Detlev Van Ravenswaay/Science Photo Library **116** Sputnik/Science Photo Library **117 l** Smithsonian Institution, National Air and Space Museum **117 r** Bettmann/Getty Images **118** NASA **119** NASA **120 l** Sputnik/Science Photo Library **120 r** NASA **121** NASA **122** NASA **123 l** Smithsonian Institution, National Air and Space Museum, Photo: USAF **123 r** Smithsonian Institution, National Air and Space Museum, Photo: NASA **124 l** ITAR-TASS News Agency/Alamy Stock Photo **124 r** Sputnik/Alamy Stock Photo **125 t** Sputnik/Science Photo Library **125 bl** Sputnik/Science Photo Library **125 br** ITAR-TASS News Agency/Alamy Stock Photo **126 l** Sovfoto/UIG via Getty Images **126 r** Ralph Crane/The LIFE Picture Collection/Getty Images **127 l** NASA/Science Photo Library **127 r** NASA **128–9** NASA **130 t** Interfoto/Alamy Stock Photo **130 b** Sputnik/Science Photo Library **131 l** Smithsonian Institution, National Air and Space Museum, Transferred from NASA-Jet Propulsion Laboratory **131 r** NASA/Glenn Research Center **132** NASA **133 l** NASA **133 r** NASA **134** Paul Schutzer/The LIFE Picture Collection/Getty Images **135 l** Bettmann/Getty Images **135 r** NASA **136** A.Sokolov & A.Leonov/ASAP/Science Photo Library **137 tl** NASA/Langley Research Center **137 tr** NASA **137 b** NASA/Science Photo Library **138** Sputnik/Science Photo Library **139 t** NASA/Science Photo Library **139 bl** National Reconnaissance Office/Science Photo Library **139 br** NASA **140** Smithsonian Institution, National Air and Space Museum **141 l** Smithsonian Institution, National Air and Space Museum **141 r** NASA **142** Time Life Pictures/NASA/The LIFE Picture Collection/Getty Images **143 l** NASA/Science Photo Library

143 r Archive PL/Alamy Stock Photo **144 t** Pictorial Press Ltd/Alamy Stock Photo **144 b** Moviestore collection Ltd/Alamy Stock Photo **145 t** © Arthur C. Clarke books: 2001: A Space Odyssey, Orbit, 2001; 2010: Odyssey Two, Granada Publishing Ltd, 1982; 2061: Odyssey Three, Del Rey, 1987; 3001: The Final Odyssey, Del Rey, 1997 **145 b** Smithsonian Institution, National Air and Space Museum, Photo: Foto Schikola **146 t** NASA **146 b** NASA **147 l** NASA **147 r** Smithsonian Institution, National Air and Space Museum, Transferred from NASA - Manned Spacecraft Cente **148** NASA **149 l** NASA **149 r** Smithsonian Institution, National Air and Space Museum, Transferred from the National Aeronautics and Space Administration **150** NASA **151 l** PhotoQuest/Getty Images **151 r** NASA **152** Science History Images/Alamy Stock Photo **153** NASA Photo/Alamy Stock Photo **154 t** NASA **154 b** NASA **155 t** Science History Images/Alamy Stock Photo **155 bl** NASA **155 br** Smithsonian Institution, National Air and Space Museum, Photo: Eric Long **156** NASA **157 t** AF archive/Alamy Stock Photo **157 bl** NASA **157 br** NASA **158** NASA **159** NASA **160** NASA **161 l** Sputnik/Science Photo Library **161 r** Detlev Van Ravenswaay/Science Photo Library **162–3** NASA Archive/Alamy Stock Photo **164–5** NASA/JPL **166** Patrick Landmann/Science Photo Library **167 l** Smithsonian Institution, National Air and Space Museum **167 r** NASA **168** The Asahi Shimbun via Getty Images **170 t** Luo Yangyang/VCG via Getty Images **170 b** Rolls Press/Popperfoto/Getty Images **171** Smithsonian Institution, National Air and Space Museum, Transferred from the National Aeronautics and Space Administration **172 t** NASA **172 b** NASA **173** NASA **174 tl** Smithsonian Institution, National Air and Space Museum, Transferred from the Smithsonian Astrophysical Observatory **174 tr** Smithsonian Institution, National Air and Space Museum **175** NASA **176–7** NASA **178 l** Entertainment Pictures/Alamy Stock Photo **178 r** Detlev Van Ravenswaay/Science Photo Library **179 l** NASA/JPL-Caltech **179 r** NASA/JPL/USGS **181** European Space Agency/Aeos Medialab/Science Photo Library **182 l** Keystone Pictures USA/Alamy Stock Photo **182 r** European Space Agency/Science Photo Library **183 l** ESA **183 r** NASA **184 t** NASA **184 b** NASA **185** CNES/Arianespace - D. Ducros/European Space Agency/Science Photo Library **186** Smithsonian Institution, National Air and Space Museum **187** NASA/Science Photo Library **188 tl** NASA **188 tr** Smithsonian Institution, National Air and Space Museum, Photo: Eric Long **188 b** NASA **189 t** NASA **189 b** NASA **190–1** NASA Photo/Alamy Stock Photo **192** China Great Wall Industry Corporation/Science Photo Library **193 t** Bettmann/Getty Images **193 b** Geoeye/Science Photo Library **194 r** Xinhua/Alamy Stock Photo **195** VCG via Getty Images **196 t** The French National Centre for Space Studies (CNES) **196 b** Space prime/Alamy Stock Photo **197 t** NASA **197 br** Sputnik/Science Photo Library **198** NASA **199 t** The Asahi Shimbun via Getty Images **199 b** NASA **200** Bettmann/Getty Images **202** NASA **203 l** NASA **203 r** NASA **204 b** NASA **205 r** Smithsonian Institution, National Air and Space Museum, Transferred from NASA Johnson Space Center **206–7** NASA/Dembinsky Photo Associates/Alamy Stock Photo **208 t** Sovfoto/UIG via Getty Images **208 b** NASA **209 t** Sputnik/Alamy Stock Photo **209 b** Smithsonian Institution, National Air and Space Museum, From the collection of Mr. Robin John Burrows, space enthusiast since reading "Sands of Mars" at age ten and E. LaVerne Johnson **210 l** Bettman/Getty Images **210 r** For Alan/Alamy Stock Photo **211 l** NASA **211 r** NASA **212 l** Smithsonian Institution, National Air and Space Museum **212 r** NASA **213** NASA **214** NASA Photo/Alamy Stock Photo **215** NASA **216 t** **216 b** NASA **217 l** Smithsonian Institution, National Air and Space Museum, Transferred from Boeing, Rocketdyne **217 r** NASA **218** NASA **219** NASA **220** NASA **221 l** NASA **221 r** NASA **222** NASA **223 l** NASA **223 r** Smithsonian Institution, National Air and Space Museum, Transfer from NASA **224** NASA **225 t** NASA **225 b** NASA **226 l** NASA Photo/Alamy Stock Photo **226 r** NASA **227 t** NASA **227 b** NASA **228** Sputnik/Alamy Stock Photo **229 l** Smithsonian Institution, National Air and Space Museum, Copyright Glavkosmos/SCC, 1989 **229 r** Sputnik/Science Photo Library **230** NASA **231** NASA **232** NASA **234** NASA **235** NASA **236 l** NASA **236 r** NASA **237** NASA **238** NASA **239 l** NASA **239 r** NASA **240** NASA **241** NASA **242** Collection Christophel/Alamy Stock Photo **243 l** Smithsonian Institution, National Air and Space Museum **243 r** Smithsonian Institution, National Air and Space Museum **244** Smithsonian Institution, National Air and Space Museum **245 t** NASA/Science Photo Library **245 b** NASA **246–7** NASA/JPL/USGS **248** Detlev Van Ravenswaay/Science Photo Library **249 l** Detlev Van Ravenswaay/Science Photo Library **249 r** Science History Images/Alamy Stock Photo **250 l** SSPL/Getty Images **250 r** Science Photo Library **251** Detlev Van Ravenswaay/Science Photo Library **252** Photo **12**/Alamy Stock Photo **253** BFA/Alamy Stock Photo **254 l** Chronicle/Alamy Stock Photo **254 r** Pictorial Press Ltd/Alamy Stock Photo **255 l** AF archive/Alamy Stock Photo **255 r** Warner Brothers/Getty Images **256 t** Smithsonian Institution, National Air and Space Museum, Transferred from NASA, Marshall Space Flight Center **256 b** NASA **257 l** NASA **257 r** Detlev Van Ravenswaay/Science Photo Library **258 l** Sputnik/Science Photo Library **258 r** Sputnik/Alamy Stock Photo **259** Sputnik/Science Photo Library **260** Sputnik/Science Photo Library **262** Smithsonian Institution, National Air and Space Museum **263** NASA/JPL-Caltech/Dan Goods **264 l** Smithsonian Institution, National Air and Space Museum, Transferred from the National Aeronautics and Space Administration **264 r** NASA **265 l** NASA/JPL **265 r** Smithsonian Institution, National Air and Space Museum, Transferred from NASA Jet Propulsion Laboratory **266 l** NASA/JPL **266 r** NASA/JPL-Caltech/University of Arizona **267** NASA **268 t** NASA/JPL/MSSS **268 c** NASA/JPL **268 b** NASA/JPL **269** NASA **270 t** Sputnik/Science Photo Library **270 b** NASA **271** NASA **272** NASA **273 t** Science History Images/Alamy Stock Photo **273 c** NASA Image Collection/Alamy Stock Photo **273 b** NASA/JPL/University of Arizona **275 tl** NASA/JPL/MSSS **275 tr** NASA/JPL/MSSS **275 b** NASA/JPL/Lockheed Martin **276** NASA/JPL/Cornell University **277** NASA/JPL **278** NASA/JPL-Caltech/Cornell **279** NASA/JPL-Caltech/Cornell University **280 t** Pallava Bagla/Corbis via Getty Images **280 b** NASA **281 l** JPL/NASA/ESA **281 r** European Space Agency/Science Photo Library **282** NASA/JPL-Caltech/MSSS **283** NASA/JPL-Caltech/MSSS **284 l** NASA **284 r** NASA **285 l** NASA **285 r** Smithsonian Institution, National Air and Space Museum, Transferred from NASA, Johnson Space Center **286 t** United Archives GmbH/Alamy Stock Photo **286 b** Pictorial Press Ltd/Alamy Stock Photo **287 l** NASA **287 r** AF archive/Alamy Stock Photo **288** NASA **289 t** NASA **289 b** NASA **290–1** NASA **292 b** NASA **293** NASA/Johns Hopkins University Applied Physics Laboratory/Southwest Research Institute/Lunar and Planetary Institute **294** NASA **296 l** NASA **296 r** Smithsonian Institution, National Air and Space Museum, Transferred from National Aeronautics and Space Administration **297 l** NASA **297 r** Smithsonian Institution, National Air and Space Museum, Transferred from NASA, Ames Research Center **298** NASA/JPL **300** NASA/JPL **301** Smithsonian Institution, National Air and Space Museum **303** NASA/JPL-Caltech **304 t** NASA/Copyright ESA/Rosetta/Philae/CIVA **304 b** NASA/JPL/JHUAPL **305 l** NASA/JPL **305 r** NASA **306** Joe Tucciarone/Science Photo Library **307** David Parker/Science Photo Library **308** NASA **309** NASA/Copyright M. Ahmetvaleev **311 t** Sputnik/Science Photo Library **311 bl** NASA **311 br** NASA/ESA/Giotto Project **313 l** NASA/JPL/DLR **313 r** NASA, ESA, H. Weaver and E. Smith (STScI) and J. Trauger and R. Evans (NASA's Jet Propulsion Laboratory) **314** NASA **315** NASA/JPL-Caltech **316** NASA/JPL/Space Science Institute **317** NASA **318** NASA/JPL-Caltech **319** NASA/JPL-Caltech/Space Science Institute **321 l** Smithsonian Institution, National Air and Space Museum, NASA/JPL/ESA **321 r** Smithsonian Institution, National Air and Space Museum **322** NASA/JPL/University of Arizona **323 t** ESA/NASA/JPL-Caltech/University of Arizona/USGS **323 bl** NASA/JPL-Caltech/ASI/Cornell **323 br** NASA/JPL-Caltech **324** Smithsonian Institution, National Air and Space Museum, Gift of John Hopkins University, Applied Physics Laboratory **326** NASA **327 l** NASA **327 r** Science Photo Library **328** Smithsonian Institution, National Air and Space Museum, Transferred from NASA **329** NASA/JPL-Caltech **330** NASA **332 l** NASA **332 r** NASA **333** NASA **334** NASA/CXC **335** NASA **336** NASA/JPL **337 l** NASA/JPL-Caltech **337 c** NASA/JPL-Caltech **337 r** NASA/JPL-Caltech **338** NASA **339** NASA **340 t** NASA **340 b** NASA/JPL-Caltech **341** NASA/Ames/JPL-Caltech **342** Oliver Gerhard/Alamy Stock Photo **343 l** Phil Degginger/Science Photo Library **343 r** NASA **344 l** NASA **344 r** NASA **345** Dr Seth Shostak/Science Photo Library **346–7** NASA/Clouds AO/SEArch **348** NASA **350 l** NASA **350 r** NASA **351 l** Blue Origin/Alamy Stock Photo **351 r** Chuck Bigger/Alamy Stock Photo **352** NASA **353 t** NASA **353 bl** NASA **353 br** **354** NASA/Langley Research Center **355 l** dpa picture alliance archive/Alamy Stock Photo **355 r** NASA **356** NASA **357 l** NASA **357 r** Smithsonian Institution, National Air and Space Museum, Transferred from NASA Langley Research Center **358 l** NASA **358 r** NASA **359 l** NASA **359 tr** NASA **359 br** NASA **360** NASA **361** Richard Bizley/Science Photo Library **362 t** dpa picture alliance/Alamy Stock Photo **363** NASA **364 l** NASA **364 r** NASA **365 l** NASA **365 r** NASA **366** NASA **367 l** NASA **367 r** NASA **368** NASA **369** Smithsonian Institution, National Air and Space Museum **370** NASA **371** NASA **372** Detlev Van Ravenswaay/Science Photo Library **373 l** Stocktrek Images, Inc./Alamy Stock Photo **373 r** NASA **374** Stocktrek Images, Inc./Alamy Stock Photo **375** NASA **376** Victor Habbick Visions/Science Photo Library **377 l** NASA **377 r** NASA **378** NASA **379 tl** NASA **379 tr** NASA **379 b** AF archive/Alamy Stock Photo **380** NASA **381 t** NASA/JPL-Caltech **381 b** Chronicle/Alamy Stock Photo **382** Julian Baum/Science Photo Library **383** NASA **384** NASA **385** NASA **386 t** NASA/JPL-Caltech/SwRI/MSSS/Gerald Eichstadt/Sean Doran **386 b** NASA/JPL-Caltech **387 l** NASA **387 r** NASA **388–89** NASA/JPL-Caltech/MSSS

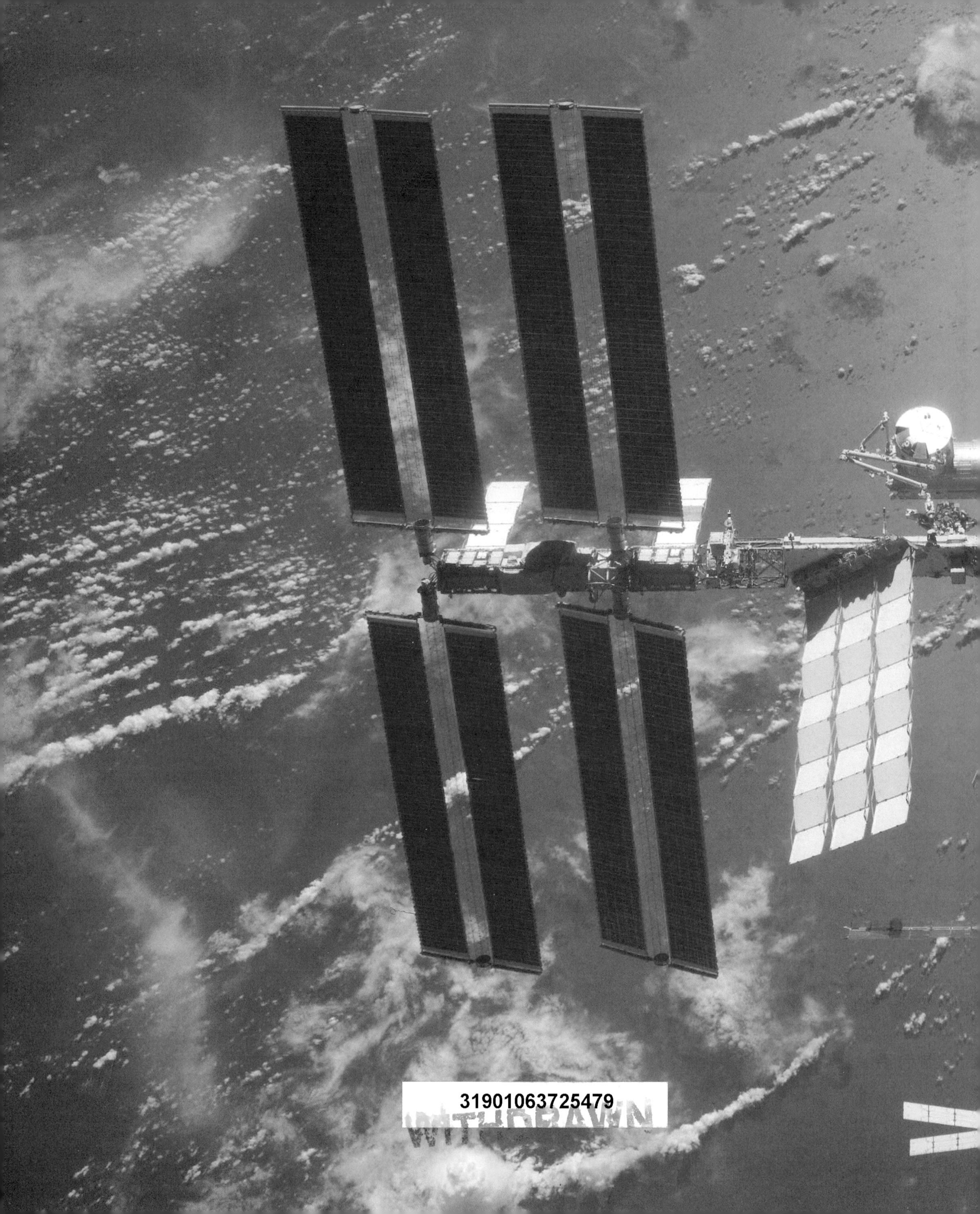